Stephen James O'Meara's
Observing the Night Sky with Binoculars

Month by month, constellation by constellation, star by star, object by object, Stephen James O'Meara takes readers on a celestial journey to many of the most prominent stars and constellations visible from mid-northern latitudes.

Filled with interesting anecdotes about the stars and constellations and their intriguing histories, this book is both a useful guide for amateur astronomers, and a great first-time reference for those just starting out. After describing a constellation's mythology, readers are guided in locating and identifying its brightest stars in the sky, as well as any other bright targets of interest – colorful stars, double or multiple stars, star clusters and asterisms, nebulae, galaxies, variable stars, and more.

This book will help beginning stargazers become familiar with the stars and constellations visible from their backyards, and explore the brightest and best stars, nebulae, and clusters visible through inexpensive, handheld binoculars.

STEPHEN JAMES O'MEARA has spent much of his career on the editorial staff of *Sky & Telescope*, and is a columnist and contributing editor for *Astronomy* magazine. He is an award-winning visual observer. His remarkable skills continually reset the standard of quality for other visual observers, and he was the first to sight Halley's Comet on its return in 1985. The International Astronomical Union named asteroid 3637 O'Meara in his honor. Steve is the recipient of the prestigious Lone Stargazer Award (2001) and the Omega Centauri Award (1994) for "his efforts in advancing astronomy through observation, writing, and promotion, and for sharing his love of the sky." He has also been awarded the Caroline Herschel Award for his pre Voyager visual discovery of the spokes in Saturn's B-ring and for being the first to determine visually the rotation period of Uranus. Steve is also a contract videographer for *National Geographic* Digital Motion, and a contract photographer for *National Geographic* Image Collection.

Stephen James O'Meara's

Observing the Night Sky with Binoculars

A Simple Guide to the Heavens

CAMBRIDGE UNIVERSITY PRESS
Cambridge, New York, Melbourne, Madrid, Cape Town, Singapore, São Paulo, Delhi

Cambridge University Press
The Edinburgh Building, Cambridge CB2 8RU, UK

Published in the United States of America by Cambridge University Press, New York

www.cambridge.org
Information on this title: www.cambridge.org/9780521721707

First published 2008

Printed in the United Kingdom at the University Press, Cambridge

A catalog record for this publication is available from the British Library

Library of Congress Cataloging in Publication data
O'Meara, Stephen James, 1956–
Stephen James O'Meara's observing the night sky with binoculars : a simple guide to the heavens / Stephen James O'Meara.
p. cm.
Includes index.
ISBN 978-0-521-72170-7 (pbk.)
1. Stars – Observers' manuals. 2. Constellations – Observers' manuals. 3. Astronomy – Observers' manuals. 4. Binoculars. I. Title.
QB63.O64 2008
523.8 – dc22 2008027706

ISBN 978-0-521-72170-7 paperback

To Donna,
The love of my life
To Milky Way, Miranda Piewacket, and Pele,
You are my starlight
And to Daisy Duke Such a Joy
My little bit of Heaven here on Earth

Contents

Preface

The night sky is full of wonder. It has a history as old as human thought, and a scope that is truly infinite. Seeing a star-filled sky from a dark location can inspire a lifetime of passion. And what the eye alone sees as an infinite expanse is suddenly and magnificently magnified in binoculars. The view awakens the mind to the humbling reality that the heavens are a vast and intangible wilderness with a quiet spirit. There is so much to see, so much to explore, so much to experience, that knowing where to begin and where to look can present a challenge. Now consider that the stars move as the Earth turns, and our perspective changes as our planet orbits the Sun. In a way, trying to befriend a star is like trying to grab hold of the golden ring on a merry-go-round. That's why I created this book, to help you make sense of the night sky while guiding you to some of the brightest and most attractive deep-sky wonders visible through handheld binoculars.

Handheld binoculars are ones that can be held comfortably in the hand – such as 7 × 50 and 10 × 50 binoculars; these are, in fact, the binoculars that virtually all credible sources overwhelmingly recommend to beginners, because they have wide fields of view, the best eye relief, and are comfortable to hold; they are also within the price range of most beginning observers. Expensive, quality optics *do* make a great difference in the view. That said, the binoculars I used to make the observations in this book were both inexpensive: one is a pair of old 7 × 50s that I bought at a yard sale for $5; the other is a pair of Meade 10 × 50s that I purchased at Radio Shack for $20. While the objects in this book have been selected with common optics in mind, they can also be enjoyed by owners of large binoculars, which will reveal them with greater ease and show them with greater clarity. The point is you do not have to spend a lot of money to enjoy the stars, the enjoyment comes from within.

My approach to helping you get started is different. The book opens by introducing you to the Big Dipper, the most recognizable star pattern in the night sky. I then help you to explore many of the naked-eye and binocular wonders within it. In doing so, you will learn many of the terms used in amateur astronomy and discover ways to measure sky distances, determine direction. You will also learn about star names, stellar magnitudes, spectral types, and the variety of deep-sky objects accessible through binoculars. This section also includes several useful tips on binocular observing. Using the Big Dipper as a learning center has several advantages. Its stars never set from mid-northern latitudes, they always circle the heavens in the north, and the stars are bright and easy to see, even under city lights.

What follows is a month-by-month journey to many of the most prominent stars and constellations visible from mid-northern latitudes within two hours of the time when they are highest in the sky. I encourage you to start with the opening chapter on the Big Dipper, which will introduce you to just about all the general and repetitive terms you'll encounter throughout the book. After that you can start your observing program at any given month in any given season. Once you begin your journey, try to follow the progression I have laid out in the book. I have carefully planned a very methodical approach to learning the stars, with repetition leading to familiarity.

Each section opens with an introduction that sets the celestial stage. It then goes on to preview each constellation individually, introducing you to the constellation's mythology, how to locate and identify its stars in the sky, and how to find any bright targets of interest – colorful stars, double or multiple stars, star clusters and asterisms, nebulae, galaxies, variable stars, and more. Most of the objects are bright and should be easy to see under a dark sky, though occasionally, I toss out a challenge. One thing is certain. As the months go by and your observing skills sharpen, you'll find it easier and easier to see fainter and fainter objects.

Much of the data concerning the stars and their planets was gleaned from "Stars," a wonderful website (http://www.astro.uiuc.edu/~kaler/sow/sowlist.html) created by Jim Kaler, Professor Emeritus of Astronomy, University of Illinois. Variable star data is mainly from the American Association of Variable Star Observers. I'd like to thank Simon Mitton and Vince Higgs at Cambridge University Press for encouraging the creation of this book. This book also owes a lot to the loving support of my wife Donna. Finally, a deep bow to my copy-editor Zoë Lewin, whose curiosity for all things beautiful in the night sky, and her attention to detail, made this a better book. I, of course, take full responsibility for any slips of the tongue.

Stephen James O'Meara
Volcano, Hawaii
July 2007

If the stars should appear one night in a thousand years, how would men believe and adore; and preserve for many generations the remembrance of the city of God which had been shown! But every night come out these envoys of beauty, and light the universe with their admonishing smile.

– Emerson, from *Nature*

A Dipper full of wonder

There is no better way to start our journey than by admiring the Big Dipper (also known as the Plough or Wagon), the most famous star pattern in the world. Spending your first nights with this easily seen star pattern is also an excellent way to learn the basics of observational astronomy. Doing so will teach you how to use your binoculars to full advantage; besides, if you're just beginning in this hobby, you'll need to know the basics, which are the building blocks of a strong foundation. I've always been a proponent of learning by doing. So let's get started. But before we do, understand that there's no need to rush. The stars, just like our friends, will always be there for you, night after night, month after month, year after year. It's the beauty of longevity.

Let's start our journey with some understanding. First, the Big Dipper is not a constellation. It is an *asterism* in the constellation Ursa Major, the Great Bear. An asterism is a familiar pattern of stars that forms part of one or more of the 88 officially recognized *constellations* (see Appendix A). The bowl of the Big Dipper lies in the body of the Bear, while the Dipper's handle is the Bear's tail – at least that's how the early Greek and Roman stargazers saw it; other cultures, such as the Cherokee Indians of the southern Appalachian Mountains, saw the three handle stars as three hunters in pursuit of a bear.

How the Bear got its tail

Anyone who has seen a bear knows that its tail is nothing but a short stub. So why is Ursa Major's tail so long? The answer, though a bit of a stretch, comes from the mythological tale of Callisto – one of several chaste maidens who cared for Diana (Artemis),[1] the virgin goddess of the Moon. As fate would have it, Callisto caught the ever-roaming eye of Jupiter (Zeus), the thunderbolt-wielding King of the Gods. After a roll in the hay, the maiden bore Jupiter a son, who she named Arcas (Arctos).

Jupiter never could keep a secret from Juno (Hera), his wife and twin sister, who became enraged with jealousy. When Juno sought out and found Callisto, she grabbed the nymph by her long golden locks and threw her to the ground, screaming, "Curse your beauty! I'll make sure that no one will desire you ever again!" With those words, Juno raised her arms. Lightning flashed, and thunder pealed. Powerless to defend herself, Callisto watched in horror as Juno, red with rage, transformed her into a giant bear. First Callisto's arms turned hairy, then her legs. Her sensitive jaw became crooked. Her gentle lips peeled back to reveal sharp teeth. When the deed was done, Callisto fled into the woods on all fours. She became terrified of her new situation, because, although Callisto looked like a beast, she had retained her human thoughts and heart.

One day, many years later, a young hunter encountered a foraging bear and startled it. Out of surprise, the mighty bear stood on its hind legs, opened wide its paws, and let out a terrifying roar. The hunter jumped back. The bear charged. But just as the hunter raised his spear to lance the beast, Jupiter appeared on the scene in time to stop the killing. Yes, the bear was Callisto, who, forgetting her appearance, was racing toward her long-lost son to embrace him!

To prevent any future tragedy, Jupiter turned Arcas into a bear, so he could forever enjoy the company of his mother. The great god then grabbed the two bears by their tails and swirled them high over his head. As he swirled the massive beasts, their tails stretched to great length before Jupiter finally flung them into the heavens. Today we see Callisto and Arcas, as Ursa Major and Ursa Minor, the Great and Little Bear, respectively.

But the story is not over. When Juno discovered what Jupiter had done, she was outraged. How could her husband place that adulteress and her spawn in the night sky

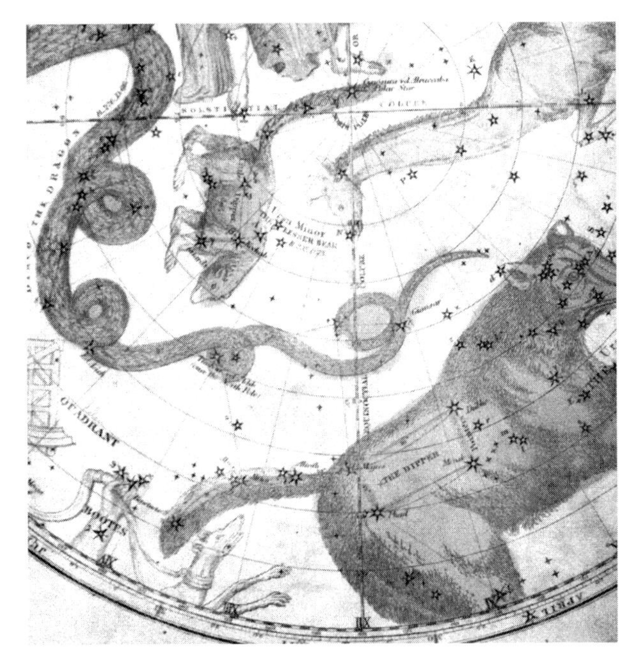

[1] The names in parentheses are the Greek counterparts of the Roman or Latin names.

so near the heavenly pole? Seeking comfort, she went to her foster parents who ruled the seas and complained: "Why should anyone fear offending Jupiter when he rewards them for causing me displeasure?" She beseeched her relatives to grant her one request: that the bears never set foot in her parents' sacred waters, as do the other stars on their nightly courses; instead, she wanted the bears to circle the pole, night after night, without time to rest. And so it is today that observers from mid-northern latitudes never see Ursa Major or Ursa Minor set below the horizon; they are *circumpolar* constellations – ones that circle the north celestial pole without ever setting, which is why finding the Big Dipper is always a good way to start your skywatching experience. Given a low horizon, you can see it at any time of the year or night.

How to find the Big Dipper

To find the Big Dipper, you'll need to get your bearings. Go outside when twilight ends and allow at least 15 minutes for your eyes to adapt to the growing darkness. As you wait, determine which way is north; use a compass or the compass setting on your Global Positioning Service (GPS), if necessary. When you face north, your left arm will point west, your right arm will point east, and south will lie directly behind you. Find a tree, house, or other landmark at these cardinal points, and record them in your observing log. When outside observing, be sure to use a red flashlight to record your notes; the eye's rod cells, which are responsible for our keen night vision, are insensitive to red light.

The landmarks you select will serve as guides to certain stars or constellations mentioned later in this book. For instance, if I were to tell you to look for a bright star halfway up the northeastern sky, you will know how to position your body (say, halfway between an oak tree in the north and a neighbor's house in the east) before looking up. That said, your first challenge is to find Polaris, the North Star, which will serve as your faithful guiding light. Be warned, despite popular myth, the North Star is not the brightest star in the night sky; it ranks 48th. But it is solitary and obvious; it also shines with a yellow light, so it's not difficult to detect.

The height of the North Star above your horizon (its *altitude*) is the same as your latitude on Earth. (Latitude is measured from 0° at the equator to 90° at the pole.) If you lived on the North Pole (latitude 90°), the North Star would be at the *zenith* (the point in the sky directly overhead; altitude 90°). The reason for this coincidence is that the North Star happens to be within 1° of the point where Earth's imaginary axis of rotation intersects the dome of the sky.[2]

If you lived on the equator (latitude 0°), the North Star would sit on the north horizon (altitude 0°). Likewise, if, you live in New York City (40° latitude), the North Star is 40° above the north horizon (40° altitude). How high is 40°? Draw an imaginary line from the north horizon to the point overhead (the sky's *zenith*). That line spans an angular distance of 90°. Forty degrees is almost one-half of the way from the horizon to the zenith. Now hold an upright fist at arm's length and look at it with one eye closed. The amount of sky covered by the fist is about 10°.

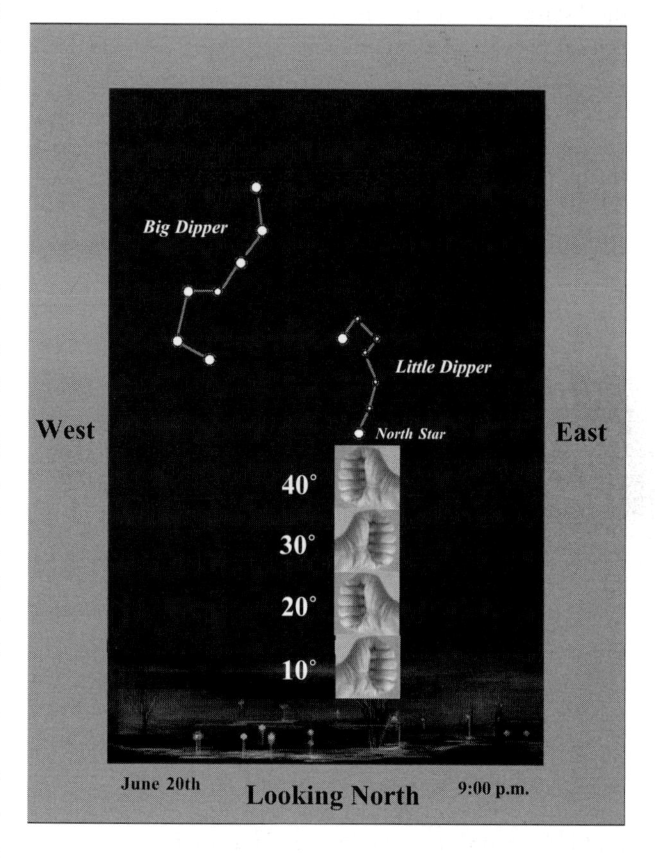

For an observer living in New York City to find the North Star, he or she would face north, place the base of their upright fist on the horizon line, make a fist with his or her other hand, and place it on top of the first fist (like one potato, two potato). Four fists equals about 40°. The North Star should be sitting on the top fist.

Now let's look for the Big Dipper. Since the Dipper is circumpolar, you can go out at any time of the night, at any time of the year to find it. If you go outside and observe the Big Dipper and the North Star at, say, 9:00 pm on April 1, then observe them again on the same night three hours later, you will see that the Dipper has rotated counterclockwise around the north celestial pole

[2] It is only by chance that Polaris is so close to the north celestial pole. It was not always our pole star and it will not always be our pole star. As the Earth spins on its axis, gravitational tugs by the Sun and Moon cause our planet to wobble like a top. Since one wobble takes 26,000 years to complete, Earth's imaginary axis slowly precesses over the years, transcribing an invisible circular path in the northern sky. Right now, that point on the circle just happens to bring it close to Polaris, our present day North Star. In AD 14,000, the bright star Vega will be our pole star. This polar precession is also responsible for the dawning of new zodiacal ages, such as the Age of Aquarius.

(from east to west). The pole star will not have moved perceptibly. That's because the Earth spins on its axis once every 24 hours, which causes us to see the entire sky turn like a giant wheel at a speed of about 15° per hour. Just as the Sun rises and sets each day so too do the stars (except for the circumpolar constellations). Since the North Star is very close to the point where Earth's imaginary axis intercepts the celestial sphere, it stays relatively fixed as all the other stars rotate around it. Imagine an umbrella dappled with dots representing the stars. If you spin the axis of the umbrella, the axis will remain fixed while all the "stars" on the umbrella turn around the axis. This eternal parade happens in the sky night after night, year after year.

If you were standing at the North Pole, Earth's axis would intersect the sky at a point directly overhead (a point we call the north celestial pole). No stars would rise and set because they are moving parallel to the horizon. If you were at the equator, the north celestial pole would be on the horizon, so all the stars would rise and set. For an observer halfway between the pole and the equator, the north celestial pole is halfway (45°) above the horizon. Therefore, any star within 45° of the north celestial pole is circumpolar, meaning that as the Earth turns, it will remain in view all night and never set.

If the Earth did not orbit the Sun, this eternal parade would always start and end at the same point in the sky. In other words, the Big Dipper would be in the same position in the sky on April 1 at 9:00 pm as it would be on December 1 at 9:00 pm. But the Earth does orbit the Sun – once every $365\frac{1}{4}$ days. As a result, the part of the sky that we see each night changes. Each night any given star will rise four minutes earlier than it did on the previous night. The change adds up. A star that rises in the east at 9:00 pm on April 1 will be high in the south at 9:00 pm on July 1. At 9:00 pm on October 1, the star will be setting in the west. Although the circumpolar constellations never rise or set, we still see them perform a yearly, counterclockwise march around the north celestial pole. Look at the chart on this page. The Big Dipper is almost directly above the North Star on April 15 at 9:00 pm but almost directly below it on October 15 at 9:00 pm.

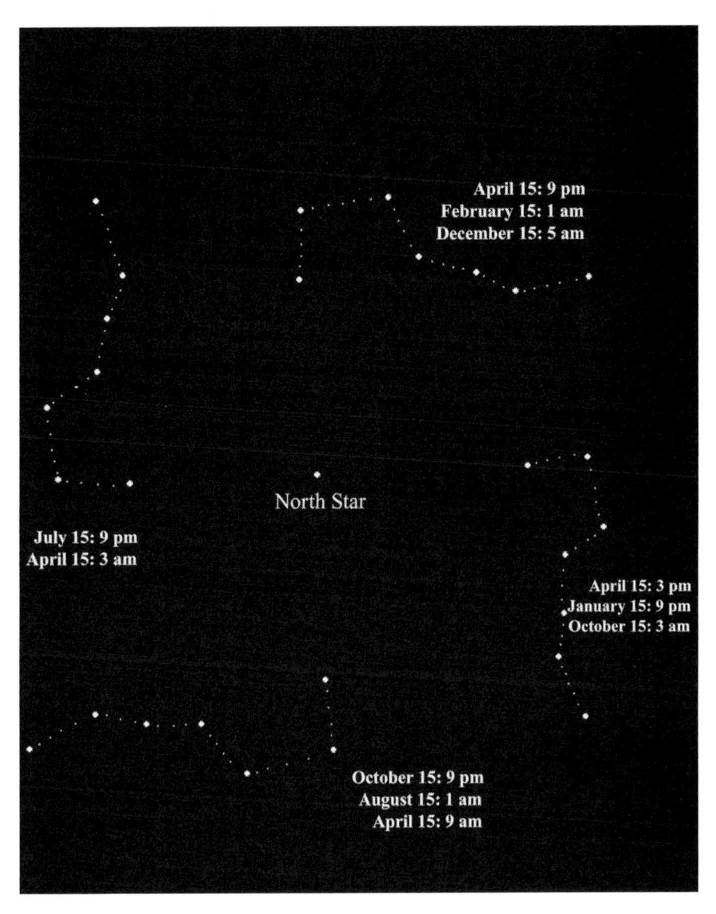

Unless it's spring, you'll need to observe from a location that offers a clear, low horizon. Otherwise, the Big Dipper may be obstructed by houses or trees. Let's assume it is April 1 at 9:00 pm. Go outside, face north, then look high overhead – high enough to get a crick in the neck. The seven stars of the Big Dipper should be prominently placed with the Dipper's bowl "pouring" its celestial waters down to the thirsty Earth (April showers bring May flowers). To confirm you have the right star pattern, hold out your hand at arm's length, close one eye, and spread your fingers; the Big Dipper should stretch fully from the tip of your thumb to the tip of your pinky. The photo on the next page illustrates this technique.

Once you're certain you have found these stars, turn your attention to Dubhe and Merak, the two stars at the end of the Big Dipper's bowl. These are the famous Pointer stars, because if you extend an imaginary line from Merak, through Dubhe away from the Bowl, they will point to the North Star. Extend your hand once again and spread your fingers. If you place the pinky of your right hand at the position of Dubhe, and twist your thumb toward the horizon, the North Star should be a short distance beyond your thumb. The North Star marks the tip of the Little Bear's tail, but we will not concern ourselves with the rest of that dim constellation right now, because it is best seen in the autumn.

Sky measures

The Big Dipper should also be three fists above the North Star. Of course, the amount of sky your fist covers will depend on the size of your hand and the length of your arm. For instance, I find my fist covers a little more than 10°, and that I'm more comfortable using just my four fingers.

You can determine more precisely how your fingers and hands measure up by using the stars of the Big Dipper to help you visualize apparent sky distances. Hold up a fist against the bowl of the Big Dipper. It should fit conveniently inside the stars that mark the open part of the bowl, which are separated by 10°. Your hand with outstretched fingers spans about 22°, which, again, is nearly the full length of the Big Dipper – from tip of the bowl to tip of the handle. To make life easier, most sources round that number off to 20° (which I've done in this book), though I find my outstretched hand covers about 25°(see the photo at above). So measuring sky distances with your hands is a rough sport but one that at least gets you in the ballpark. That said, the distance between Dubhe and Merak is only 5° – about the width of two or three fingers held at arm's length; you decide. Now take your binoculars and hold them up to the Pointer stars in the Big Dipper's bowl. Most binoculars give a field of view around 7°, so the pointer stars should fit nicely in the binocular field, being near opposite edges. Throughout this book I have adopted a binocular field of view of 7° to help you find objects in the sky.

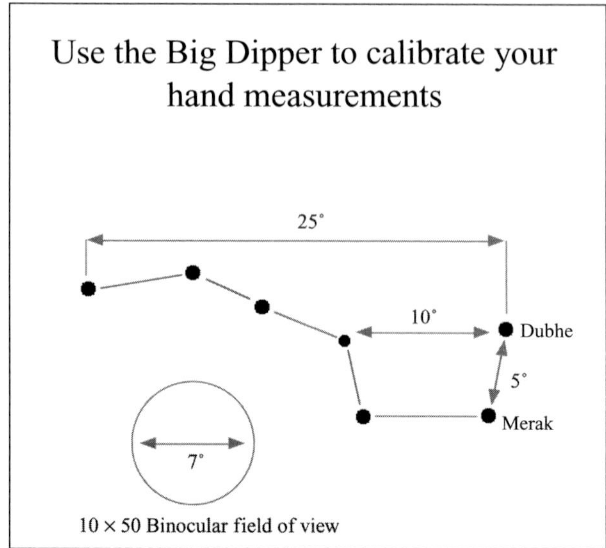

Name those stars

The chart on the next page shows the constellation of Ursa Major labeled with lower-case Greek letters and numbers. The Greek letters are a system of stellar nomenclature introduced in 1603 by Bavarian astronomer Johann Bayer, who labeled stars in each constellation according to their brightness. The most prominent star was given the letter Alpha (α); the faintest became Omega (ω).

The Greek alphabet (lower case)

α	Alpha	ι	Iota	ρ	Rho
β	Beta	κ	Kappa	σ	Sigma
γ	Gamma	λ	Lambda	τ	Tau
δ	Delta	μ	Mu	υ	Upsilon
ε	Epsilon	ν	Nu	φ	Phi
ζ	Zeta	ξ	Xi	χ	Chi
η	Eta	ο	Omicron	ψ	Psi
θ	Theta	π	Pi	ω	Omega

There are exceptions, though, such as with the stars in the Big Dipper, which are labeled in order of celestial longitude (as measured from west to east), not by brightness. So Dubhe is the Alpha (α) star of Ursa Major. Astronomers condense it all by saying that Dubhe is Alpha (α) Ursae Majoris, which is the Greek letter followed by the Latin genitive of the constellation name (see Appendix A).

Other stars have number identifications. These are *Flamsteed numbers*. Like the Greek letters, a Flamsteed number precedes the Latin genitive of the constellation: 80 Ursae Majoris (Alcor), for example. John Flamsteed was

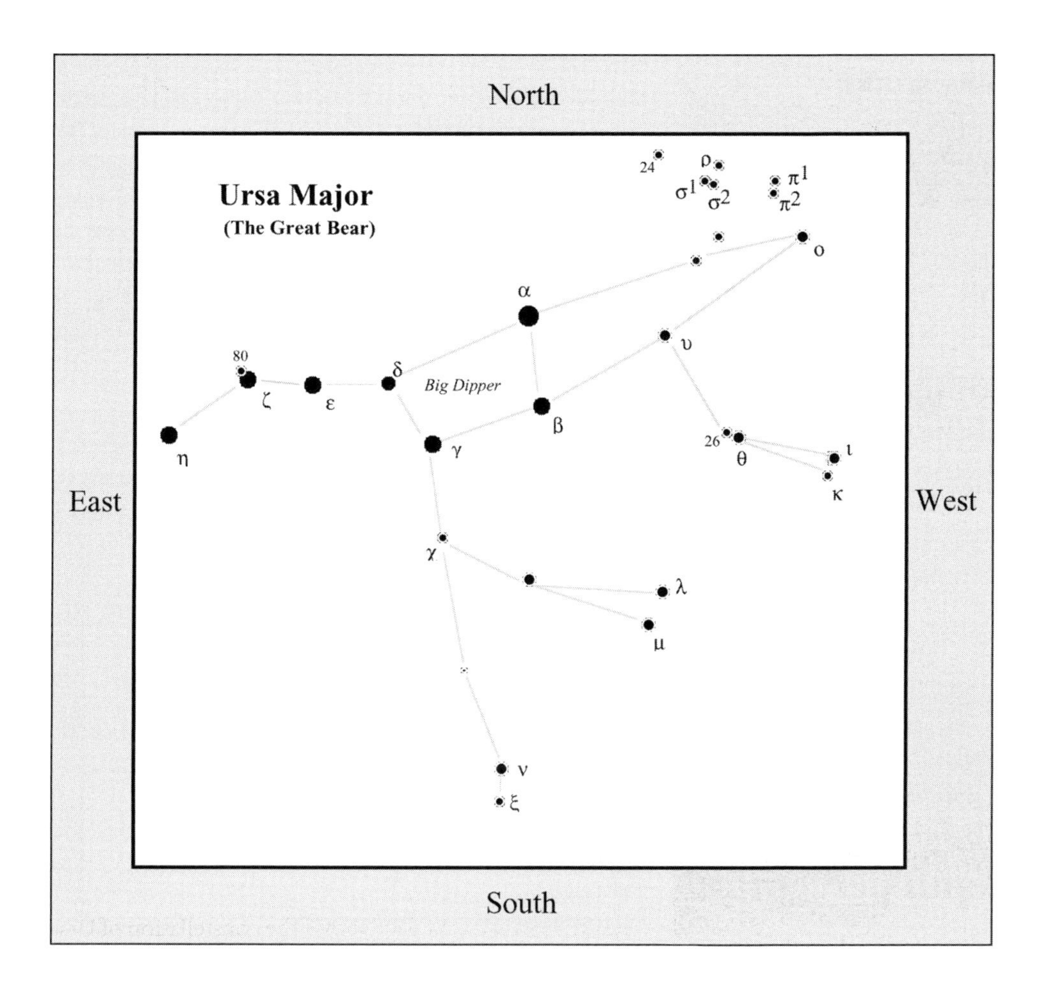

a prodigious eighteenth-century observer who dedicated 30 years of his life to measuring star positions, which he dutifully cataloged in his *Historia Coelestis Britannica* (1725) in order of their celestial longitude, again from west to east. Note that, on occasion, I have also added to the charts some stars that have an italicized lower-case letter, like *a* or *b*; these are additional guide stars (those with no Greek letter or Flamsteed designations), which you'll find described in the text as Star *a* or Star *b*, etc.

Star brightness

Look carefully at the seven stars of the Big Dipper with your unaided eyes. At first glance all of the Dipper's stars will appear to shine at about the same brightness. But notice how Delta (δ) Ursae Majoris, the star in the bowl closest to the handle, is slightly dimmer than the rest. Astronomers refer to an object's apparent brightness as its *magnitude*. The brighter an object appears, the smaller the numerical value of its apparent magnitude. On the brighter side of the magnitude scale, the values soar into the negative numbers; Sirius, for example, the brightest star in the night sky, shines at magnitude −1.5. The faintest stars visible at a glance to the unaided eye hover around 6th magnitude. The faintest stars visible *at a glance* in 10 × 50 binoculars shine at about 9th magnitude. Limiting magnitude is a *very* conservative number; your visual limit will vary depending on your location, the clarity of the atmosphere, the degree of light pollution, your binoculars, your visual acuity, the time you spend looking, and your expertise.

Mathematically, a 1st-magnitude star is 2.512 times brighter than a 2nd-magnitude star, which is 2.512 times brighter than a 3rd-magnitude star, and so on. The math works out nicely so that a star of 1st magnitude is exactly 100 times brighter than a star of 6th magnitude. Since Delta Ursae Majoris shines at 3rd magnitude, it appears about $2\frac{1}{2}$ times fainter than the other stars in the Big Dipper, which are all 2nd magnitude. (Think of magnitude as "class," where a first-class star is brighter than a second-class star, and so on.)

The chart on the next page shows some star magnitudes (placed in parentheses) in and around the Big Dipper's handle. The magnitudes have been rounded off. Learn to discern the difference between these magnitudes, because throughout the book I will refer to stars by their rounded-off magnitudes, telling you to look, say, for a 4th-magnitude, or an 8th-magnitude, star. Having a clear idea of how bright or faint a star will appear with your unaided eyes or through binoculars will help you in your searches.

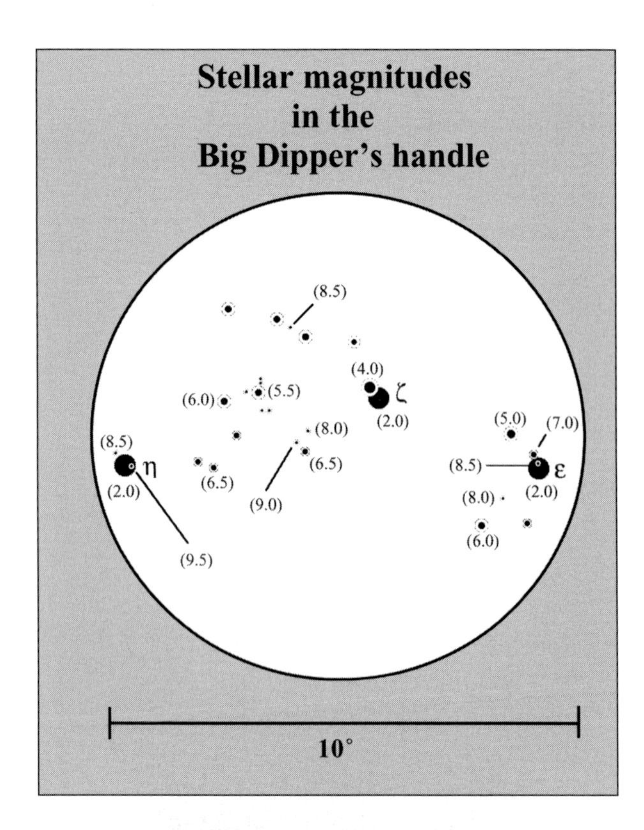

Distance and depth perception

Look at the chart below. It gives the distances to the stars in the Big Dipper as measured in *light years* (*ly*). A light year is the distance light will travel in one year. At a rate of 186,000 miles per second, light will travel 6 trillion miles in a year, so 1 light year equals 6 trillion miles. Notice how the five central stars of the Big Dipper – Beta (β), Gamma (γ), Delta (δ), Epsilon (ε), and Zeta (ζ) – all lie at similar distances (roughly 80 ly). That's because they are the brightest members of the Ursa Major Moving Group – a loosely bound *open star cluster* that has, in addition to the five stars of the Big Dipper mentioned above, about a half dozen fainter members splashed across a volume of space that measures 18 by 30 light years.

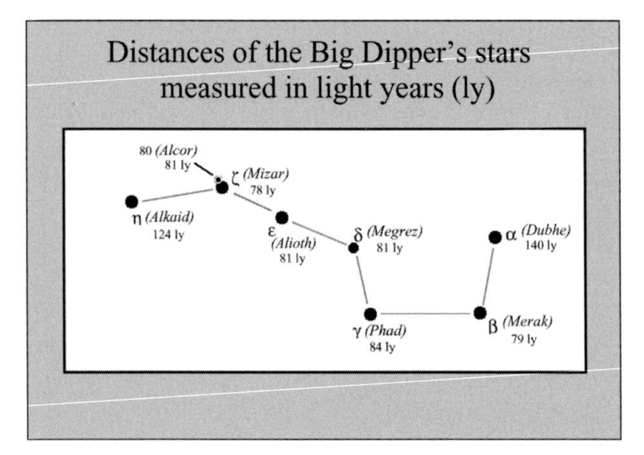

Alpha (α) Ursae Majoris (Dubhe) and Eta (η) Ursae Majoris (Alkaid) are not members of the Moving Group, being 140 and 124 light years distant, respectively. They are moving in their own separate paths. Take a moment now and try to perceive these different distances. It's hard to do because we see the sky as a two-dimensional sphere, but the night sky really is a vast and deep sea of space. (Try to imagine the stars, as bioluminescent deep-sea creatures, at different depths.) Come back to Earth 100,000 years from now, and the Big Dipper will no longer appear as it does today. As Dubhe and Alkaid move on their own separate ways, the familiar Dipper pattern will change to what looks more like a scoop than the Plough so familiar to British observers.

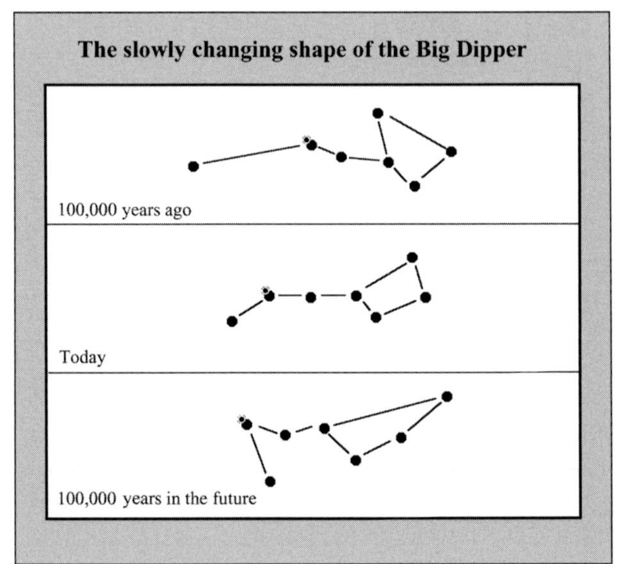

Another member of the Ursa Major Moving Group can be seen with the naked eye. Look at Zeta (ζ) Ursae Majoris (Mizar), the second star from the tip of the Bear's tail. Do you see its 4th-magnitude companion immediately to the northeast? If so, you've seen the lightweight Rider (Alcor) of the ancient Arabian Horse (Mizar). You've also seen your first optical *double star* and perhaps your first physical *binary star* as well.

An optical *double star* is a pair of stars that, to the eye, appear very close together on the celestial sphere. Some double stars will look single to the unaided eye but appear as two close objects through binoculars. Although the two stars look like physical companions, they're actually many light years apart and just happen to appear in nearly the same line of sight; imagine two ships a mile apart passing one another on a distant horizon.

Now imagine that the closer the ships appear to get, the more difficult they become to see as two objects. But if you were to look at the ships with binoculars, they would once again be separated by a respectable distance. In time, they would once again blend. But if you looked at them through a telescope, they would split once again and so on. And so it is with the night sky and the chance alignment of stars. Some pairs of stars are easily seen with the unaided eye, others require either binoculars or a telescope to detect.

If two close stars are physically related, meaning that they lie at the same distance and orbit one another around a common center of gravity, then we call it a *binary star*

system. Since Alcor and Mizar are part of the Ursa Major Moving Group, they are, in fact, relatively close, being separated by only three light years. Although it's uncertain as to whether the two stars can be physically attracted at that distance, it is possible that they may be weakly associated, taking at least 750,000 years to orbit one another!

If you have a telescope, you'll see that Mizar splits again, having another faint, companion nearby. In fact, astronomers with powerful telescopes have found that each of these two components is again double! So Mizar is not a double star but a quartet of stars, a double-double. It becomes a *quintuple* star if you include Alcor.

So far, we've only encountered angular sky measures on the order of degrees. But they also divide into smaller units called arc minutes (′) and arc seconds (″): 1° is $\frac{1}{360}$ of a circle; 1 arc minute (1′) is $\frac{1}{60}$ of a degree, and 1 arc second (1″) is $\frac{1}{60}$ of an arc minute. Your pinky held at arm's length covers about 1° of sky; half a pinky, then is equal to 30 arc minutes (30′). The full Moon measures 30′ across, as does the Sun. The unaided eye can resolve objects that are separated by about 4′ and greater, while a pair of 10 × 50 binoculars can split double stars on the range of 34 arc seconds (34″) and greater, depending on your expertise or whether you use a tripod.

Look at Alcor and Mizar with your unaided eyes. These stars are separated by a respectable 14 arc minutes (14′), which is about half the apparent diameter of the Moon or Sun. Now use your binoculars to look for an 8th-magnitude star almost midway between them and a bit south. This is *Sidus Ludoviciana*. While surveying the heavens with a small telescope on December 2, 1722, J. G. Liebknecht, a German theologian and mathematician, chanced upon this star. After making some crude and inaccurate measurements of the star's position, he convinced himself that the object had moved. He then wrote and distributed a pamphlet announcing his discovery of a new planet, which he named after his sovereign, the Landgrave Ludwig of Hessen-Darmstadt. Liebknecht's contemporaries were not convinced, and they quickly proved that *Sidus Ludoviciana* is a fixed star. Furthermore, the noted German philosopher L. P. Thule distributed his own pamphlet criticizing Liebknecht's claim, adding that it was hardly necessary for Liebknecht to announce every telescopic star as new and to give it a special name.

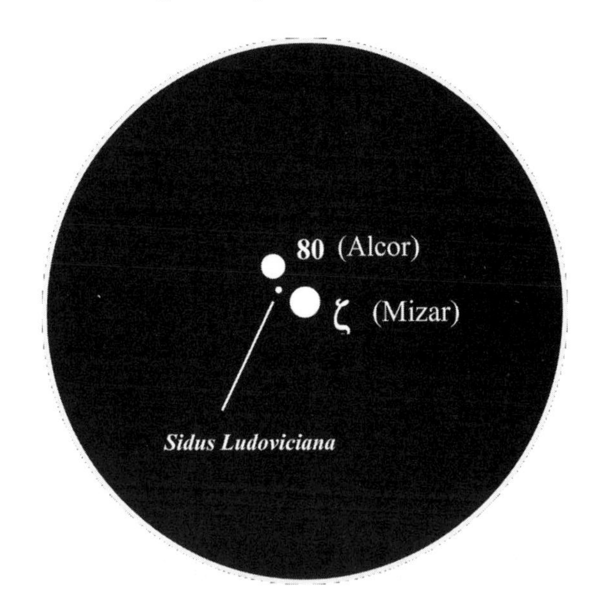

Averted vision

If you cannot see *Sidus Ludoviciana* using direct vision, try looking out of the corner of your eye. The star should snap clearly into view. This observing technique, called averted, or peripheral, vision, is used by all visual astronomers. Averted vision places the object we want to study on the eye's night-sensitive (low-light) rod cells, which line the outer surface of the retina (the layer of cells at the back of the eye). When you use direct vision, the star's light falls on the retina's fovea (central portion), which is lined with day-sensitive (bright-light) cone cells, so it does not appear as prominent.

The more you practice using averted vision, the easier it becomes to see fainter and fainter objects. If you still can't see the star, don't worry. Try again on another night. Your eyes might be fatigued, or the sky conditions might not be optimal. Note, however, that sometimes when objects are bright enough, and close enough, *direct* vision will help you to resolve them best. Be sure to record the details of each observation in your observing log. Throughout the book, I will remind you when to use averted vision, especially if the object being discussed is faint.

Your retina may also have a specific spot that's highly sensitive to dim light. You can find that spot by trial and error. Once that visual "hot spot" is located, you will know how to position your head to best see dim objects. I find, for instance, that if I direct my gaze at the 4:00 position angle from the object, I have the best chance of seeing it clearly. Note, however, that the eye has a blind

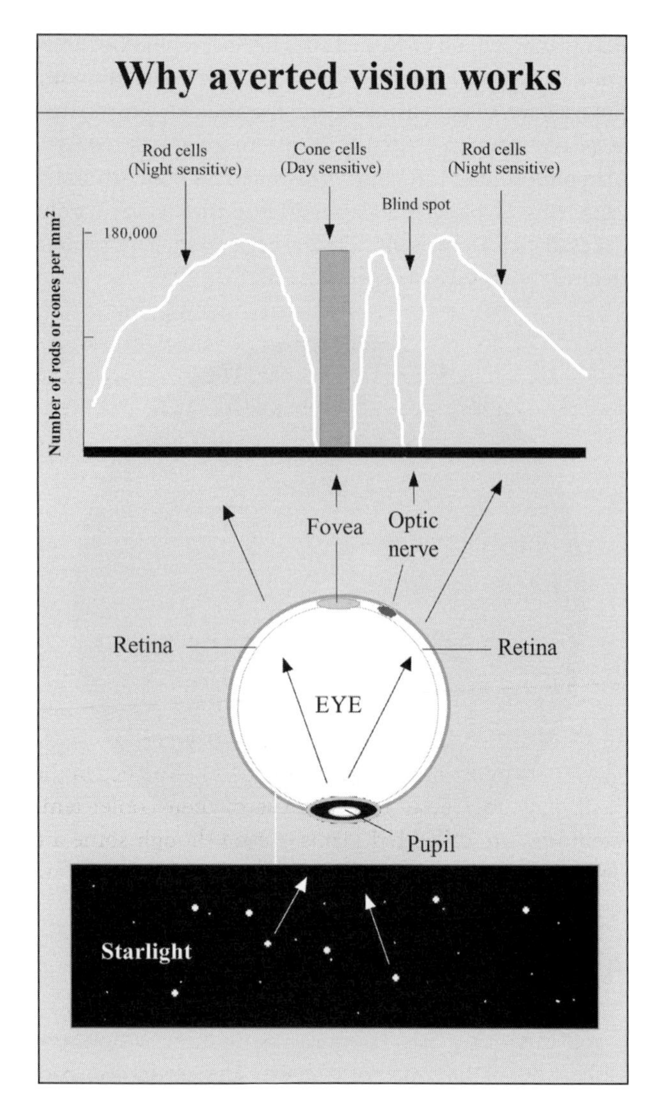

spot at the position of the optic nerve; if starlight hits that spot, it will disappear. It takes practice to know just how far away from the object you have to look in the direction of your visual "hot spot" to get the maximum benefit. Observing, like any sport, requires practice if you want to be good at the game.

Star color

One of the many enjoyable aspects of binocular observing is looking for color differences among the stars. For this you want to use direct vision, so that the star's light falls directly onto the eye's color-sensitive, cone cells. While most stars look white at a glance, a careful survey of the sky will show that some have a bluish tint (like glacial ice), others look more yellow (like our Sun); still others can appear orange or red. The color differences you see might be negligible or subtle at first, but they do become more apparent with experience. It's a skill (or perception) that grows with time.

Start by looking at Eta (η) Ursae Majoris (Alkaid), the star at the end of the Big Dipper's handle. Now move star by star along the Big Dipper until you reach Alpha (α) Ursae Majoris (Dubhe). Two things should be immediately apparent: All the stars in the Big Dipper have a whitish, or bluish-white hue, except for Dubhe, which has a distinct golden patina. Actually Dubhe is a double treat. Not only is the star colorful, but it has a striking (but physically unrelated), 7th-magnitude binocular companion about 5 arc minutes (5′) away, roughly in the direction of Beta (β) Ursae Majoris.

A star's color depends on the temperature of its surface gases. Think of molten metal when it cools: the color gradually shifts from blue-white, to yellow, to orange, to red, before it turns completely black. Each color in this *spectrum* (rainbow) corresponds to a specific *wavelength* of light, which emits a certain amount of heat energy. Like the hottest metals, the hottest stars shine with a bluish-white hue, while the coolest stars have a reddish tint. (Thus the saying "Red-giant stars aren't so hot.")

Ironically, in the visual arts, warmer colors have the coolest thermal properties, and vice versa, so don't be confused. The visual warmth of color refers to the appearance of an object as if its surface were being warmed by the rich golden shades of sunset, or by the orange glow of a distant fire; cool colors on the other hand refer to those that remind one of snow or ice. So, while Dubhe has a warmer visual hue than Alkaid, Dubhe's physical temperature is actually much lower, although still hot by Earth standards.

Using the color spectrum as a guide, astronomers have created a classification scheme for stars based largely on their surface temperatures. The scheme uses seven letters to represent the seven main spectral types. Ranging from hot to cold, they are *O, B, A, F, G, K, M*. (One favorite mnemonic for memorizing the spectral types of stars is *Oh, Be A Fine Guy/Gal, Kiss Me.*) For precision, astronomers have further subdivided the spectral types into 10 subclasses (which range from 0 to 9). If you know a star's spectral type, then you'll know its temperature range. A star's temperature is measured in *Kelvin* (K), which begins at absolute zero (−273 degrees Celsius, or −459 degrees Fahrenheit). The chart below shows the

The classification of stars by spectral type

Spectral type	Color range	Temperature (K)	Examples in Ursa Major (UMa)
O	Blue	31,500–50,000	
B	Blue-white	10,000–31,500	Eta (η) UMa
A	White	7,500–10,000	Beta (β), Gamma (γ), Delta (δ), Epsilon (ε), Zeta (ζ) UMa
F	Yellow-white	6,000–7,500	Theta (θ) UMa
G	Yellow	5,300–6,000	Xi (ξ) UMa
K	Orange	3,800–5,300	Alpha (α) UMa
M	Orange-red	2,100–3,800	Iota (ι) UMa

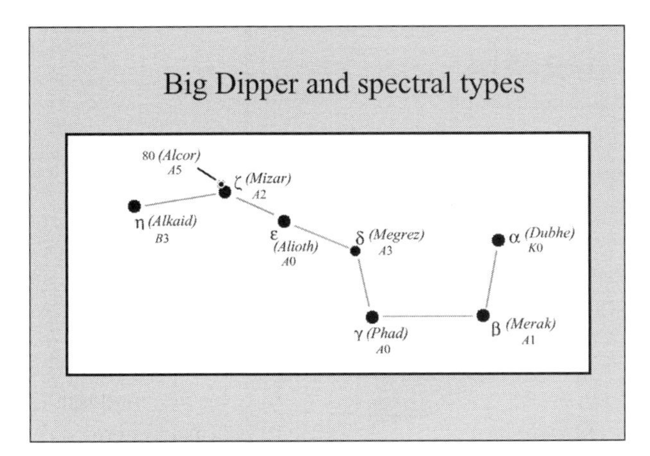

main spectral types, their color, temperature range, and some examples of stars in these classes in Ursa Major. Our Sun, for comparison is a type G2 star, and the North Star is type F7, so both shine with a distinct yellowish hue.

The colors we see depend on the sensitivity of our eyes to certain wavelengths of light and atmospheric conditions, so any description is highly subjective. Some astronomers claim that the cells in our eyes cannot detect color in stars. True or not, do not let the science of color distract you from the pleasure of seeing – real or imaginary – color. One of the most popular pastimes in summer observing events, for instance, is to have different observers look at the colorful double star Albireo in Cygnus, the Swan, then share what colors they see. Few people ever agree, and the shades they describe can be illuminating and funny. In fact, you will notice throughout this book, that I am quite literally "colorful" in my descriptions of star color, saying a star is "sunflower yellow" or "hot-sauce orange." You won't necessarily see the colors I describe, but you should record whatever color *you* see. To record the colors of stars, I match what I see to the named colors on ColorPlace® paint swatches from Wal*Mart®.

Dubhe and the life of a star

More than a star's surface temperature, a star's color also tells us a little something about its size and age. All stars are born in vast clouds of dust and gas (a *nebula*[3]) under the agency of gravity. As matter condenses, it forms a body that fuses hydrogen atoms into helium at the core. When it does, a star is born, throwing off its signature heat and light. Of course, how brightly a star appears to us on Earth also depends on its distance. A flashlight beam looks brighter when it is closer, for instance. So why does Dubhe, which is 1.5 times farther away from the Ursa Major Moving Group, shine more brightly than the stars in the group?

In the early twentieth century, two astronomers (Ejnar Hertzsprung and Henry Norris Russell) created what's now known as the Hertzsprung–Russell (H–R) diagram – a clever way to make sense of our two-dimensional sky. They found that if you plot a star's temperature against its absolute magnitude (the apparent magnitude of a star if it were magically placed at a distance of 10 parsecs [32.6 light years]), 90 percent of all the visible stars line up in a curved diagonal line. They called this line the "main sequence," and it's a family portrait, of sorts.

Look at the main sequence on the H–R diagram on page 10. Notice how that for every increase in brightness, the star's size and temperature also rise. The lower-right end of the main sequence is dominated by cool, red-dwarf stars, which are only about 1/10th as big as the Sun; except for a few exceptions, these stars are too faint to be seen in simple binoculars. At the upper-left end of the main sequence, we find the contrary – hot, bright stars some 10 to 20 times larger than our Sun; most of the stars in the Big Dipper congregate here. The Sun, our closest star, lies in the middle of the main sequence.

Despite the variety of star sizes on the main sequence, astronomers consider them all (including our Sun) to be dwarfs. The reason becomes clear if you look at the giant stars to the upper right of the main sequence in the H–R diagram. These stars, because of their cooler temperatures, are called red giants – even though some are yellow giants and some are orange giants. Red giants are typically about 25 times the size of the Sun and hundreds of times more luminous. With a surface temperature of 4,500 K, Dubhe is a type *K* (orange) giant 30 times larger than the Sun and 300 times more luminous. That's why, even though it's farther away than the Ursa Major Moving Group, Dubhe shines more brightly than any star in it.

To see a red giant star is to see the ultimate fate of our Sun. After living its life on the main sequence for 10 billion years, a star like our Sun will evolve off the main sequence to become a red giant star. (Don't worry, our Sun is only about halfway through its life cycle, so it has a healthy 5 billion years left to burn.) When a star reaches the red-giant phase, it has exhausted, or nearly exhausted, its nuclear fuels and is fusing hydrogen into helium in a layer of gas outside the core. This new energy burn causes the star to swell by a factor of 10 or more; when the Sun reaches this stage, its outer atmosphere will envelop the Earth. Dubhe is now in this core, helium-burning stage, signaling to us that its death is near.

Eventually, a red giant's core becomes so hot that it fuses helium into carbon and oxygen, causing the star to swell even more. In many cases, strong stellar winds push the star's outer shells out beyond the star's gravitational hold. As more and more shells expand outward, the star appears to vary in brightness as viewed from Earth. Red-giant stars in this stage are called Mira stars,[4] or long-period-variables, with an amplitude range of 2.5 to 11 magnitudes over a period of 80 to 1,000 days.

[3] Any bright nebulae in Ursa Major have all but faded from view, but we will encounter many fine examples in the rich summer Milky Way.

[4] Mira stars are named for the prototype star in the constellation of Cetus known as Mira.

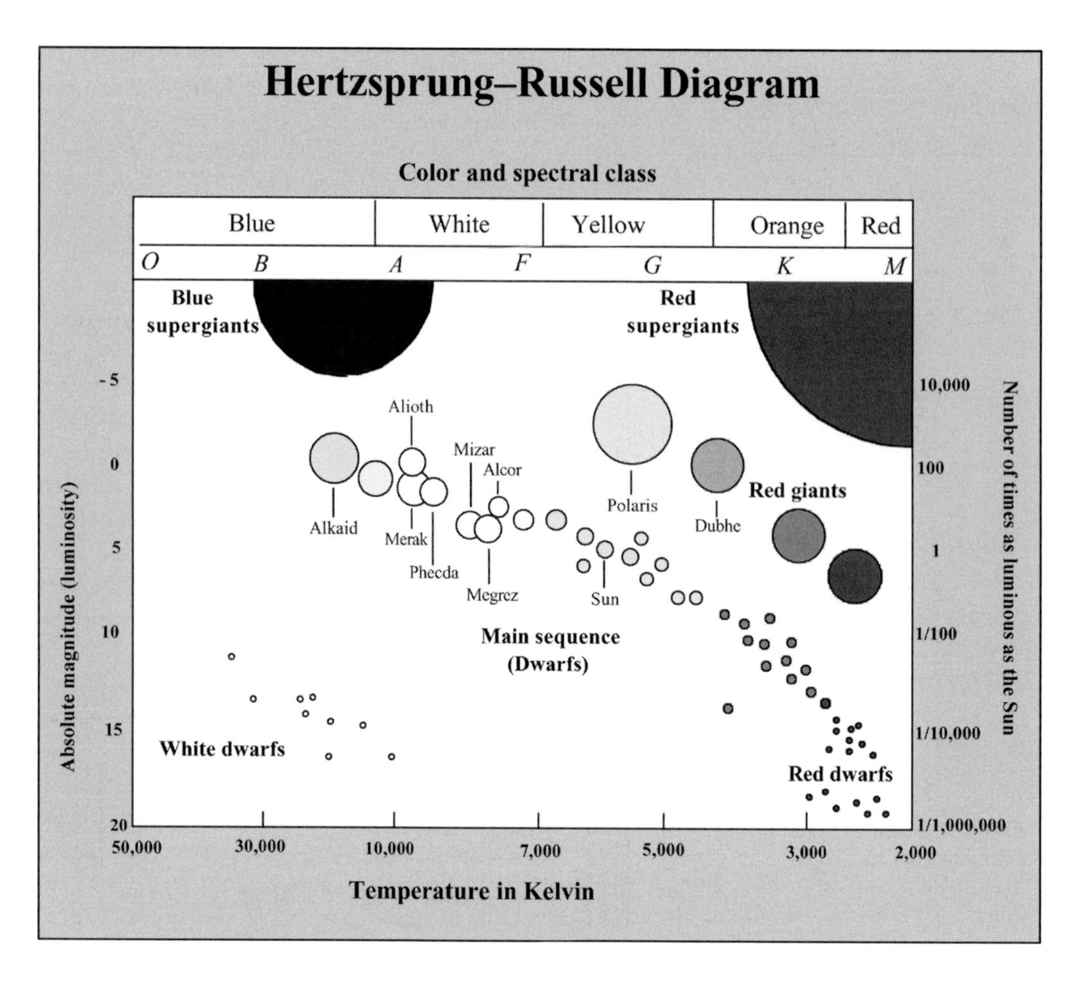
Hertzsprung–Russell Diagram
Color and spectral class
Blue
White
Yellow
Orange
Red
O
B
A
F
G
K
M
Blue supergiants
Red supergiants
Alioth
Mizar
Alcor
Alkaid
Merak
Phecda
Megrez
Polaris
Sun
Dubhe
Red giants
Main sequence (Dwarfs)
White dwarfs
Red dwarfs
Absolute magnitude (luminosity)
-5
0
5
10
15
20
10,000
100
1
1/100
1/10,000
1/1,000,000
Number of times as luminous as the Sun
50,000
30,000
10,000
7,000
5,000
3,000
2,000
Temperature in Kelvin

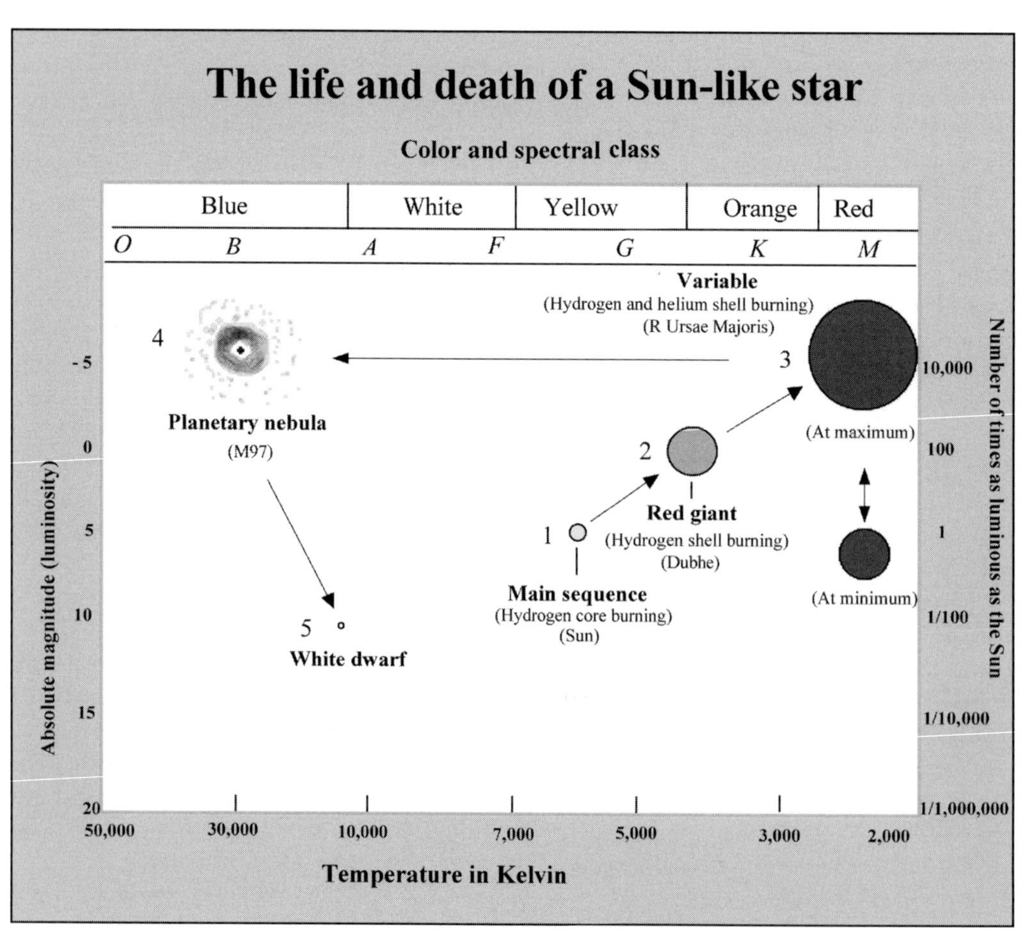
The life and death of a Sun-like star
Color and spectral class
Blue
White
Yellow
Orange
Red
O
B
A
F
G
K
M
Variable
(Hydrogen and helium shell burning)
(R Ursae Majoris)
4
3
Planetary nebula
(M97)
2
(At maximum)
Red giant
(Hydrogen shell burning)
(Dubhe)
1
Main sequence
(Hydrogen core burning)
(Sun)
(At minimum)
5
White dwarf
Absolute magnitude (luminosity)
-5
0
5
10
15
20
10,000
100
1
1/100
1/10,000
1/1,000,000
Number of times as luminous as the Sun
50,000
30,000
10,000
7,000
5,000
3,000
2,000
Temperature in Kelvin

Although a long-period variable is still called a red-giant star (because of its large size and color), it has moved to the left of the Red Giant Branch of the H–R diagram. It will remain in this region, which is known as the Asymptotic (parallel) Giant Branch, for the final 200,000 to 1 million years of its life.

Your first Mira variable star

Ursa Major has several long-period variables within its borders, but one Mira-type star is not only within binocular range, but also only about one binocular field northwest of Dubhe. It's called R Ursae Majoris[5] and its light varies between magnitude 6.6 (easy to see in binoculars) to 14.3 (difficult to see in a telescope) every 301.62 days.

Now it's time to test your hunting skills. Use the chart at right to find the field of R Ursae Majoris. Remember, the star varies in brightness, so it may not be visible on the first day you look. If the star is not visible, you should continue to check the field every few weeks until you catch sight of it. Seeing a variable star swell into view after weeks, or even months, of invisibility, can be a joyous occasion; it's like spotting your first breaching whale after its long winter migration. You can start to record the star's ever-changing brightness, week by week. Of course, if the star is visible, you can start to record its magnitude right away.

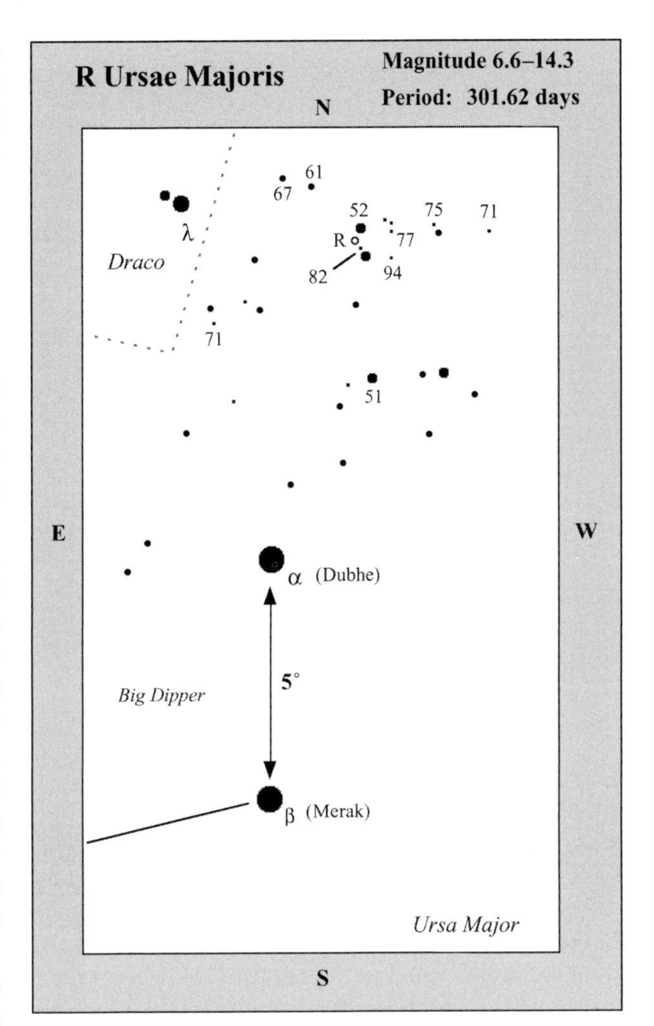

Use the chart at right, which has been adapted from one by the American Association of Variable Star Observers (www.aavso.org) to estimate the variable's brightness. To make a magnitude estimate, compare the light of the variable star to that of two or more nearby stars that do not vary in brightness. The magnitudes of these comparison stars are given with points omitted, so they won't be mistaken for stars; thus, a star of magnitude 6.7 appears as 67. You may need to interpolate. For instance, if you estimate the brightness of R Ursae Majoris is halfway between the 6.7-magnitude comparison star and the 7.5-magnitude comparison star, then your estimate is magnitude 7.1. You can further refine your estimate by comparing it with the 7.1-magnitude comparison star. When you have finished, record the date and time of the observation as well as the magnitude. For consistency, it is always best to look at the star and the comparison star so that they are parallel to the long axis of your eye; this may involve turning your head as you look through your binoculars.

If you continue this project for weeks, you can plot the star's brightness over time. Of course, making a complete light curve of this particular variable star is impossible in binoculars, but you will encounter other variables in other constellations, that can be followed through an entire cycle of variability.

The final phase

When all the helium in a red giant's core has fused into carbon, the solar furnace burns no more. In just a few thousand years (the blink of an astronomical eye), the star's core will all but implode under the force of gravity, shrinking amazingly until it is about the size of the Earth. This remnant core, called a white-dwarf star, will be a whopping 1 million degrees hot and so dense that a teaspoon of its matter will weigh a ton!

During this final core collapse, the star's outer layers also fall inward, but they ignite before they reach the core, and the star "goes out" with a bang – ejecting 40 percent of its mass into space as layers of expanding luminescent gas called a planetary nebula. The name has nothing to do with planets, except that the great eighteenth-century observer, Sir William Herschel, coined the term because, through his telescope, these objects appeared round in form and resembled the green, gas-giant planet Uranus, which he discovered. University of Washington astronomer Bruce Balick says that he likes to think of

[5] In the early nineteenth century few variable stars were known, so it seemed reasonable to use the letters of the Roman alphabet, starting from the letter R and ending with Z. When more variables were discovered they continue with RR . . . RZ, then use SS . . . SZ, TT . . . TZ and so on until ZZ.

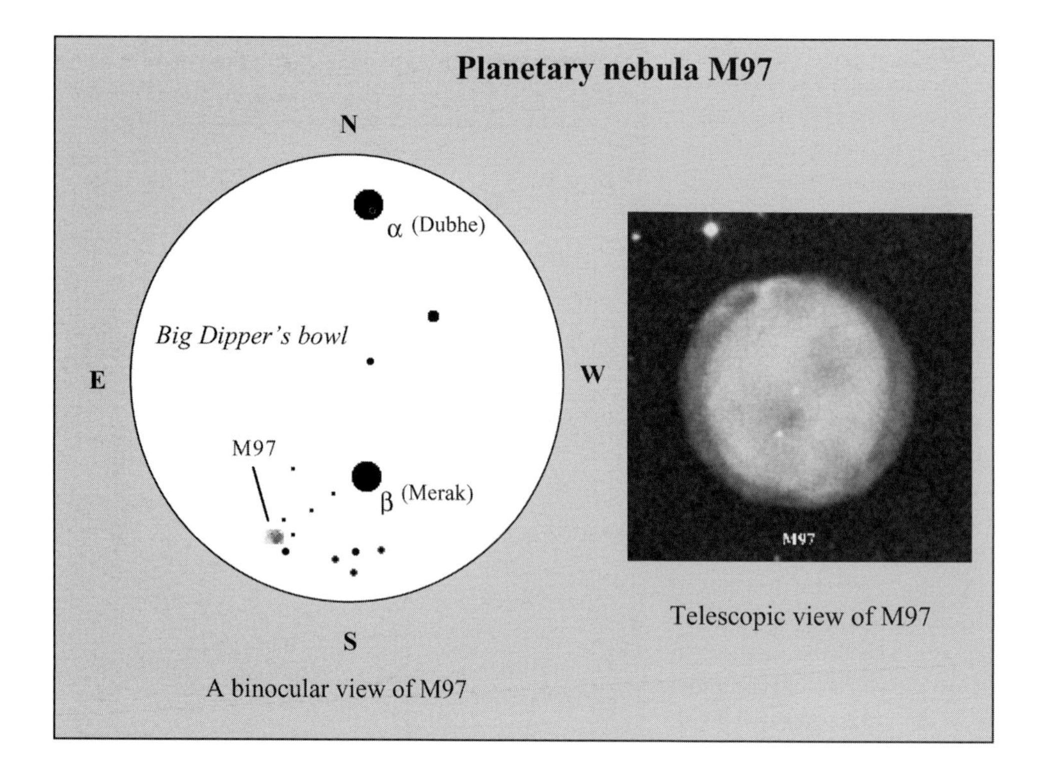

a planetary nebula as a cloud of smoke which escapes from a burning log as it collapses and crumbles into embers. Amazingly, becoming a planetary nebula is the fate reserved for 95 percent of all stars in our galaxy. "If you lived for 100 million years instead of just a century," Balick says, "then planetary nebulae would be the twinkling ornaments of the Milky Way."

If you want truly to test your vision, a famous planetary nebula, called the Owl Nebula, lies just outside the bowl of the Big Dipper. To see this roughly 10th-magnitude object you will need to be under a very dark sky and have at least 10 × 50 binoculars to see it "well." I place well in quotes because you will also need to use averted vision to see it. When it comes to observing deep-sky objects like nebulae, "magnitude" is deceiving. In most cases, deep-sky objects do not appear as point sources. Instead, their light is spread across a specific area of sky. A 10th-magnitude nebula, then, will appear dimmer than a 10th-magnitude star, because the light is no longer concentrated but diffused over a greater area of sky. Imagine how the concentration of light differs when you use the different settings of a flashlight with an adjustable beam. The wider the beam, the less intense the beam appears. This dimming effect is intensified under less-than-perfect sky conditions.

Use the chart above to locate the Owl Nebula's field, which is about 2° southeast of Beta (β) Ursae Majoris (Merak) – so the nebula and that bright star are in the same binocular field. Look for a soft, round glow about 3 arc minutes (3′) in diameter. Remember, use averted vision. If you can, be sure to view this planetary nebula through a telescope, which may show the Owl's two dark eyes.

Other types of deep-sky objects

The Owl Nebula is also known as M97 – the 97th object cataloged by the French astronomer Charles Messier (1730–1817), who created the first extensive catalog of nebulae and star clusters discovered mostly by him and his contemporary Pierre Mechain. M objects, or Messier objects, are the most popular deep-sky objects visible from mid-northern latitudes; many can be seen in binoculars, and you will be introduced to the brightest and best as we tour the binocular sky. The list includes other planetary nebulae, diffuse nebulae, open star clusters, globular star clusters, and galaxies.[6]

Before we describe these other objects, let's first look at our Milky Way. Except for a few galaxies that can be seen with the unaided eye, everything we see in the night sky belongs to our galaxy, the Milky Way galaxy, of which the Sun and its family of planets are a part. One common question is how can we be in the Milky Way galaxy yet see it as a misty river "out there" – as if we were in a jet looking down at the Mississippi River. The answer is that the milky path we see in the night sky is only a part of our galaxy. Once astronomers discovered that relationship, they decided to name our galaxy after that beautiful luminous pathway in the sky.

Because we are in the Milky Way, it is hard to know exactly what our galaxy looks like. It's like trying to imagine the shape of a forest from a spot deep within. But

[6] *Deep-Sky Companions: The Messier Objects*, Stephen James O'Meara (Cambridge, Cambridge University Press; Cambridge, MA, Sky Publishing, 2000).

What our Milky Way galaxy might look like seen face on (left) and edge on (right).

years of study have led us to believe that ours is a spiral galaxy about 100,000 light years across and about 2,500 light years thick. It contains some 200 to 400 billion stars, which are distributed across the galaxy's wheel-shaped disk and central, football-shaped hub (or bulge). At the heart of the bulge, which measures 13,000 light years thick, is a dense nucleus that spans only 10 light years across. The wheel-shaped galaxy is surrounded by a spherical halo (about 130,000 light years across) of older stars and interstellar matter that contains some 200 or so globular star clusters (described on page 14) which are strongly concentrated toward the galaxy's nucleus. Finally, these luminous parts of our galaxy are surrounded by a mysterious dark halo, comprised of dark matter, whose form is unknown. The photo illustrations at upper right show what we believe our galaxy would look like if we could view it from above and from the side.

But how do we explain the appearance of the Milky Way in the night sky? In a sense, it is a visual illusion. The Milky Way forms a complete band around the night sky. It gives the illusion that we are in the center of a wheel (or the mental illusion that we are at the center of the universe). But we are not. Our Sun is roughly 30,000 light years from the galactic center (and about 14 light years above the central plane of the disk) in a spiral arm known as the Orion Arm. The Orion Arm is sandwiched between two larger arms: The outer Perseus Arm and the inner Sagittarius Arm. This is why, when we look at the Milky Way in the direction of Sagittarius, we see it suddenly blossom in width, because we are looking in the direction of the galactic bulge. We do not see the Milky Way slicing through Ursa Major, because when we look at it, we are not looking into the plane of the galaxy but high above it.

The direction we look above the galaxy's plane to see the North Star and the Big Dipper

The two photo illustrations at left on the next page show our home in the galaxy and how we see the Milky Way. Now let's look at a few deep-sky objects within our galaxy. A *bright nebula* is a luminous, interstellar cloud of dust and

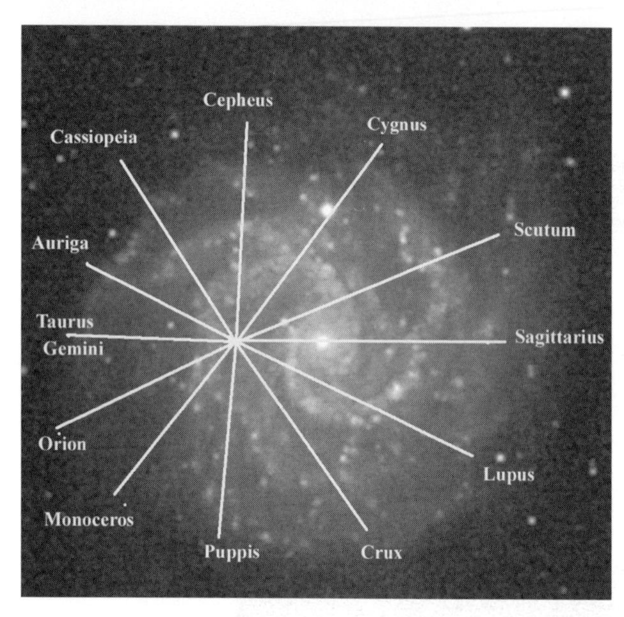

gas that either gives off its own light (emission) or shines by reflecting the light of nearby stars (reflection). There are no bright, diffuse nebulae in Ursa Major, but we will see many in the summer night sky.

Open star clusters are loose and irregularly shaped collections of dozens or hundreds of young stars that travel mostly in the thin disk of stars, dust, and gas that comprises the galaxy's plane. They occupy a volume of space typically less than 50 light years across, are loosely held together by gravity, and are fated to disperse over a period of several hundred million years. The Ursa Major Moving Group is an open cluster that lies high outside the galaxy's plane; it looks so big and loose because it is the closest star cluster to our Sun. We will encounter many smaller, dimmer, and more spectacular examples of open star clusters throughout this book.

Globular star clusters are orbs of ancient starlight, some 10 to 14 billion years old, that reside in the Milky Way's halo tens of thousands of light years distant. Many lie far above or below the plane of our galaxy (in its halo). All contain from tens of thousands to millions of stars, and measure from 100 to 300 light years across. There are no globular star clusters in Ursa Major but we will find several fine examples of them, especially in the summer night sky.

Our final deep-sky category, *galaxies*, takes us beyond our Milky Way system and into the realm of intergalactic space. Galaxies, also called island universes, are vast citadels of stars, gas, and dust, into which most of the visible matter of the universe is concentrated. To see them, we must look through the stars in our Milky Way, then across millions of light years of empty space. They are literally starships sailing the vast ocean of space. Galaxies range in size from small dwarfs measuring only a few hundred light years across with just a few million suns, through normal galaxies like our own Milky

Way with a few hundred billion stars, to giant ellipticals spanning over hundreds of thousands of light years and containing several trillion stars. There are three basic types: spirals (normal, barred, or mixed), ellipticals, and irregular. Lenticular galaxies are those midway in form between a spiral and an elliptical. We see the various types of galaxies in a variety of orientations – from face on to edge on, to, in the case of some ellipticals, end on.

Ursa Major has some fine examples of galaxies seen at various angles. The brightest, Messier 81 (M81) is a bit of a challenge to find, but well worth the effort. Use the chart on page 16 to find Alpha (α) Ursae Majoris. Now look about 10° (one fist width) northwest for 4.5-magnitude 24 Ursae Majoris, which marks one of the Bear's ears. You can confirm 24 Ursae Majoris in your binoculars by looking about $4\frac{1}{2}$ southwest for a beautiful, 1°-wide acute triangle of three roughly 5th-magnitude suns: Rho (ρ), Omicron^2 (o^2), and Omicron^1 (o^1) Ursae Majoris. Center 24 Ursae Majoris in your binoculars, then look 2° east–southeast for the diffuse oval glow of M81. The galaxy shines at 7th magnitude but its light is spread over an area that measures $27' \times 14'$. Still, skilled observers, like Brent Archinal of Flagstaff, Arizona, have detected it with the unaided eye under very dark sky conditions.

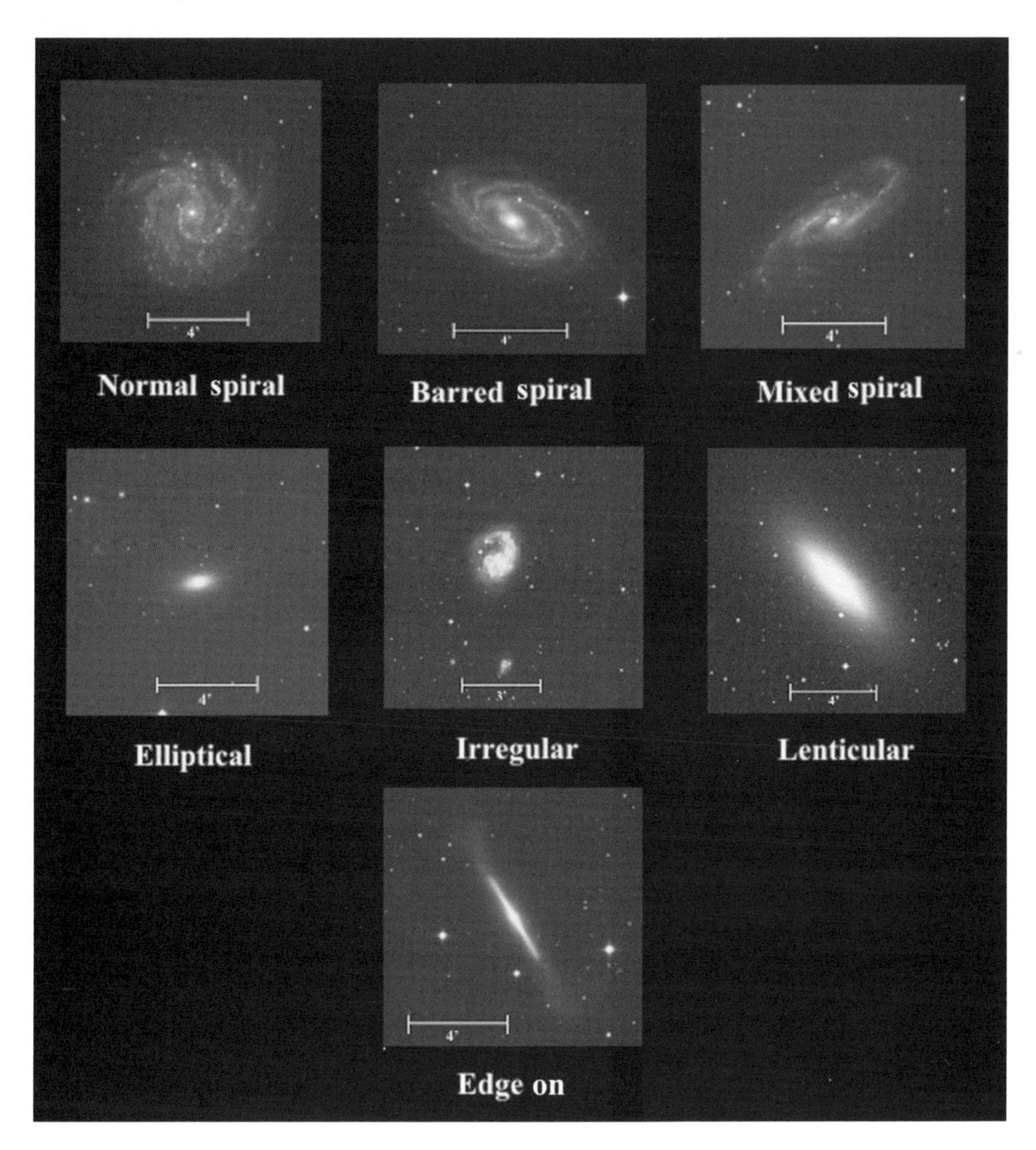

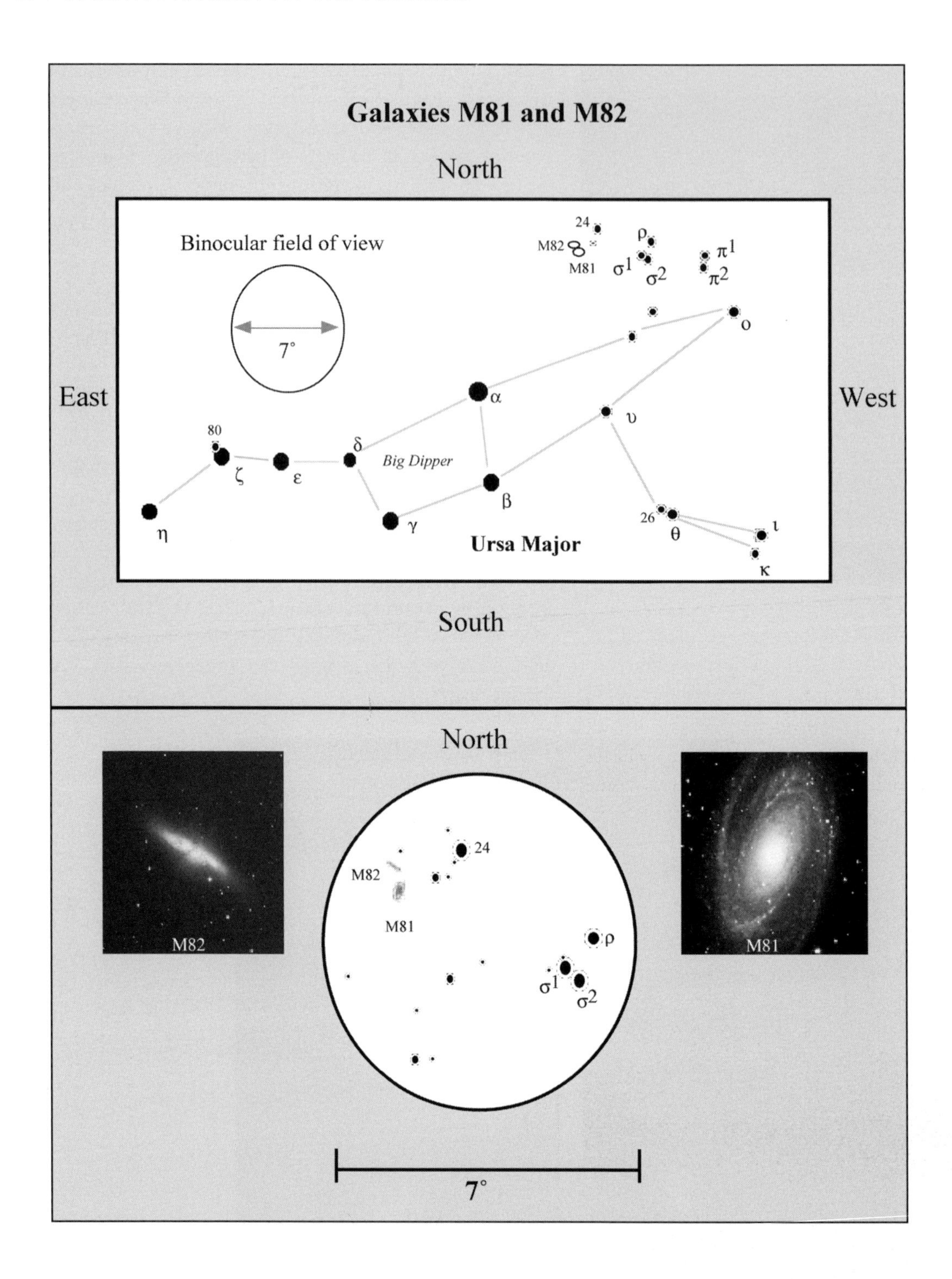

With your binoculars, you'll just be seeing the galaxy's disk of unresolved starlight, so don't be fooled by the photograph shown here. To the eye, the galaxy will look like a small ghostly glow that gradually gets brighter toward the center. But you are glimpsing starlight that has traveled a distance of 12 million light years to reach your eye. The light we see from that distant object left its host galaxy during the Miocene epoch, a period of time when 100 or so species of Old World apes began roaming the Earth. Messier 81 is among the nearest and brightest spiral galaxies, which we see inclined about 40° to the plane of the sky.

Stars aren't the only celestial bodies that interact, or come in pairs. Messier 81 is gravitationally interacting with Messier 82 (M82) – a classic irregular galaxy 150,000 light years away. In telescopes, M81 and M82 look like two dancers on a stage, both of which deserve applause and attention. But in binoculars, M82 is a much more difficult object to see; it shines at magnitude 8.4 and measures 11′ × 4′ in size, so it is a very faint phantom of light that's long and thin like a cigar. Can you see it?

To see the beautiful face-on spiral galaxy, Messier 101 (M101), first locate Zeta (ζ) Ursae Majoris (Mizar) in the Big Dipper's handle. Now use your binoculars and the chart on page 17 to follow the $3\frac{1}{2}$°-long path of four 5th-magnitude stars (81, 83, 84, and 86 Ursae Majoris). These stars are set up like stellar flagstones, leading us to our 8th-magnitude target. Messier 101 lies only $1\frac{1}{2}$°

northeast of 86 Ursae Majoris. It covers an area of sky nearly 30′ across – the same apparent size as the full Moon – and is a very nice round glow.

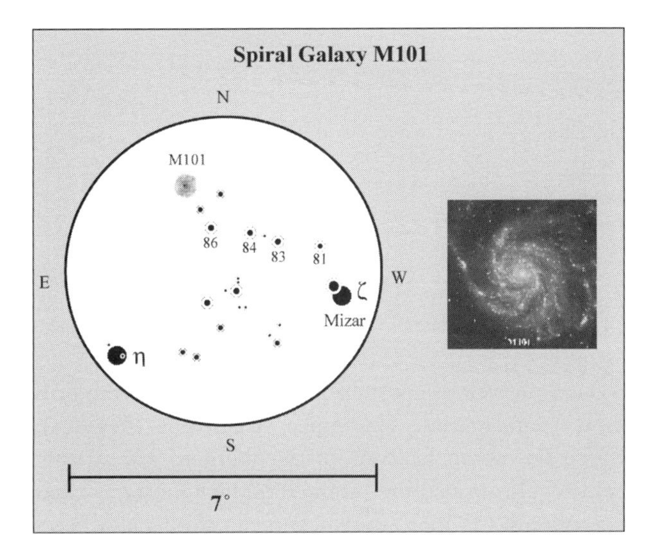

Messier 81 and M82 are part of a small cluster of galaxies some 12 million miles from the Local Group of Galaxies, which includes our Milky Way. Messier 101 is part of another group of galaxies 17.5 million light years distant. The Local Group lies at the edge of the Virgo Supercluster of galaxies, which includes a vast system of galaxies (some 2,500 members) in Virgo and Coma Berenices some 50 million light years distant. While the Virgo Supercluster totally dominates our corner of the universe, the universe is replete with other galaxy clusters, some of which have bright members visible in binoculars.

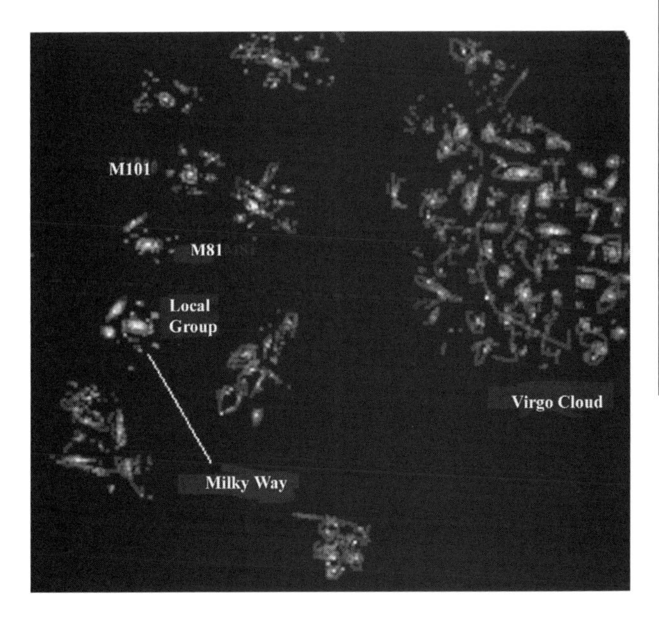

Extrasolar planets

We end our brief tour of the universe with an exercise for the imagination. Use the star chart below to find 47 Ursae Majoris near the rear foot of the Bear. Then look at it with your binoculars. This unassuming 5th-magnitude stars lies about 46 light years from the Sun, so if you are 46 years of age, you're looking at light burning from that star in the year you were born. But that's not what's interesting. In 1996, astronomers discovered a Jupiter-like planet orbiting this Sun-like star with indications of an even larger planet in an outer orbit. The Jupiter-like planet orbits the star every three years and lies at a distance that places it in an orbit that, were it our Sun, would be between the orbits of Mars and Jupiter in the main asteroid belt. Although you cannot see this planet with your binoculars, perhaps you can with your mind's eye.

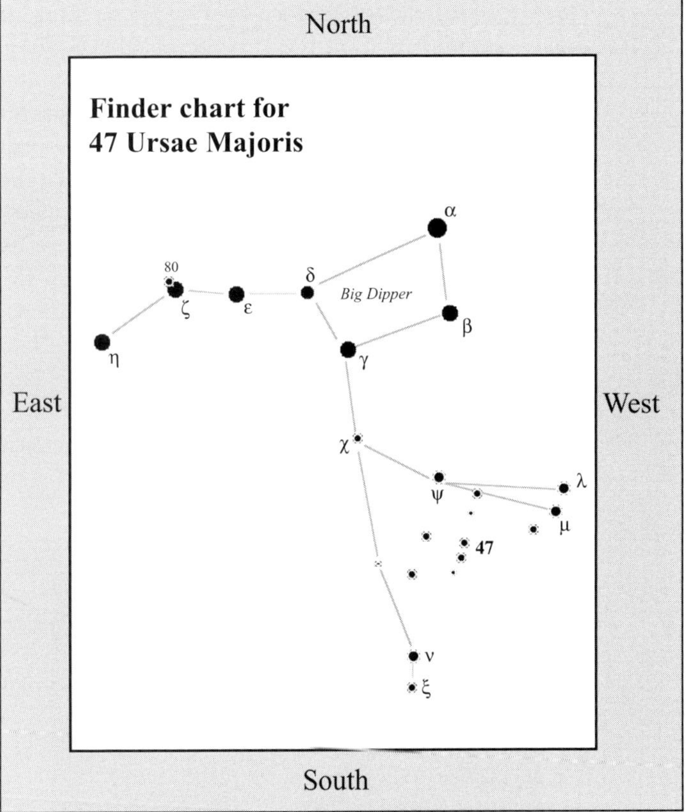

CHAPTER 2

The spring stars

1 April

Everywhere about us they are glowing,
Some like stars, to tell us Spring is born;
Henry Wadsworth Longfellow, *Flowers* (1839)

Flowers, Longfellow's "tender wishes, blossoming at night," accompany us in our first month of stargazing. Spring has arrived. The long winter nights are over. And though the air is crisp, it's also filled with the secret scent of budding flowers and fresh earth. (Actually, the chill of the night is a friend to stargazers, who know that the words "clear" and "cold" go hand in glove.) Under foot, the soil feels both fragile and firm. Overhead, the stars scintillate with a certain purity. That's because, for the most part, when we gaze upon the spring constellations, we are looking high above the plane of the Milky Way, through a thin veneer of stars. In the north, the Big Dipper is high overhead. But we need to turn our attention to the south, where Leo, the Lion, is prowling the Serengeti of the high heavens.

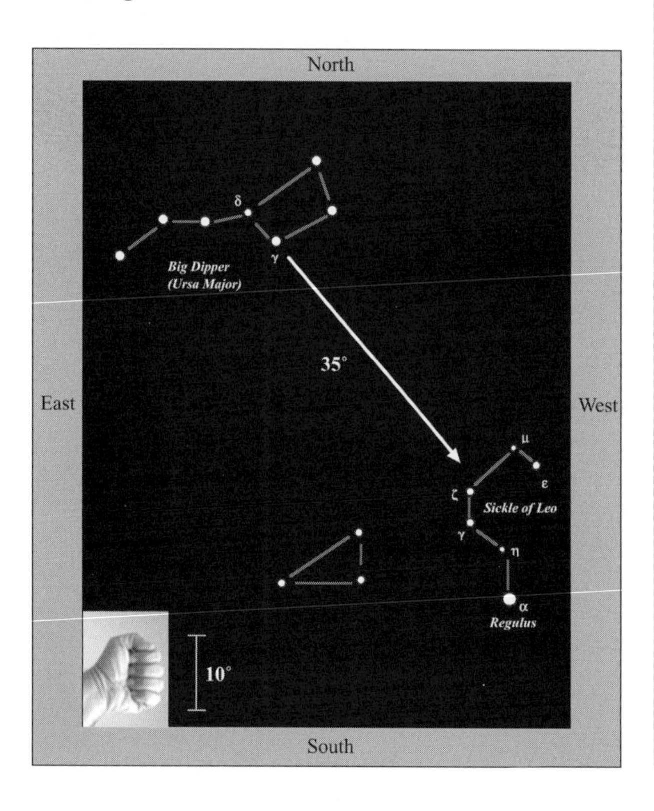

Leo, the Lion

Leo is a marvelous constellation, punctuated by the brilliant, 1st-magnitude star, Alpha (α) Leonis (Regulus), which lies about halfway up the southern sky at 9:00 pm. Regulus is also the brightest and southernmost star in a roughly 12°-long asterism known as the Sickle of Leo, which also looks like a backward question mark (see the chart at lower left). The asterism – which is comprised of Alpha, Eta (η), Gamma (γ), Zeta (ζ), Mu (μ), and Epsilon (ε) Leonis – should frame your fist held at arm's length. Confirm the Sickle by imagining a line drawn from Delta (δ) Ursae Majoris, through Gamma (γ) Ursae Majoris, and extending it three and a half fists to the southwest, where it should intersect the Sickle, whose blade faces to the west.

Important Note: If you cannot make sense of these stars, consider that a planet might be interfering with the view. Leo is one of the sky's 12 *zodiacal constellations*. So every now and again, the Moon or a bright planet will sail past its paws like a playful mouse. The photograph below, for instance, shows the Sickle of Leo rising in the east with the planet Saturn.

The zodiac is a band around the sky about 18° wide, centered on the ecliptic, in which the Sun,

Moon, and planets move. The band is divided into 12 signs of the zodiac, each 30° long, that were named by the ancient Greeks after the constellations that used to occupy these positions; "zodiac" means "circle of animals." The zodiacal constellations are Aries the Ram, Taurus the Bull, Gemini the Twins, Cancer the Crab, Leo the Lion, Virgo the Virgin, Libra the Scales, Scorpius the Scorpion, Sagittarius the Archer, Capricornus the Sea Goat, Aquarius the Water Bearer, and Pisces the Fish; only Libra is inanimate, and that's because it was a later addition, being made from the claws of the Scorpion to even out the formerly 11 zodiacal constellations. Since the planets still sail through the boundaries of all these constellations, they can disrupt their appearances and confuse beginning stargazers. Be sure to check your newspapers or astronomy magazines for information on the appearance and location of the planets.

Constellations of the Zodiac

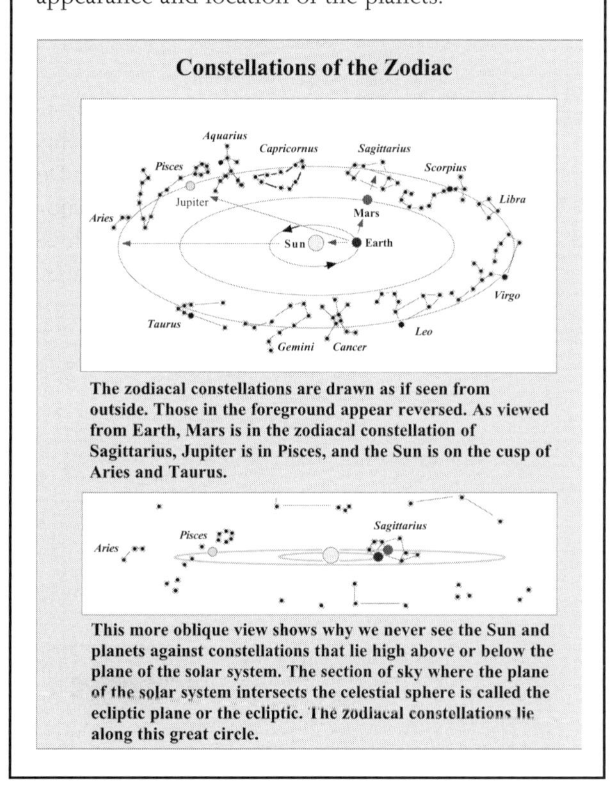

The zodiacal constellations are drawn as if seen from outside. Those in the foreground appear reversed. As viewed from Earth, Mars is in the zodiacal constellation of Sagittarius, Jupiter is in Pisces, and the Sun is on the cusp of Aries and Taurus.

This more oblique view shows why we never see the Sun and planets against constellations that lie high above or below the plane of the solar system. The section of sky where the plane of the solar system intersects the celestial sphere is called the ecliptic plane or the ecliptic. The zodiacal constellations lie along this great circle.

To see the Lion (see the chart on page 20), imagine the Sickle as the beast's thick and hairy mane, and Regulus as its icy heart. Now look about 15° east for a roughly 10°-wide triangle of 2nd-magnitude stars: Delta (δ), Theta (θ), and Beta (β) Leonis; Delta and Theta Leonis lie in the Lion's flank, while Beta Leonis (Denebola) marks the tip of the Lion's tail. Seen in this way, the Lion appears to be lying on its belly in the stellar grass, facing west. In fact, this view may have once represented the Sphinx of ancient Egypt. The Sphinx was Egypt's earliest colossal royal statue. During the New Kingdom (circa 1550–1069 BC) it was regarded as a manifestation of the Sun god. To these ancient people, Leo was the summer House of the Sun and was most likely worshiped because the annual life-sustaining rise of the Nile coincided with the Sun's entrance into the constellation. In his geography of Egypt, Pliny the Elder notes that "The Nile begins to increase at the next new moon after the summer solstice. . . . it is at its greatest height while the sun is passing through Leo." The image below shows the Lion as the alabaster Sphinx at Memphis. Below it is a more modern interpretation – the Papillon.

The Greek representation of Leo includes dim hind legs stretching to the south and front paws reaching forward; so the Lion is on the pounce.

Actually, in classical mythology, Leo is the one being pounced upon. That's because it had the misfortune of being the mighty Lion of Nemea – the first of twelve successive labors Hercules (Heracles), the mythical Strongman (see page 48), had to overcome to survive the wrath of Juno. (Hercules was one of Jupiter's many illegitimate sons.) Leo was believed to have come down from the Moon as a shooting star, which fell to Earth in the town of Nemea near Corinth. There, in a bleak valley littered with the bleached bones of human prey, the beast terrorized the inhabitants of the small village. The supernatural animal lived high in a valley wall, in a dark recess with two entrances, and was all but untouchable. From his lofty perch, Leo had complete command of all he saw. The Lion was fearless. It possessed a coat as thick as

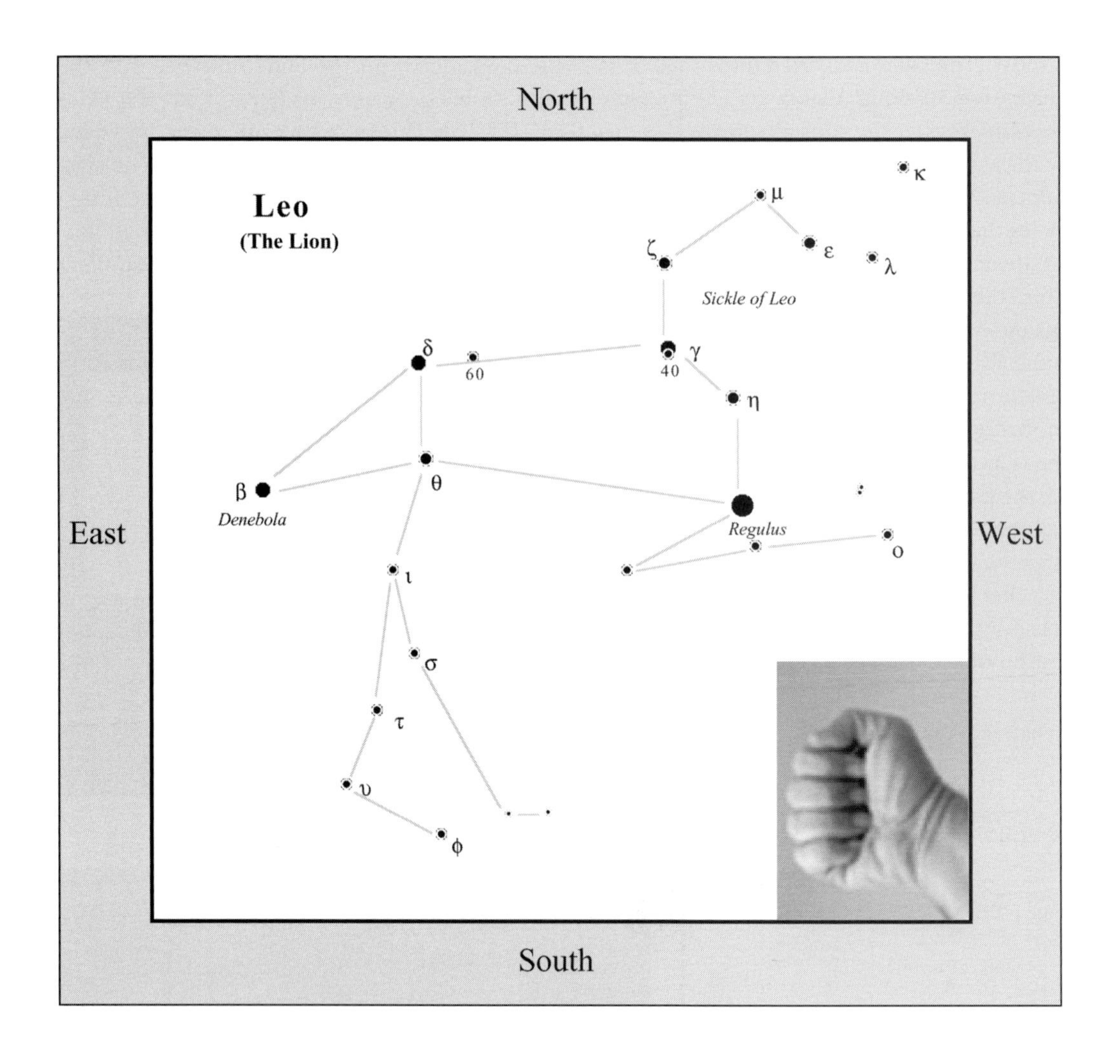

armor – so thick that no mortal weapon could penetrate it, not even the knobby club of Hercules. But the Strongman outwitted the Lion by entering the cave at nightfall. Once safely in, the Strongman blocked his entrance with an enormous boulder, then sneaked up on the sleeping cat. Carefully and noiselessly, Hercules wrapped his muscular arms around the animal's dense mane and squeezed the Lion's neck until he heard the beast take his final breath. Upon seeing Hercules' success, Jupiter honored the Lion by returning it to his home in the sky, where we see it full of life and vigor – a fitting symbol of the recuperative power of spring.

Let's start our binocular journey of this constellation by raising our glasses to radiant Regulus – the Lion's Alpha star, which is only 77 light years distant and marks the Lion's heart. Regulus means "the little king," a title conferred upon it by Nicolaus Copernicus (1473–1543) because skywatchers from different cultures regarded it as one of the leading stars in the sky for some 2,000 years. Leo itself is one of the oldest constellations, having been known as a Lion by the peoples of ancient India and Egypt. We see its form in the celebrated zodiac in the Egyptian Temple of Dedera. Back then, the Sun slipped into this constellation during the intense heat of summer and became a prognosticator of weather, as the Greek poet Aratus (315 – 240 BC) informs us:

> Most scorching is the chariot of the Sun,
> And waving spikes no longer hide the furrows
> When he begins to travel with the Lion.

Today, especially as seen through binoculars, regal Regulus shines with a dazzling icy light. Indeed, the Lion's heart is a main-sequence star that shines 140 times brighter than the Sun – at a white-hot temperature of 12,000 Kelvin. But the star's a weight-watcher's nightmare, suffering from an extreme battle of the bulge; Regulus's equatorial diameter is 32 percent greater than its polar diameter – a symptom of the star's extremely rapid rate of rotation. Unlike the Sun, which turns on its axis once a month, Regulus does the same in 15.9 hours! If Regulus spun just 10 percent faster this egg-shaped ball of gas might fly apart.

If you're up for a challenge, and you own 10 × 50 or larger binoculars, go outside in the twilight and look at Regulus when it's just beginning to shine. If the atmosphere is stable, and Regulus looks sharp, you might glimpse its 8th-magnitude companion (not shown in the charts) about 3′ to the northwest. Next to blazing Regulus, the companion looks extremely impish and requires averted vision to see well; it becomes almost impossible to see when darkness falls, because the glare from Regulus overpowers it. If you do not succeed on your first night, try again on another. Now imagine that that dim orange dwarf star orbits Regulus with a period of at least 130,000 years, at a distance that's at least 100 times that of Pluto from our Sun. From a planet orbiting that orange dwarf (which itself has a dim telescopic companion), Regulus would shine in the night sky with the light of six of our Moons; it would also be visible as a pale dot in broad daylight.

When you're done admiring the little "lion king," turn your binoculars to 2nd-magnitude Gamma (γ) Leonis (Algieba) in the Lion's mane. Although the true beauty of this star is revealed in a telescope (it is a close and colorful double), binoculars will show the star burning with a deep yellow hue; the contrast between it and the vestal purity of Regulus is not only sharp but delightful. Like Alpha (α) Ursae Majoris (Dubhe), Gamma Leonis is a K-type (orange) giant with a temperature of 4,400 Kelvin. It lies at a distance of 126 light years, making it slightly closer to us than Dubhe. Now compare the colors of these two K-type stars. Do they look the same? Remember what I said about color sensitivity and the eye in Chapter 1? Return your gaze to Gamma Leonis and look 20′ to the south–southwest for 5th-magnitude 40 Leonis. This F-type star is 69 light years distant, making it an optical companion, not a true physical binary. I see 40 Leonis shining with a slightly greenish tint. This might be a trick of the mind – the contrasting color to lemony Gamma – but it's curious to see . . . or imagine.

Speaking of imagining, Gamma Leonis is very near the radiant point of the famous Leonid meteor shower. The shower occurs each November as our planet plows into the dusty debris shed from Comet 55P/Tempel-Tuttle whose orbit happens to intersect Earth's at that point. The shower activity usually peaks every November 16–17. That's when we can see some 10 to 20 meteors per hour. But every 33 years, more dramatic displays can occur. In 1966, for instance, observers across the central plains of the United States saw about 5,000 of these "shooting stars" in a 20-minute period. Occasionally, the meteors truly storm. As was witnessed in 1833 and 1836, the night sky turned into a blizzard of shooting stars. During such storms it is possible to see the meteors shooting out from a point near Gamma Leonis like falling snow seen

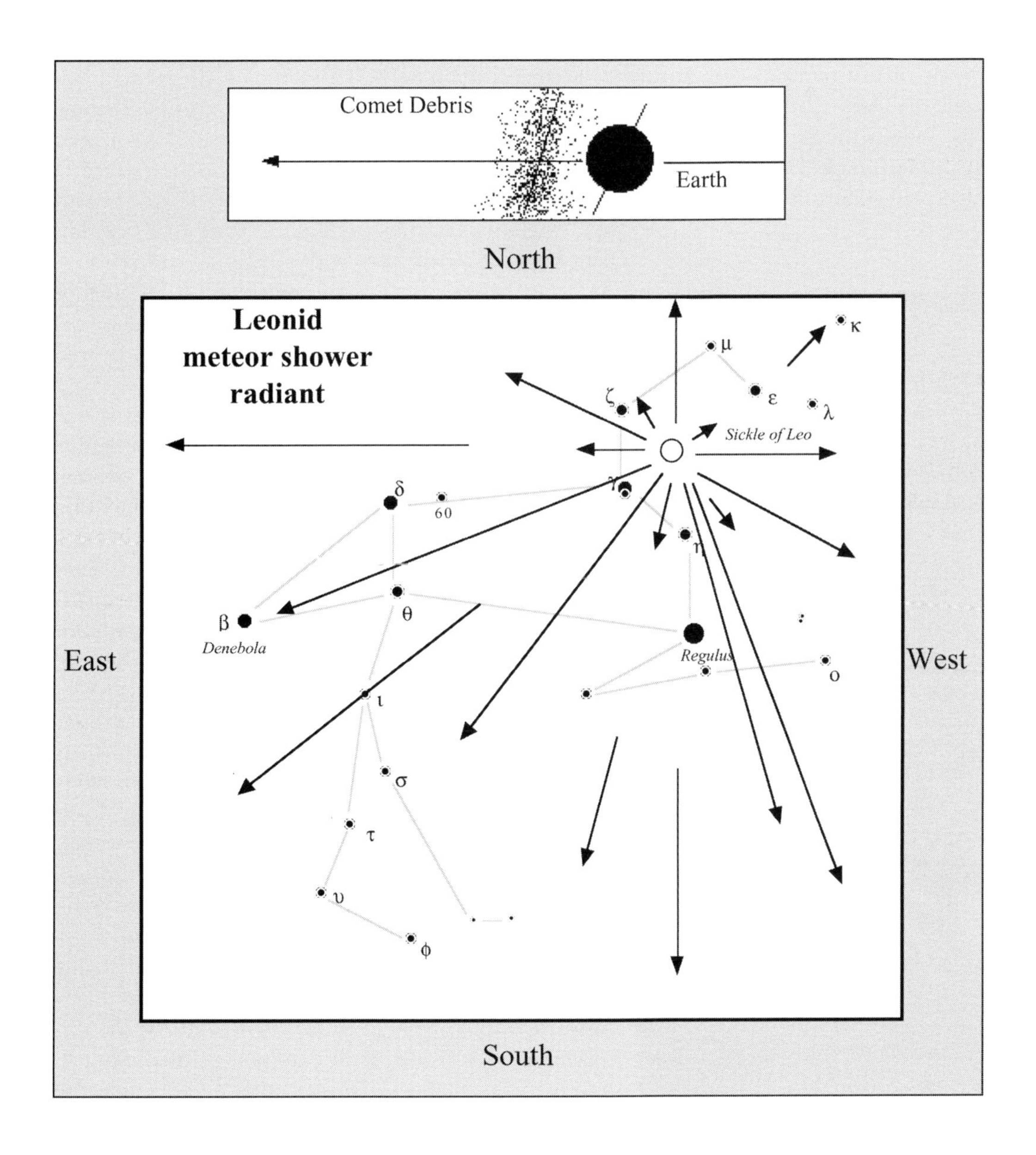

through the windshield of a moving car – only in this case, the moving car is the Earth plowing through miles and miles of "comet sweat!"

Although the Leonids did not storm for visual observers in 1999, or in any of the years surrounding it – in other words, no one saw the car of the Earth driving through a blizzard of meteors – we were treated to many memorable years of dramatic Leonid activity. The most memorable activity for me occurred in the predawn hours of November 16, 1998 (my birthday). I was not planning to observe that morning. My wife, Donna, was away in Boston, and I was in bed with our late Pomeranian, Pele, curled up on Donna's pillow. But I awakened at 5:00 am to the soundless flashing of light. When I looked out the bedroom window, the back yard suddenly lit up as if by lightning. Just as my mind screamed, "What was that!," I saw a brilliant fireball fall from the heavens causing the yard to light up once again. Quickly I grabbed Pele, wrapped her up in a blanket, and dashed outside. In the next 45 minutes, I saw 30 dazzling fireballs and four other meteors raining down from the Sickle of Leo, which was high overhead. With each falling star came a brilliant visual report that bathed the night in an eerie glow (see the photo illustration below). The magic of that morning was in the drama of the unexpected; the meteor activity on the following morning, the morning of the predicted peak, did not measure up to the previous night's spectacle. It was a valuable reminder that prediction is not an exact science, and that we as observers should always expect the unexpected.

From Gamma Leonis, look about half a binocular field north for 3rd-magnitude Zeta (ζ) Leonis (Adhafera), which, in Arabic, means "the lock of hair." It's uncertain whether the intent was to name this star for the Lion's mane, or if it was somehow misplaced, since it may also refer to the lock of hair belonging to neighboring Coma Berenices (see page 29); such accidents were common years ago. Adhafera has a clean yellow patina that contrasts well with its two rust-colored, 6th-magnitude companions: 35 Leonis about 5′ to the northwest and 39 Leonis about 20′ to the southeast. This attractive coupling is yet another chance alignment. Adhafera, a rare yellow-white F-type giant, lies 260 light years away; 35 Leonis (another F-type star) is 100 light years distant; and 39 Leonis is 26 light years closer. If you look less than 10′ northeast of 39 Leonis, you'll see yet another companion – a colorless 7.5-magnitude star of unknown distance. Interestingly, while I see 35 Leonis as being rust colored, it is, in fact, a G-type star with the same surface temperature and color as our Sun's; this star is, in fact, just beginning its evolution to the giant state.

Second-magnitude Beta (β) Leonis (Denebola), the Lion's bright tail star has a 6th-magnitude optical companion some 19′ to the south–southwest. But if you own a pair of 10 × 50 binoculars, challenge yourself to see its dim, 8th-magnitude optical companion, about 4′ south of Denebola's burning white skirt. I find that that dim sun pops into view only with relaxed averted vision – in other words, don't strain your eyes in an attempt to see it. Just relax your averted gaze and breathe normally as you concentrate your attention on the desired location.

Denebola is a fascinating star. This *A*-type main-sequence dwarf 36 light years distant appears to be surrounded by a cloud of dust – the same kind, we suspect, that formed the planets in our Solar System. If true, Denebola may have planets already orbiting it, though we have no evidence to support that contention. Still, imagine the possibility. If a planet inhabited by intelligent beings is orbiting that star, we could, theoretically, communicate with them in the span of a human lifetime. By the way, if you know anyone who's 36 years old, showing him or her Denebola on the night of their birth would be a great present, since the light that reaches their eyes will have left that star around the day of their birth. Likewise, as a purely thought-provoking exercise, if you could then somehow magically transport this birthday friend to Denebola, they could look back at Earth through an equally magical telescope and watch themselves being born.

From Beta, swing about one binocular field to the west and center 3rd-magnitude Theta (θ) Leonis (Chertan). Now look about 2° to the south–southeast for 5th-magnitude 73 Leonis. If you are under a dark sky and can brace your binoculars so that they do not jiggle, use your averted vision to look less than 1° east–southeast of

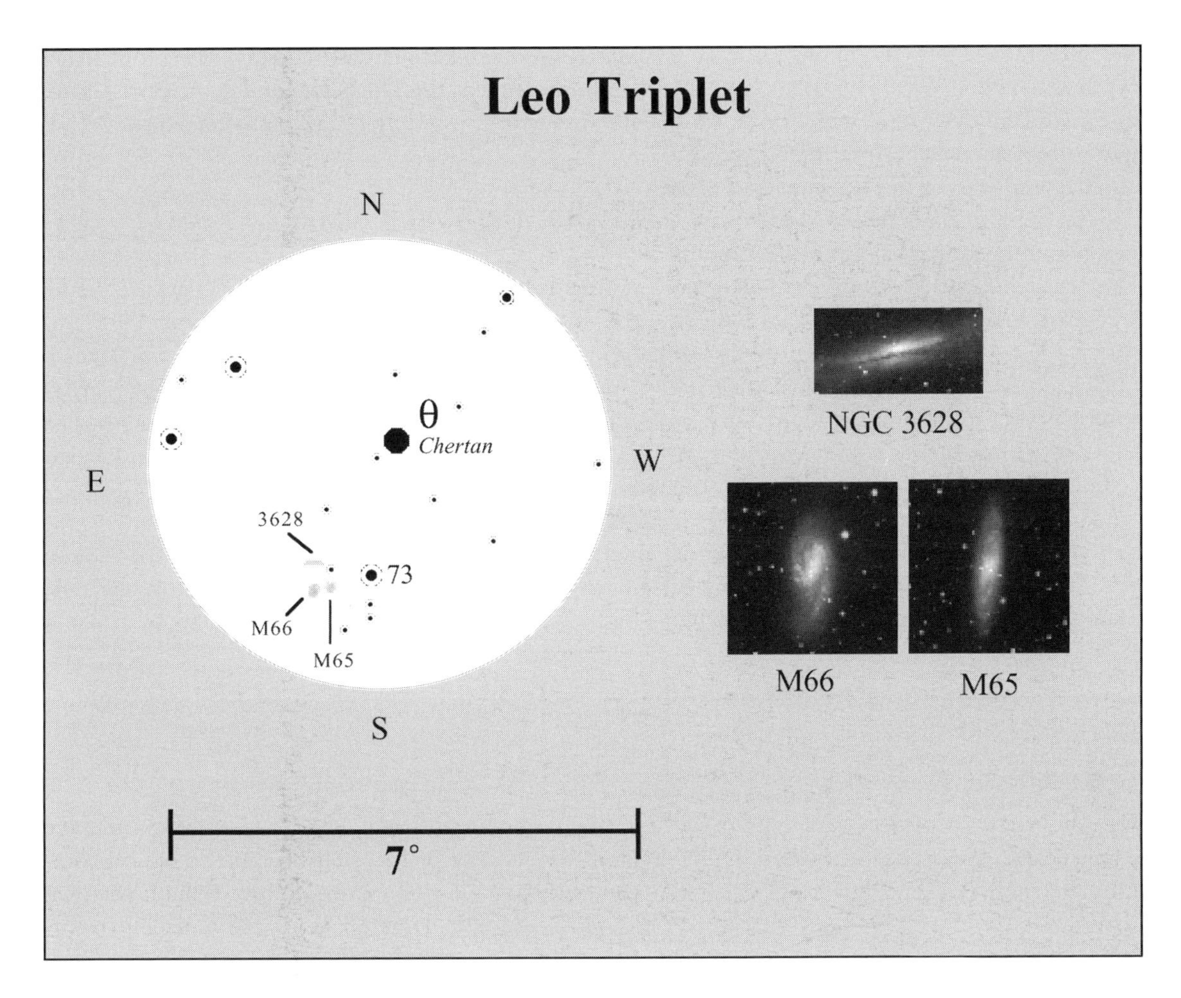

Theta for the ghostly forms of two 9th-magnitude Messier galaxies: M65 and M66. They should appear as two slightly oval glows less than 10′ across. I find M66 more apparent, probably because M65 hugs a 7th-magnitude star immediately to its north. These spiral wonders belong to the Leo Spur of galaxies – an outlying branch of galaxies some 20 million light years distant between us and the great Virgo Cluster of Galaxies. If you have a telescope, you should look for NGC 3628 – the 3,628th entry in the *New General Catalogue* (NGC) of clusters and nebulae – and the third member of what's known as the Leo Triplet: NGC 3628 is a mysterious edge-on sliver of dim light, 1° due east of 73 Leonis.

Use the chart on page 20 to locate 5th-magnitude Tau (τ) Leonis, which is a little more than 10° south–southeast of 73 Leonis. This G-type sun looks rose-petal pink and is accompanied by a fleck of ashen light (a 7th-magnitude optical companion) immediately to the south; the companion becomes more apparent with averted vision. Binocular guru Phil Harrington of New York recommends you place Tau Leonis toward the upper right corner of your field and carefully look around for three more faint doubles, which, when seen together, form an asterism that he calls the Double Cross, "because all four pairs collectively fall into an X pattern." Give it a try.

Return to Regulus. Now look less than one binocular field due west for the pair of 6th-magnitude stars 18 and 19 Leonis, whose magnitudes are 5.8 and 6.4, respectively (labeled as 58 and 64 on the chart). The Mira-type variable R Leonis should be visible immediately southeast of 19 Leonis. It forms the east–northeast apex of a tiny 5′-wide equilateral triangle with two dim binocular stars.

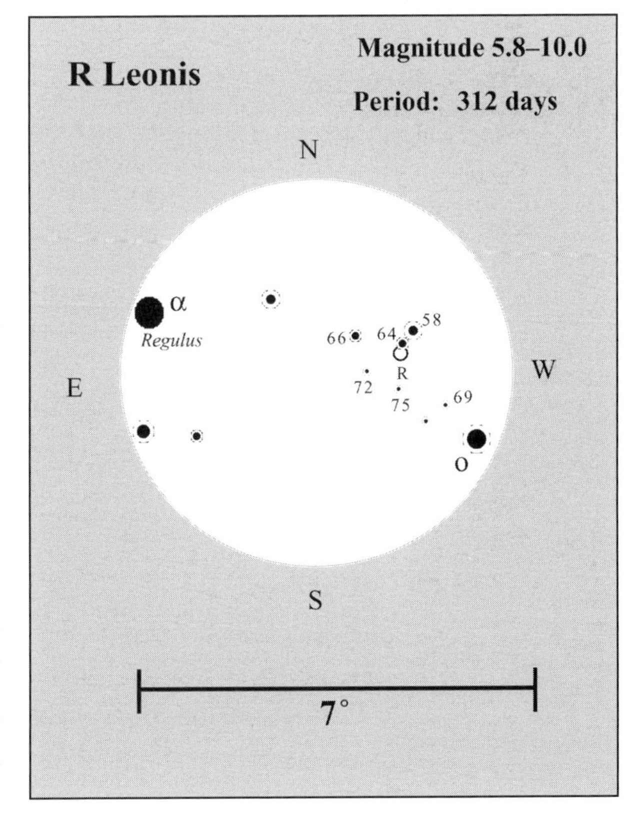

R Leonis ranges in brightness from magnitude 5.8 to 10.0 every 312 days. If you own 10 × 50 binoculars, you can watch the star rise and fall through its entire cycle, from maximum light to minimum light, over the course of a year. R Leonis is indeed one of the brightest and easiest long-period variables to observe and has become a favorite among beginning stargazers. But beware, as the late variable star observer Leslie Peltier writes in his 1965 autobiography *Starlight Nights*: "I feel it is my duty to warn any others who may show signs of star susceptibility that they approach the observing of variable stars with the utmost caution. It is easy to become an addict, and, as usual, the longer the indulgence is continued the more difficult it becomes to make a clean break and go back to a normal life."

By the way, if you go outside on the evening of March 1st at 8:00 pm, you will see the Sickle of Leo rising in the east, heralding the coming of spring – thus the saying March "comes in like a Lion." March goes "out like a lamb," because on the 31st of that month, Aries, the Lamb (see page 103), is setting in the west after sunset.

Hydra, the Water Snake

Return your gaze to Regulus, the heart of the Lion. Using the chart below as a reference, look about 20° (two fists) southwest for 2nd-magnitude Alpha (α) Hydrae (Alphard), the heart of Hydra, the mythical Water Snake. Alphard should be conspicuous because no other stars shine as brightly as it in the immediate vicinity. Indeed, in Arabic, Alphard loosely translates into the "Solitary One."

The star will be unmistakable in binoculars because it has a rich orange hue. The color is most dynamic under a dark sky, when the star seems to burn like a

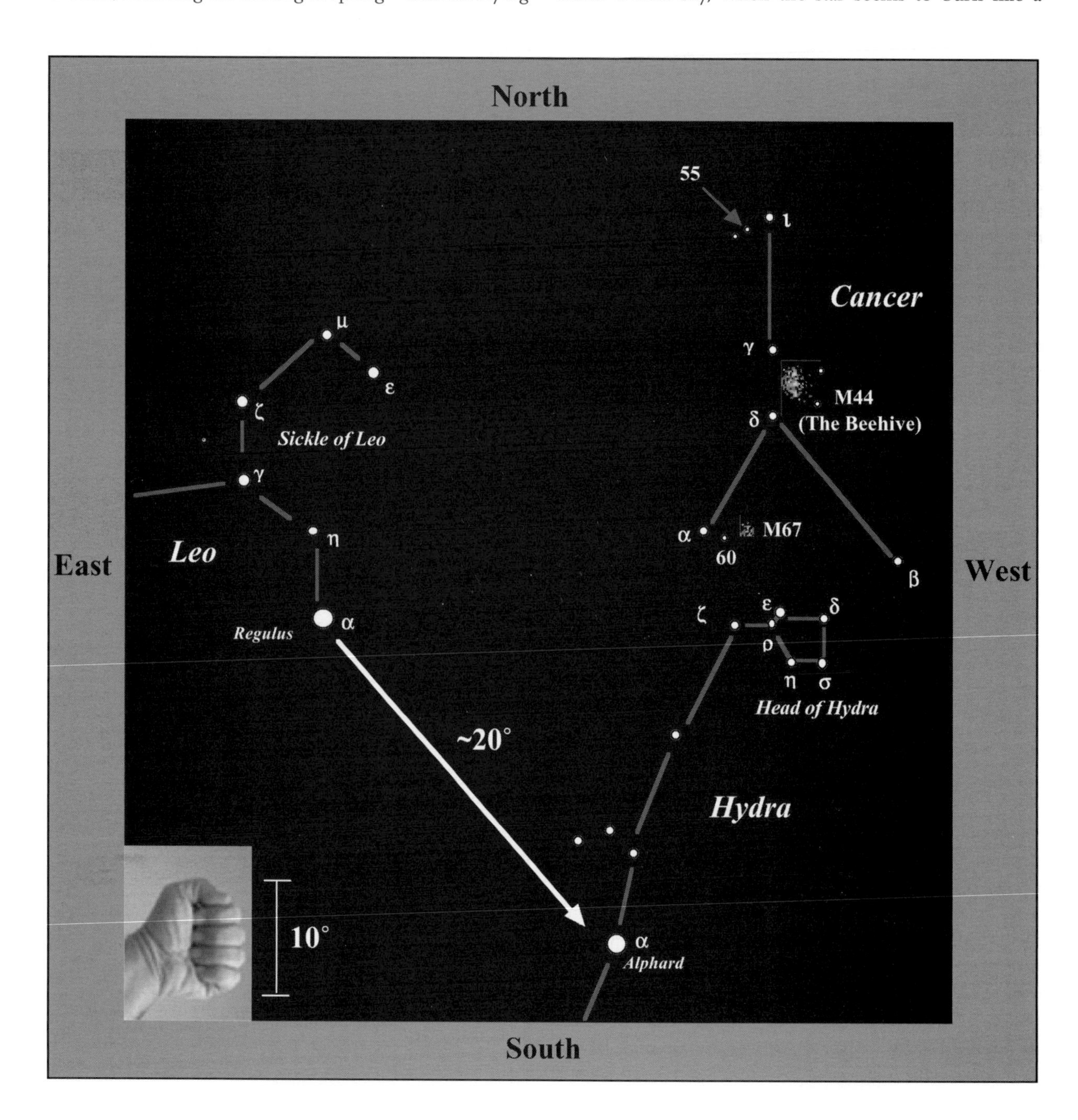

Halloween pumpkin lit by a candle from within. With a temperature of 4,000 Kelvin, Alphard is indeed an orange giant star some 40 times larger than the Sun. It lies at distance of 175 light years, making it 35 light years closer to us than Dubhe in the Big Dipper, which is not only similarly bright but also has a similar color and temperature. But Alphard's spectrum has an abundance of the metal barium. University of Illinois astronomer James Kaler notes that when Alphard was young, it had a more massive companion that died first as a dim white dwarf; in its death throes, the companion bombarded Alphard with neutrons, which formed barium and other elements in Alphard's atmosphere. So Alphard is a "barium star" – one that is itself now bloated and dying.

Speaking of dying. The slaughter of Hydra was Hercules' second labor (no rest for the wicked). After killing the Nemean Lion, Hercules jumped aboard his nephew Iolaus' chariot and the two hit the road for Argos on the northeast coast of the Peloponnesian Peninsula. There, in the village of Lerna, the great serpent Hydra was haunting the region's swamps, devouring its cattle and despoiling the forests. Hydra was no ordinary serpent; the creature had nine heads, the central one of which was immortal. Aside from the usual terror induced by such beasts on the region's inhabitants, the multi-headed Water Snake threatened the country's very livelihood. Hydra took up residence near the sacred springs and wells of Amynome, which the people of Argos needed in times of drought. Without the water, the people would die. They could also die from the stench of the beasts' foul breath or the odor rising from its tracks.

The Hydra.

When Hercules arrived, he found the beast lolling about in the swamp, the heads entertaining one another. Hercules wasted no time. He immediately confronted the beast and began wailing on its heads with his hefty club. But for every head Hercules smashed, two new ones grew in its place. Hercules retreated and called his faithful nephew for help. Together, Hercules and Iolaus plotted a way to defeat Hydra. Each time Hercules would knock off a serpent's head with his club, Iolaus would cauterize the bleeding wound with fire from his torch. With the flow of blood stopped, they figured, new heads would not grow back. They were right. As for the immortal head, after Hercules knocked it to the ground and Iolaus cauterized the bleeding stump, Hercules buried the head under a huge rock, and disposed of the beast's body.

The myth is rife with volcanic overtones. Although the supposed battle took place in Lerna (located a few miles southwest of Argos) the inspiration for Hydra itself may have been Methana[1] – an active volcano only about 20 miles east–southeast of Argos. Methana is a "multi-headed" lava-dome complex. Each lava dome can exude multiple, bulbous lobes with fiery tongue-like spines, any one of which can send deadly pyroclastic flows to surrounding regions, devastating all in their paths. At night, these flows appear as long serpentine bodies of fire. The domes themselves are swathed in vile sulfurous odors rising from snaking fissures in the earth that can not only invade rivers and despoil the countryside, but also, if released in proper concentrations, can be poisonous. What's more, volcanic domes are prone to collapse (decapitation), after which they rebuild (regeneration of new "heads").

"Multi-headed" dome complex

Snaking lava river

[1] Of course, there was also Santorini – the only large-scale natural disaster in recent geologic time which could have been known to the early Greeks; the cataclysmic eruption of Santorini volcano before the fall of Troy most likely inspired Plato's legend of Atlantis and led to the loss of the Minoan civilization. The ancient city of Argos was occupied during the time of the eruption.

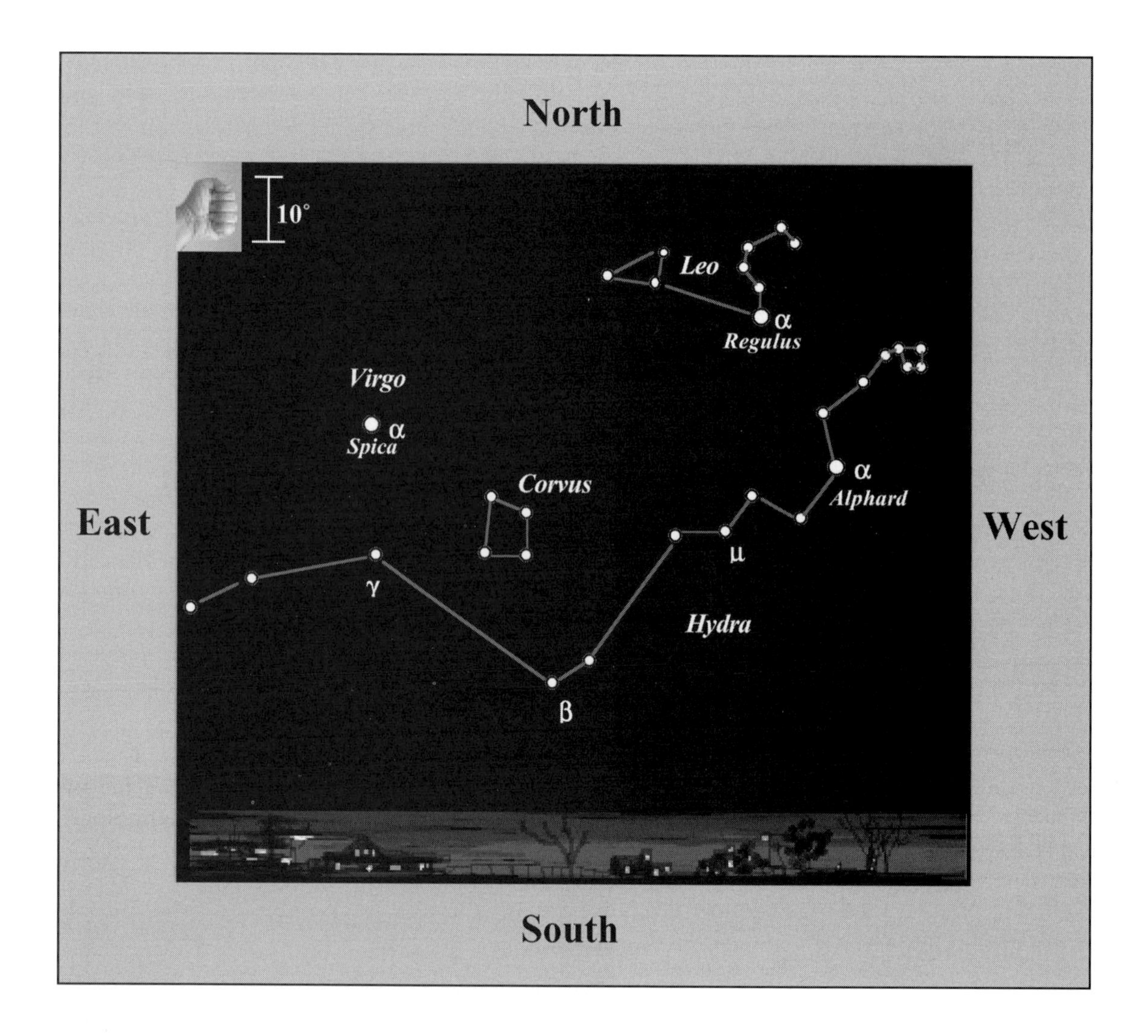

The Methana volcano consists of 10 principal domes with perhaps as many as 30 subsidiary domes or lobes. The last eruption of the Methana volcano was in 230 BC. Interestingly, the Latin author Gaius Julius Hyginus (c. 64 BC–AD 17) claimed that the Hydra possessed some 30 heads. Today, Methana, still has thermal springs and mofettes (gas exhalations), and questionable eruptive activity may have occurred in 1922, lending credence to the immortality of the mythical Hydra. During the 230 BC eruption, this volcanic monster, if you will, produced a lava flow that snaked 1,500 feet beyond the island's coast and plunged into the sea – a veritable molten Water Snake. Hydra has the distinction of being the sky's longest constellation. From nose to tail, the beast slithers across a quarter of the sky. In April, it can best be seen in full only around midnight, as shown in the chart above.

Our interest in the Water Snake is, for now, contained in the Snake's upper torso, from its heart to its head. Using the chart on page 24 as a guide, return your gaze to Alphard. Now look about 20° (two fists) northwest for a tight garland of five 4th- to 5th-magnitude suns – the Hydra's Head. The Head is a very fine naked-eye asterism, spanning about 5° of sky, so the entire Head will fit nicely in most binocular fields of view. Of particular note is Epsilon (ε) Hydrae, which is a quintuplet star system, whose combined light adds brilliance to Epsilon Hydrae, the primary star in that multiple system; only the pale yellow sheen of Epsilon can be perceived in binoculars. Compare the subtle color of Epsilon to the warmer hue of Sigma (σ) Hydrae, which is a K-type giant. The fainter a star appears in binoculars, the more difficult it will be to pick out subtle differences in star colors. But the only way to learn is to try. Zeta (ζ) Hydrae should appear similar in color to Epsilon. The other stars in the garland are a crisp white.

Cancer, the Crab

As the chart on page 24 shows, the Head of Hydra is nestled under the dim constellation, Cancer, the Crab – yet another giant creature unleashed upon Hercules by Jupiter's resentful wife to prevent him from completing his labors. As Hercules battled Hydra in the Lernaean swamp, Cancer emerged from the depths and began to sink its claws into the Strongman. Undaunted, Hercules continued to club the Hydra. Then, raising one of his mighty feet, he simply stomped on the Crab, putting an end to its annoying habits.

If you are under a dark sky, look a little more than 10° (one fist) directly above the Head of Hydra for M44, one of the sky's most glorious open star clusters. Nicknamed the Beehive, this attractive swarm of suns is visible to the unaided eyes as an elongated nebulous patch. Although the cluster shines at 3rd magnitude, that light is spread across 1° of sky, or about twice the apparent diameter of the full Moon. City or country, the cluster is a delight in binoculars, which, under the best of conditions, reveal some 80 members ranging in brightness from magnitude 6 to 10. In fact, the cluster contains about 15 stars between magnitudes 6.3 and 7.5, placing them within the range of naked-eye vision; seeing these stars is really a test for more advanced observers who have mastered the art of averted vision, are observing under dark skies, and know the importance of time and patience. On several occasions, I have consistently recorded a dozen or so cluster members without optical aid under clear dark skies at an altitude of 4,000 feet.

The Beehive cluster has been known since antiquity. Aratus and Pliny both noted that whenever the object's misty form vanished from naked-eye view, foul weather was on the way – an observation most likely related to the approach of thin, high cirrus cloud, which dims the appearance of the night sky before an impending storm. Observing from Alexandria in ancient Egypt, Ptolemy (c. AD 83–161) recorded the celestial mist as the "center of the cloud-shaped convolutions in the breast [of Cancer], called Praesepe."

Praesepe is a derivative of the Latin verb praespire (meaning to enclose), which is also the root word for presepio (nativity scene). The term appears to be linked to an early Christian church on the Esquiline Hill in Rome, known since the seventh century as Sancta Maria ad praesepe – the place, according to tradition, where the remains of the holy manger were brought for safe keeping. The Praesepe we see in the night sky, then, is the fuzzy, straw-like cloud of unresolved starlight visible to the unaided eye between the 4th-magnitude stars Gamma (γ) and Delta (δ) Cancri; seen through ancient eyes these stars were the two Aselli, or donkeys, guarding Christ's crib. The nature of the "cloud" remained a mystery until Galileo turned his tiny telescope on it in 1609 and saw a "mass of more than 40 small stars."

To my knowledge, English observer John Herschel was the first to call M44 the Beehive. In his 1833 *Treatise on Astronomy*, Herschel writes, "In the constellation Cancer, there is . . . a luminous spot, called Praesepe, or the beehive, which a very moderate telescope, – an ordinary night glass, for instance, – resolves entirely into stars." The description makes sense because in the nineteenth century, the traditional beehive, called a skep, was made of straw. So, if you're a purist, you should call the misty glow visible to the unaided eye the Praesepe (a heap of straw), and reserve the Beehive moniker for the binocular or telescopic view, which shows the individual stars as the swarming bees.

In an interesting parallel, Daphne Moore in *The Bee Book* (New York: Universe, 1976) explains that a Breton legend "tells us how the falling tears of the crucified Christ turned into bees and flew away to bring sweetness to the world."

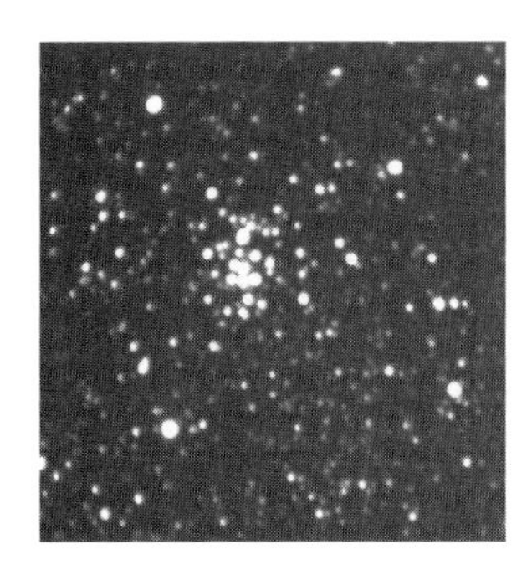

You can see the falling tears of Christ in the mind's eye. Just take the time to examine the Praesepe's shape with your unaided eyes. At first, the glow will appear round. With time and keen averted vision, however, you should see the cluster's core elongated north to south. Now relax. Take a moment to walk around while breathing regularly. When you feel refreshed, look at M44's elongated core once again, but this time see if it does not break apart into three distinct misty patches – the "falling tears of Christ." This triple aspect has been observed by sky observers since before the invention of the telescope.

Now look at the cluster again with your binoculars. This time try to relax your gaze while concentrating on the cluster's overall geometric shape. Under a dark sky, the brightest stars form the outline of a face on a wider veil of stars. The veil becomes more and more distinct the longer you look, because the face within it is bordered on all sides by dark lanes or voids of starlight, giving the veil a square shape. In keeping with the Christian overtones of the Praesepe, the veil of stars can be seen in the mind's eye as "Veronica's veil." As illustrated in the Sixth Station of the Cross, a woman named Veronica uses her veil to wipe the sweat and blood from the face of Jesus who is on his way to Calvary; the act leaves an image of his face on her veil.

The observation is a curious one because in his *Cycles of Celestial Objects*, published in the mid-1800s, Admiral William Henry Smyth describes a sighting of a similar association of stars discovered by the "Capuchin, De Rheita, who fancied he saw the napkin of S. Veronica, in 1643, with an improved telescope he had just constructed." The object, Smyth says, was found "in the sign of Leo, between the equinoctial and the zodiacal circles," which places it about 10° southeast of Regulus. Alas, no object like it has ever been seen in the area. Below is a reproduction of the imagined view. See if you do not believe M44 is not an equal or more suitable competitor for Veronica's lost veil.

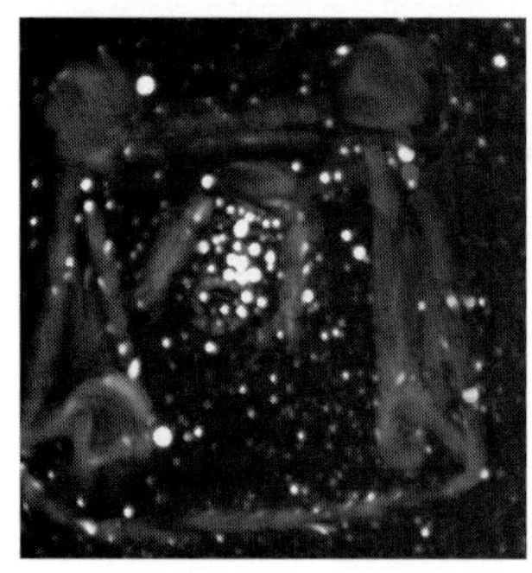

At a feeble distance of 515 light years, M44 is one of the largest, nearest, and brightest open star clusters. The age of the cluster is estimated to be about 400 million years, which means the light we are now seeing from its stars left the cluster during the Devonian Period, when the first plants blossomed on land and the earliest known sharks roamed Earth's turbulent seas.

Using the chart on page 24 as a guide, from M44, lower your gaze about 10° (one fist) southeast to 4th-magnitude Alpha (α) Cancri. Place Alpha Cancri in your binoculars and look just 2° west for the dimmer, 6th-magnitude open star cluster M67. This cluster is about 3 magnitudes dimmer and 75 percent smaller than magnificent M44. But what an illusion because M67 is 2,600 light years distant – more than five times farther away than M44. If you could bring M67 to the distance of M44, it would appear twice as large as M44 in the sky. Unlike M44, which is relatively young, M67 is a very aged cluster, being somewhere between 4 and 5 billion years old. Indeed, it is one of the oldest open star clusters known! In your binoculars, M67 should appear as a small oval haze with a bright elliptical core. The cluster's brightest stars shine around 10th magnitude, placing them near the limit of resolution in binoculars.

Take some time now to find and compare Iota (ι) and Beta (β) Cancri in your binoculars. These stars lie roughly at the same distance (300 and 290 light years, respectively) and have slightly different golden hues. Iota Cancri is a pale gold G-type giant, while Beta Cancri is a K-type orange giant that looks like a gold doubloon dipped in blood.

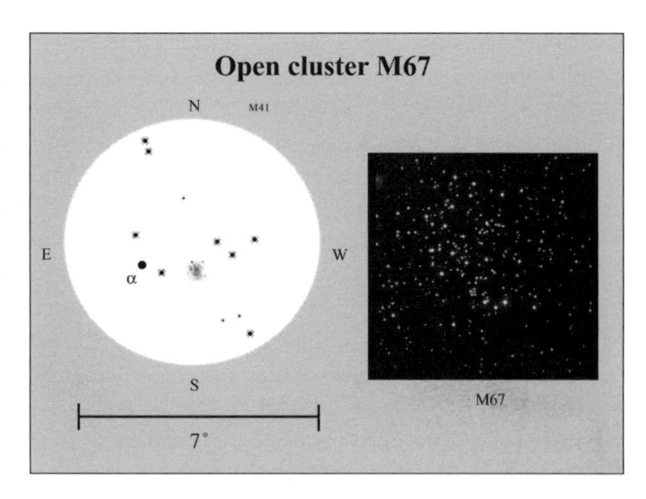

Finally, return to Iota Cancri, then use the chart on page 24 to find 6th-magnitude 55 Cancri – a G-type dwarf 41 light years distant with four suspected gas giant planets orbiting it. The largest is about four times more massive than our giant Jupiter. The smallest is a match for the planet Uranus. The solar system around 55 Cancri then, may resemble our own, with unseen terrestrial-like worlds orbiting closer to the star. It is also closest to its parent star, its orbit just 10 percent the size of that of Mercury.

2 May

The moon looks down on the lovly may
And the little star his friend and guide
Travelling together side by side
And the seven stars and charleses wain
Hangs smiling oer green woods agen
The heaven rekindles all alive
Wi light the may bees round the hive
Swarm not so thick in mornings eye
As stars do in the evening skye

John Clare (1793–1864), *May*

On the first clear evening in May, go outside around 9:00 pm and look to the south. Note that Alpha (α) Leonis (Regulus) no longer reigns high in that domain; the Lion has left his lofty perch and is now prowling the southwestern sky beyond the *meridian* – the imaginary line that passes through the north and south celestial poles and the *zenith* (the point directly overhead). Beta (β) Leonis (Denebola), the star marking the Lion's tail, should be tickling the meridian. But look a little more than a fist northeast of Denebola, and you'll see a hazy gathering of minute suns glowing like a speckled ghost. Raise your binoculars to this misty spot and you'll see that cluster of stars in its full glory. The brightest members should be in the form of an upside-down Y. This is the core of the constellation Coma Berenices (Berenice's Hair); the full extent of the lady's locks is more than twice that length. To the unaided eye, the Hair glistens like dew clinging to a cobweb under the cold light of a rising Moon.

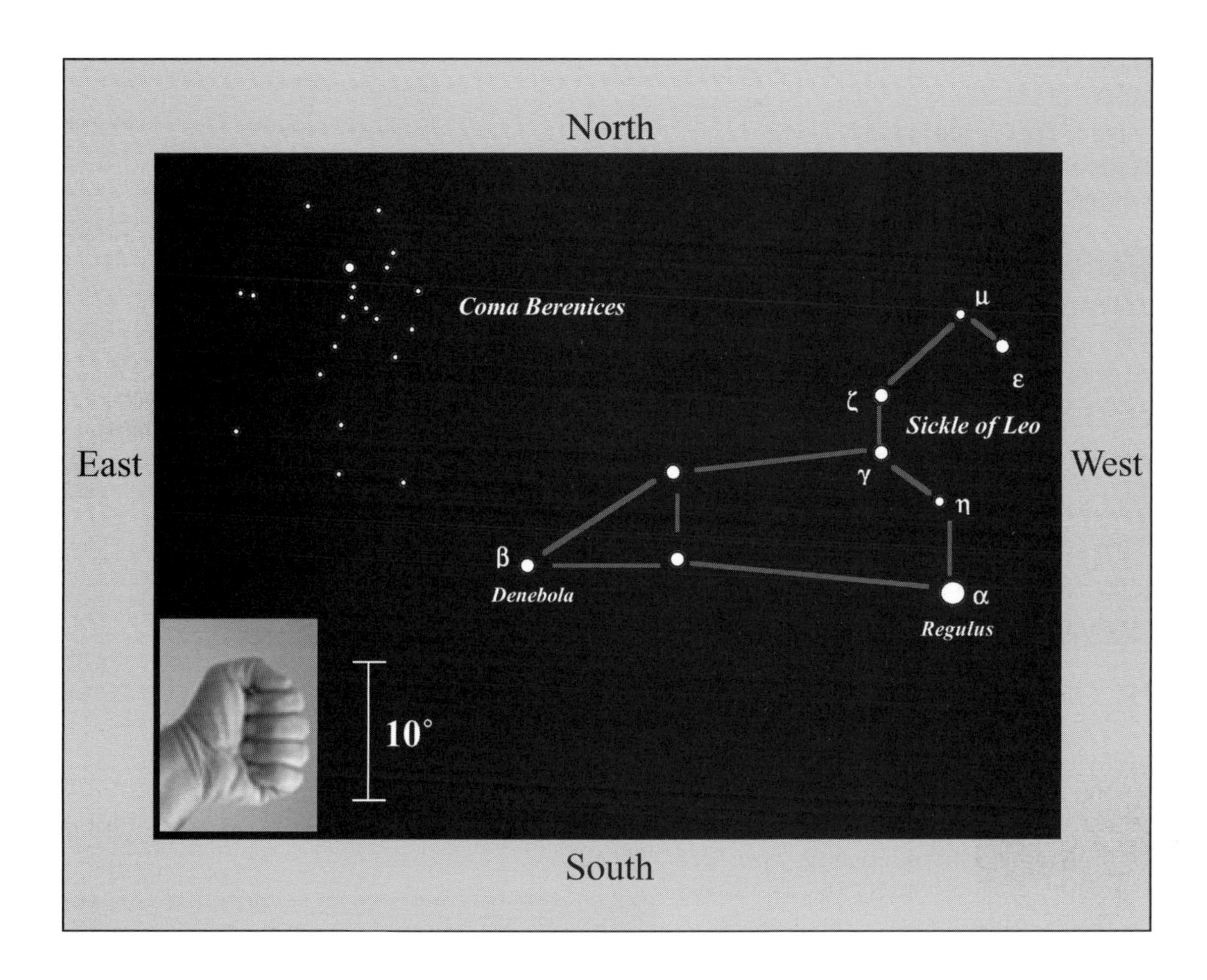

Coma Berenices, Berenice's Hair

Conon of Samos (~280–220 BC), court astronomer to Ptolemy III of Egypt, created the story of Coma Berenices after a theft. The story begins with King Ptolemy heading off to war to avenge the murder of his sister in Syria. Fearful of the outcome, his lovely sister and wife, Queen Berenice II, asks the goddess Aphrodite for help; Berenice vows that she will cut off her long amber locks and place them in Aphrodite's temple if her husband returns safely. When the Queen's wish is granted, Berenice honors her vow, cuts off her hair, and under great ceremony (but with great sadness to the King who loved her precious locks), has them placed in Aphrodite's sanctuary under the protection of his priests. But that night, under the cloak of darkness, someone steals the locks. When the King learns of the theft the next morning, he lines up his priests, draws his sword and with blood in his eyes raises the steel against the Sun. Fortunately, Conon of Samos arrives to save their necks. The court astronomer tells the King he knows where the locks are being kept and encourages him to wait until nightfall. Reluctantly, the King agrees. He sheathes his sword and waits for the ashen glow of twilight to fade. When the stars begin to show their faces, the King and Queen and all the nation's people gather in the court and wait patiently. Suddenly, Conon points to the heavens beyond the tail of Leo and cries out, "Look, it's the clustered curls of thy queen, placed in the heavens by the gods for all the world to see. Look! They glitter like a woven net, as golden as they were on Berenice's head."

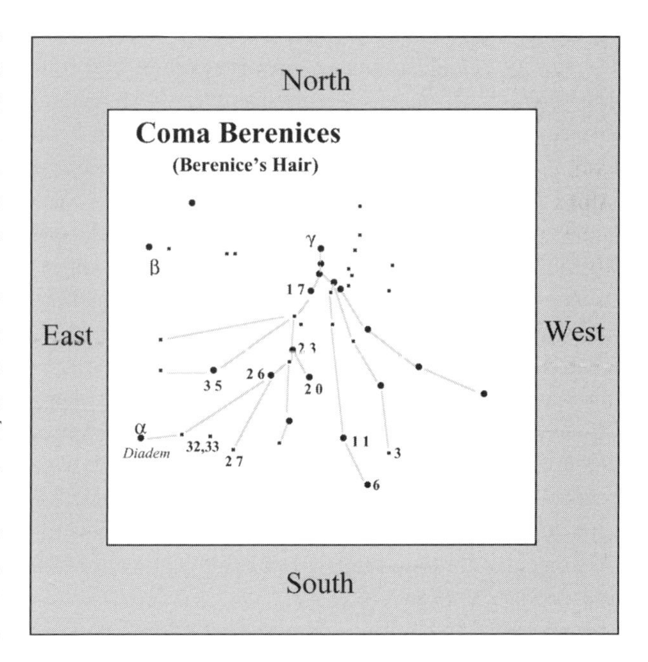

In my youth, this sparkling array of dim suns snared my attention and my imagination. I preferred to see the constellation as the silken threads of Charlotte's Web. Had I been a classicist, Coma Berenices would have become Arachne's Web. Arachne was a gifted weaver, who hanged herself after she had shamed Minerva, the Roman goddess of crafts and wisdom, with her pride. Seeing Arachne suspended by a rope caused pity to well up in Minerva's heart. As a lesson to all who dare to defy the gods, Minerva ordered Arachne to "Live!" so that she might preserve the memory of this lesson. And live she did – hanging on by a thread as a spider.

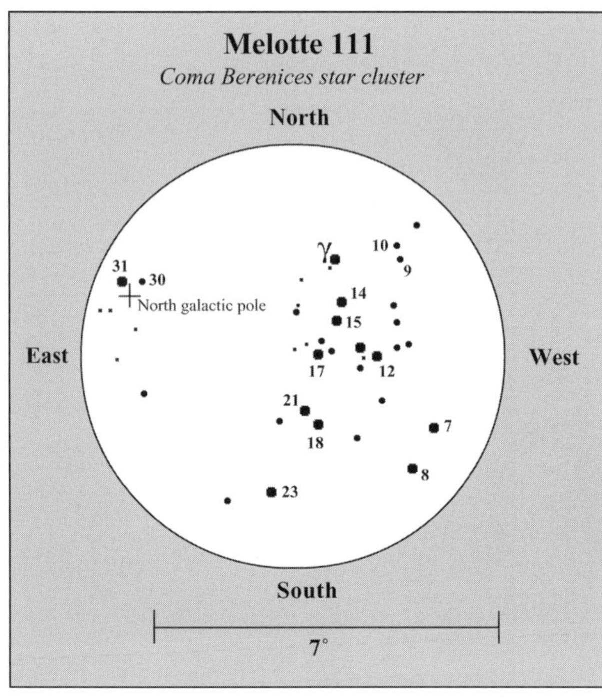

In his poetic 1888 masterpiece *Astronomy with an Opera Glass*, Garret Serviss also imagined a cobweb here: "One might think the old woman of the nursery rhyme who went to sweep the cobwebs out of the sky had skipped this corner, or else that its delicate beauty had preserved it even from her housewifely instinct."

The Y-shaped core of Coma Berenices is a marvel in binoculars. That's because it's one of the largest and brightest open star clusters in the night sky. Nearly 300 light years distant, the grouping, known as Melotte 111, is about $3\frac{1}{2}$ times farther out than the Ursa Major Moving Group and about two times closer than the bright open star cluster M44 in Cancer (see page 26). Melotte 111 is very young, having an estimated age of 500 million years, which means that its stars were born around the time when the first winged insects appeared on Earth.

The most obvious members are arranged in a rough triangle that spans 10 Moon diameters in extent, which converts to $22\frac{1}{2}$ light years in true physical extent. The cluster contains no giant stars, though the brightest members are just beginning to evolve toward the giant stage. The cluster is intrinsically only slightly brighter than the Sun, meaning that if we could place the Sun at the distance of the Coma Berenices cluster, it would shine with the brilliance of a 1st-magnitude star. Now sweep the cluster with your binoculars. Notice how Gamma (γ) Comae Berenices has a brash yellow tint, and that the star 17 Comae is a beautiful binocular double comprised of a pale-yellow primary and a soft-rust companion.

Berenice's Hair lies only about 5° west of the *north galactic pole* – the point where our galaxy's axis of rotation intersects the north celestial sphere. Through binoculars the view looks sparse, but this is an illusion, because the region is on the western fringe of the great Coma Berenices Cluster of Galaxies, a turbulent array of more than 3,000 galaxies 280 million light years distant (see the Hubble Space Telescope image below). This massive cluster of galaxies, scattered across 20 million light years of space, is the nearest one to the Local Group, which includes the Milky Way. So use your imagination and try to fathom the vast and distant cluster in your mind's eye.

Under a dark sky, you can see M64 – a much closer galaxy 19 million light years distant. Messier 64, also known as the Black Eye Galaxy, is about 1° northeast of 35 Comae Berenices (see the charts on pages 29 and 31). It shines at a modest 8.5 magnitude and spans only 10′ of sky, so it appears as a little puff of light. Through powerful telescopes, however, it is a magnificent object: here is a collision of two galaxies that has left a scarred spiral system with a spectacular dark band of absorbing dust arcing in front of the galaxy's bright nucleus like a shiner – thus the galaxy's "black eye" moniker.

Return your gaze to 35 Comae, then move about two-thirds of a binocular field to the southeast where you'll

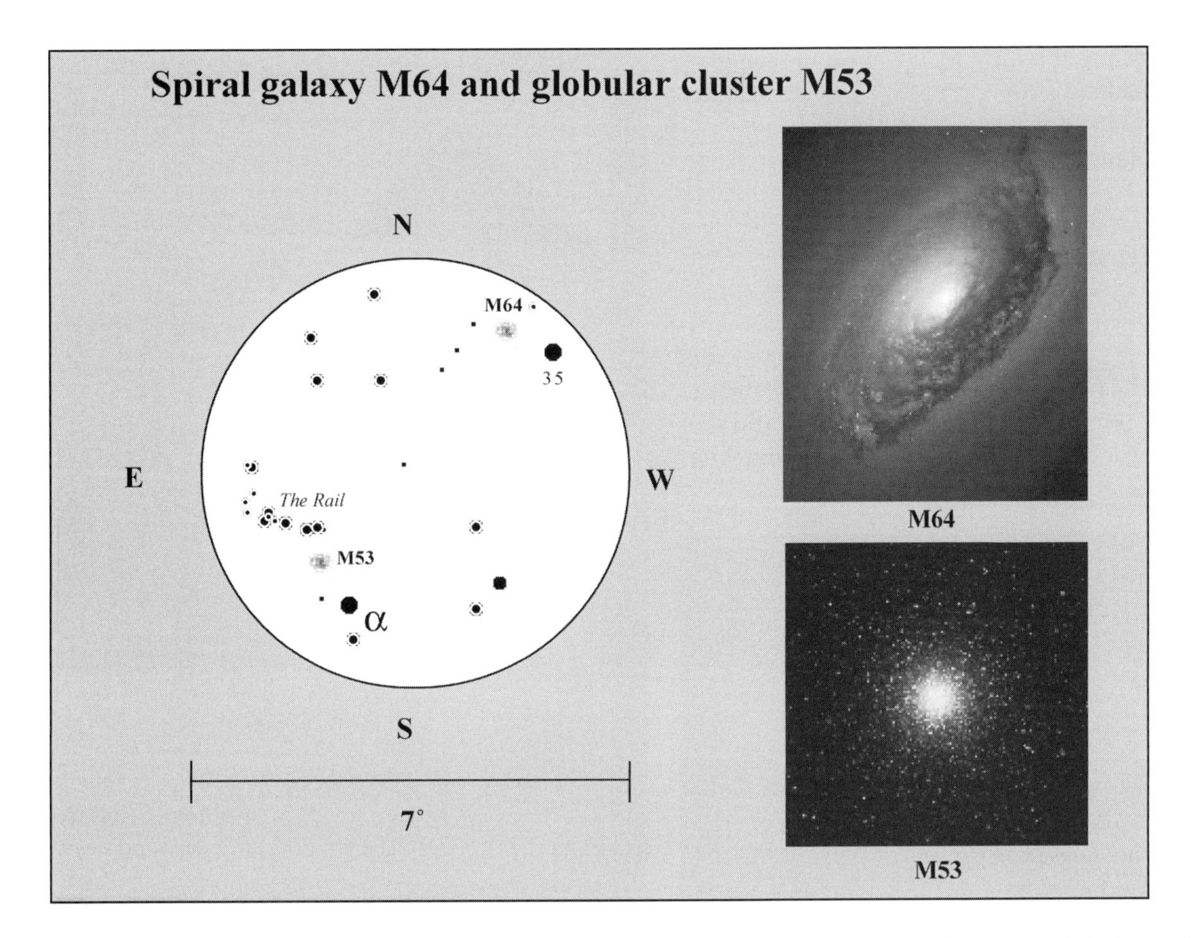

find Alpha (α) Comae Berenices (Diadem). Just about 1° north–northeast of Alpha Comae you should see another small diffuse glow – one more condensed and a bit more obvious than M64. This object is 8th-magnitude M53, a 200-light-year-wide globular star cluster 56,000 light years distant. Note that a beautiful 1°-long binocular asterism, which I call "the Rail," lies to the northeast; seeing it reminds me of one of those sleek dragsters that tore up the strips in the 1960s.

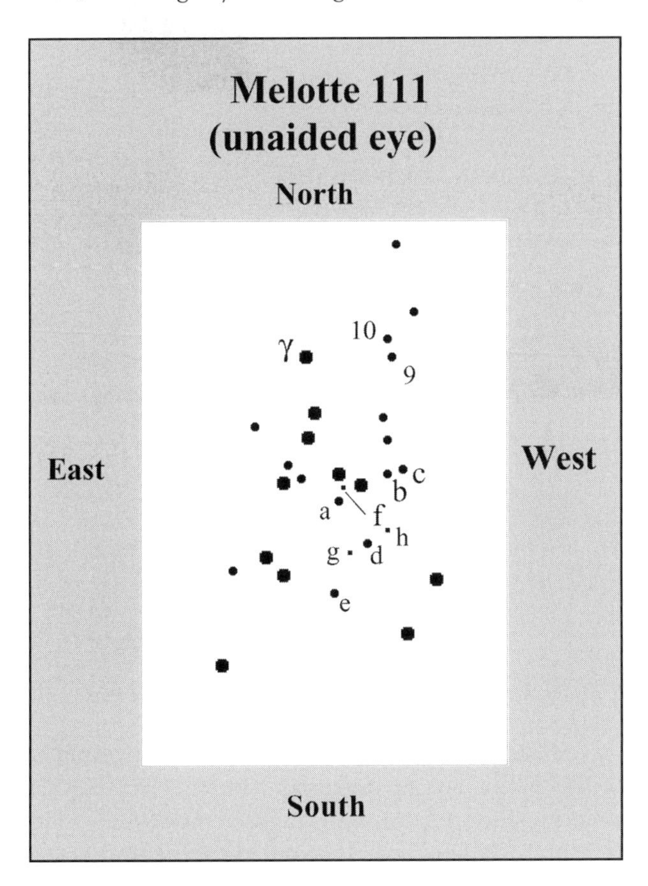

The stars 32 and 33 Comae Berenices (see the chart on page 29) form a pretty pair $4\frac{1}{2}°$ west of Alpha. And if you look at nearby 27 Comae with averted vision, you might spy its tiny companion immediately to the southwest. By the way, both Alpha and Beta (β) Comae do not belong to Melotte 111, which is nearly 4.5 times more distant. Also, while 4th-magnitude Diadem is the constellation's Alpha star, Beta Comae (which has no proper name) is just as bright and has a slightly warmer hue.

After admiring these deep-sky objects, put your binoculars aside, grab a chair (or lie on a blanket) and use the chart at left to see how many dim stars you can see in the cluster without optical aid.

The cluster contains nine stars between magnitude 6.0 and 7.4 with separations ranging from 12″ to 31″ – all within the grasp of a keen-eyed observer under a dark sky. The following magnitudes were derived from Hipparcos satellite data:

Star	Magnitude	Star	Magnitude
a	–	f	6.7
b	6.2	g	7.4
c	6.4	h	7.3
d	6.2	9	6.4
e	6.0	10	6.6

By the way, if you happen to look upon the soft glow of Coma Berenices in 2010, you will be seeing light that left the star in the year when the last legal execution for witchcraft happened: In Protestant Scotland, an elderly woman was burned at the stake after being accused of turning her daughter into a pony and riding her to a witch's sabbat. You can see the "witch" fly today in the stars of Coma Berenices: Gamma Comae is the witch's head, the triangular form of the main cluster is the witch's body. The stars 30 and 31 Comae form the witch's wind-blown cape. And the stars 4, 23, and 26 form the witch's broom, which she clutches in her hand (12 Comae). Her black cat, or familiar, is marked by two unnumbered stars near 4 Comae.

Finally, if Leo, the Lion, was indeed the Sphinx of ancient Egypt, and if the stars of Coma Berenices belong to Queen Berenice of Egypt, might we not also see the triangular-shaped core of Coma Berenices as the Great Pyramid at Giza, as seen from the rear of the Sphinx?

Canes Venatici, the Hunting Dogs

Midway between Coma Berenices and the Big Dipper's handle are the two brightest members of a relatively modern constellation: Canes Venatici, the Hunting Dogs. Polish astronomer Johannes Hevelius created the constellation in 1687 from stars that had previously been a part of Ursa Major. Don't even bother to try and see the two leaping hounds here, because the stars in this region do not follow the contours of the dogs. Instead, Hevelius seems to have placed the dogs in the voids between the stars. As a compromise, I envision the constellation's two brightest stars as marking the location of each dog, then imagine a line between them as a leash.

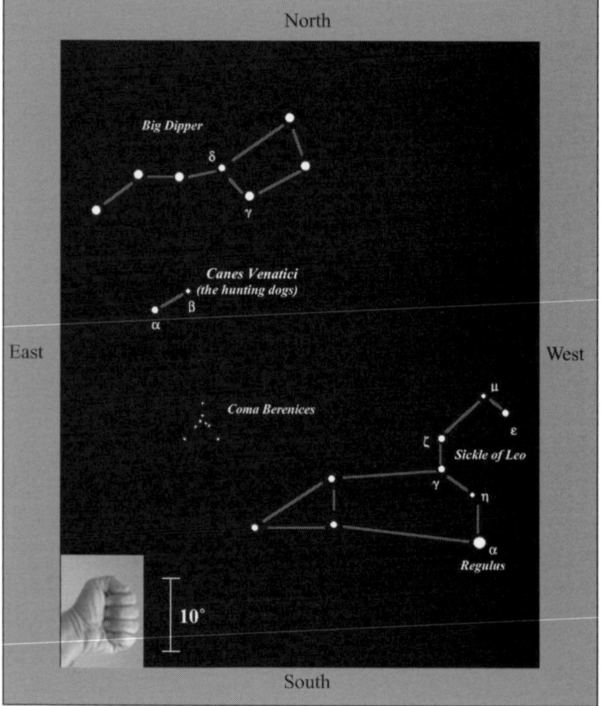

Confusing matters, Alpha (α) Canum Venaticorum is known as Cor Coroli, meaning Charles's Heart – a title honoring King Charles I of England. Apparently (and perhaps in the mind's eye) this star blossomed in brilliance on the evening of May 29, 1660, when King Charles II

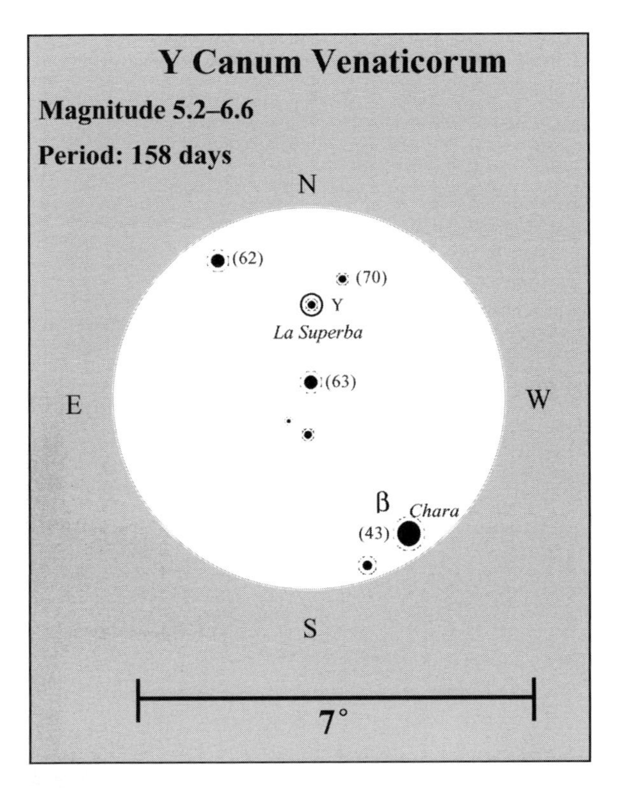

returned to London at the restoration of the monarchy. If that's the case, then shouldn't Cor Coroli honor King Charles II? The problem is that the star also has a longer name – Cor Coroli Regis Martyris – which refers to the fact that King Charles I was beheaded. Confused? Well, again I propose a compromise: You can either look upon the Alpha and Beta stars as two dogs or as two kings; which king gets the brighter star is totally up to you.

Shining at 3rd magnitude, Cor Coroli lies at a distance of 110 light years and is a white *A*-type dwarf. Through 10 × 50 binoculars, it is crystal-blue, though others claim it is pale lilac (what a lovely color!). Now look at Beta (β) Canum Venaticorum, otherwise known as Chara. Chara is Greek for "joy," a beautiful reference to the star's appearance in the fresh springtime sky. Chara is a pale lemon *G*-type star that mirrors what our Sun would look like if placed 27 light years distant. For a dramatic color shift, raise your glass one field to the north and use the chart below to look for the 710-light-year-distant semiregular variable star Y Canum Venaticorum.

Otherwise known as La Superba, Y Canum is literally one of the coolest and reddest stars visible to the unaided eyes. With a temperature of roughly 2,200 Kelvin, the star bleeds scarlet light – a color obvious in most binoculars. La Superba is a member of a rare class of red giants known as carbon stars. Most red giants and supergiants are richer in oxygen than carbon, but carbon stars have ratios of carbon to oxygen typically four to five times higher than those in normal red giants. And that excess carbon is liberally

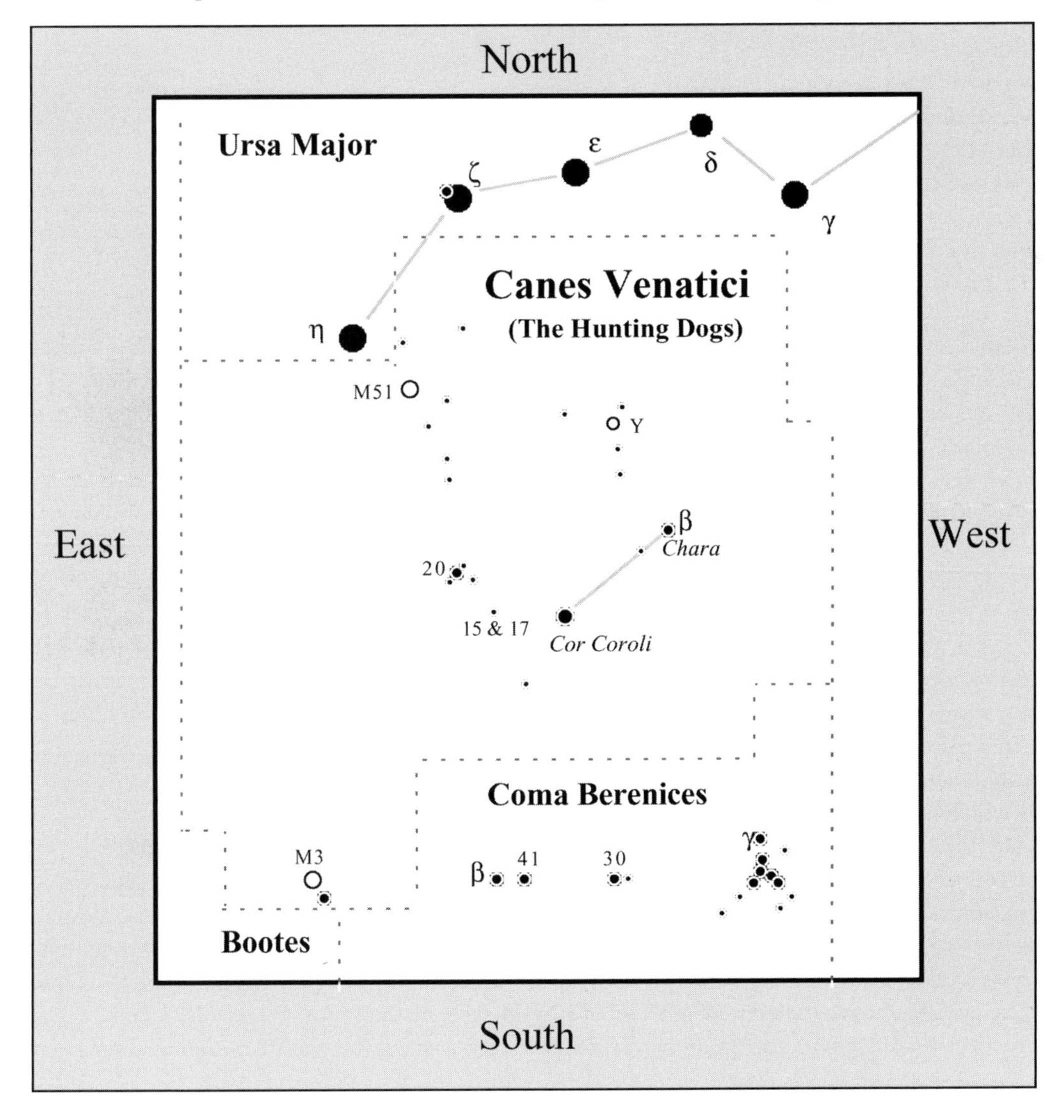

M51 and NGC 5195

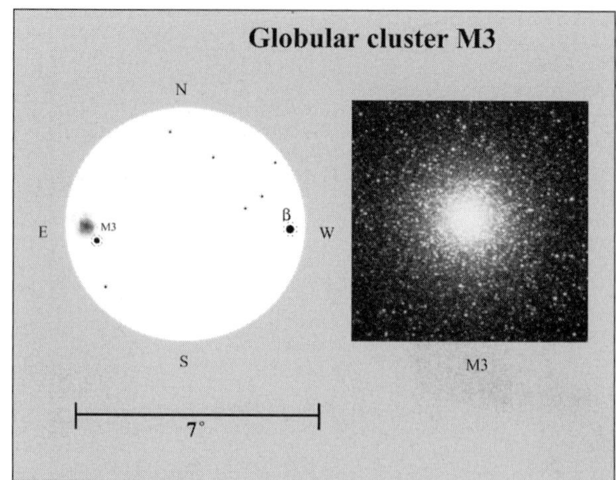

M3

lofted from the star's interior to its surface before escaping into space. La Superba is in fact surrounded by a huge detached shell of matter of its own making with a diameter of around 2.5 light years. The star varies in brightness from about magnitude 5.2 to 6.6 over a period of 158 days, though the exact period is somewhat in question.

In the chart on page 33, do you see how the dashed lines define the borders of the constellation? The modern 88 constellations all have boundaries defined by Belgian astronomer Eugene Delporte and adopted by international agreement in 1930. Such borders are arbitrary though when it comes to finding deep-sky objects. For instance, note how close the 8th-magnitude spiral galaxy M51, the famous Whirlpool Galaxy in Canes Venatici is to Eta (η) Ursae Majoris, the 2nd-magnitude star at the tip of the Big Dipper's handle. To find M51, you want to forget about constellation boundaries and just use the brightest stars nearby. To find M51, then, it's best to start at Eta Ursae Majoris. Once you center that star in your binoculars, use the chart above to get a fix on the galaxy's position $3\frac{1}{2}°$ to the southwest.

In 10×50 binoculars under a dark sky, the galaxy appears as a slightly elongated nest of diffuse light with a sharp, starlike nucleus. But M51 is really a magnificent maelstrom of some 200 billion suns 15 million light years distant. Its disk, which we see face on, spans about 50,000 light years. It travels through space with a little companion galaxy known as NGC 5195, which is visible in the photograph accompanying the chart above.

Globular star cluster M3 in Canes Venatici is another borderland object. To find it, first use your unaided eyes to locate the Y-shaped Coma Berenices star cluster, then look about 10° (one fist) east for 4th-magnitude Beta (β) Comae Berenices, which is paired with 5th-magnitude 41 Comae. Place Beta in your binoculars then use the chart at top right to locate M3 which is literally a straight shot – one binocular field – due east. The cluster is an easy catch being not only bright (6th-magnitude) but also highly concentrated at the core. Some amateur astronomers have even glimpsed M3's glow with their unaided eyes! In binoculars, the cluster has a pale-yellow sheen and swells magnificently with averted vision under a dark sky. When we look at this fantastic orb, we are seeing a half-million or so suns 32,000 light years distant, in the halo of our galaxy.

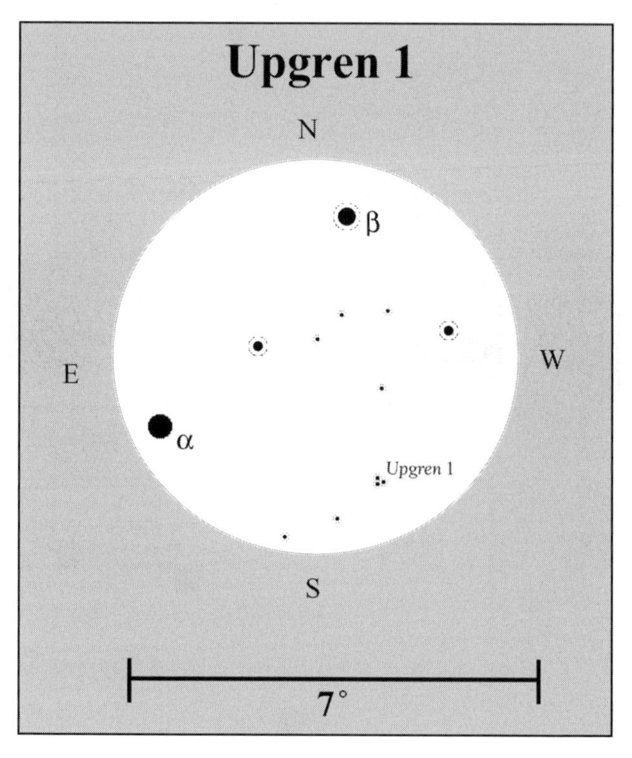

Before leaving Canes Venatici, be sure to look at the pretty gathering of little suns around 20 Canum Venaticorum, then at 15 and 17 Canum, which has a third member, 6 Canum, immediately to the north. But for a final challenge, try finding Upgren 1 – a 6th-magnitude asterism of five 7th- to 9th-magnitude stars $4\frac{1}{2}°$ west–southwest of Cor Caroli (only three stars are shown on the chart); some pretty things come in small packages.

Virgo, the Virgin

The largest constellation next to Hydra is beautiful Virgo, the Virgin, the first (and only) Lady of the Zodiac. Despite its size, Virgo is somewhat lackluster, save for its 1st-magnitude luminary Alpha (α) Virginis (Spica) – one of spring's jewels of the night. When May opens, Spica is about as far east of the meridian as equally bright

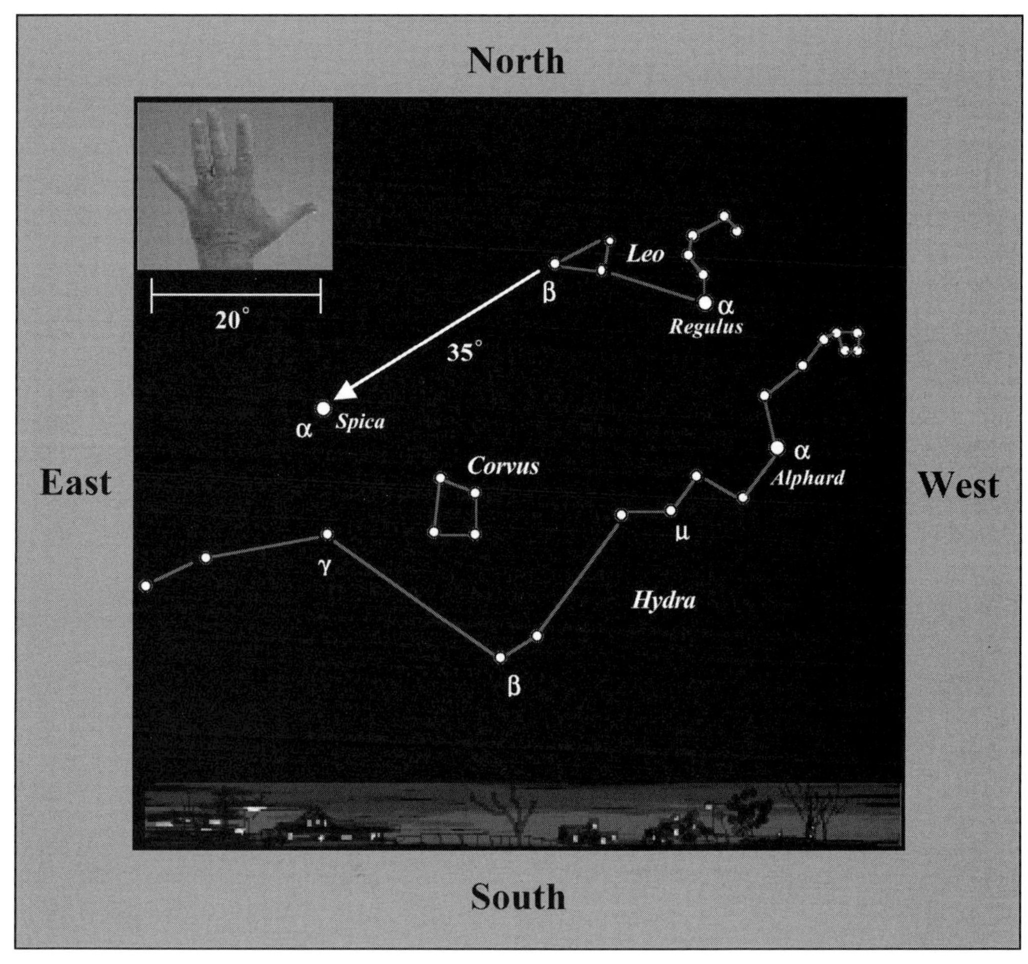
North
Leo
β
α
Regulus
20°
35°
α
Spica
Corvus
α
Alphard
East
West
μ
γ
Hydra
β
South

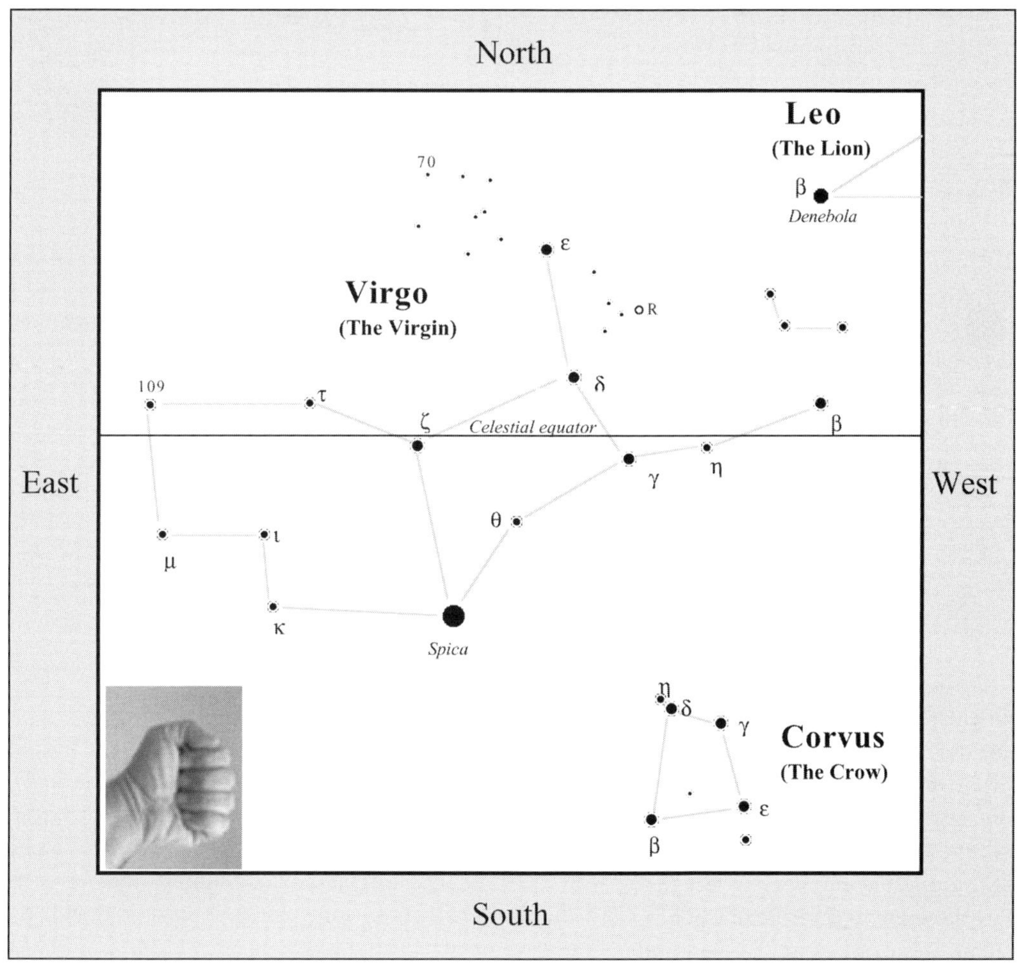
North
Leo
(The Lion)
β
Denebola
70
ε
Virgo
(The Virgin)
R
109
τ
δ
ζ
Celestial equator
β
East
γ
η
West
θ
μ
ι
κ
Spica
η
δ
γ
Corvus
(The Crow)
ε
β
South

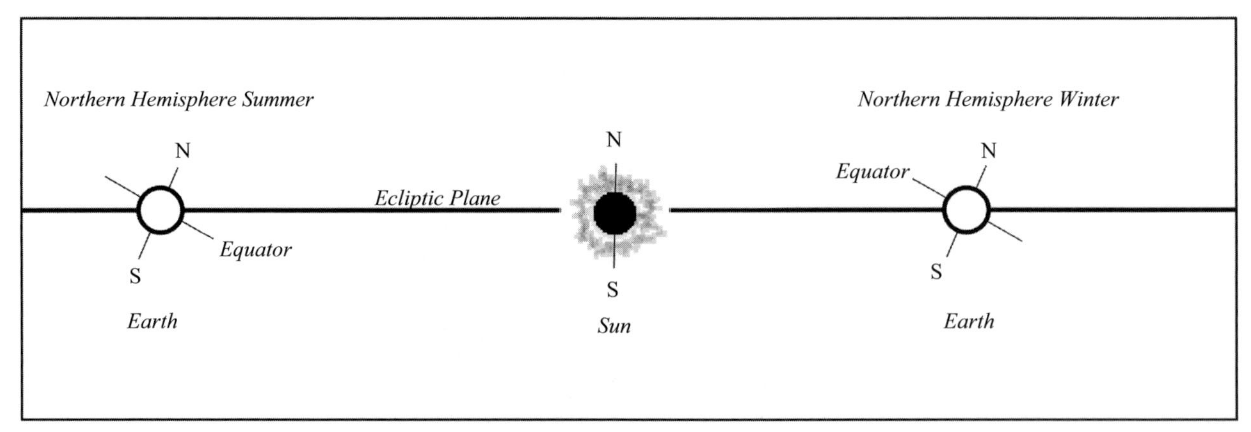

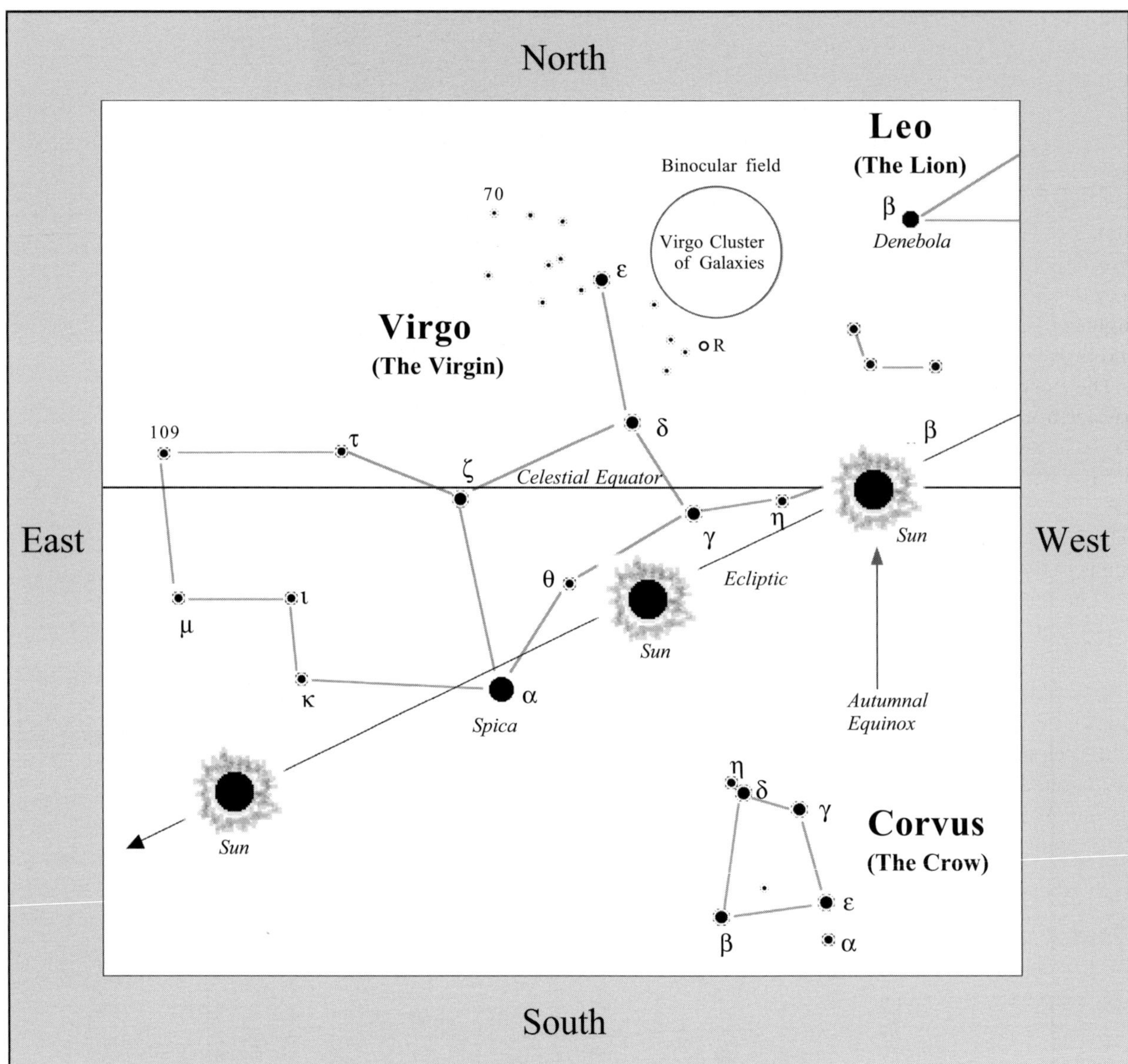

Regulus is west of it, though Spica is lower in the sky. Spica lies about 10° south of the celestial equator – the great circle that shows where Earth's equator, when projected into space, intersects the celestial sphere.

So part of Virgo lies north of the celestial equator and part of it lies south of the celestial equator – a fact that did not go unnoticed by the ancients. In Roman legends, Virgo was Proserpine, daughter of Jupiter and Ceres, and the goddess of fertility. Proserpine's tale literally has a shaky beginning. After Jupiter and his brothers buried the giant Enceladus alive under fiery Mount Etna on Sicily, the earth began to shake so violently that Pluto, God of the Underworld, believed the ground would crack and bleed unwanted sunlight into his dark domain. So Pluto flew out of the Earth in a fury, riding his dark chariot drawn by four black steeds. (The volcanic imagery is unmistakable.)

But Pluto's attention was quickly diverted by the sight of beautiful Proserpine, who was picking lilies and violets in a valley with her friends. Overtaken by lust, Pluto swooped down from the sky and kidnapped the lovely nymph. He then plunged his chariot into the River Cyane, where he opened a passage to Hades and forced Proserpine to become his bride.

Alarmed by Proserpine's disappearance, Ceres scoured the Earth for years but failed to find her daughter. As the years passed, flowers began to wilt, the land became parched, and cattle began to die. Just as life on Earth seemed near an end, Ceres discovered Proserpine's plight and implored Jupiter to intervene. Jupiter consented on one condition: that Proserpine should not have tasted any fruits of the Underworld. But the wily devil overheard Jupiter's concern and fed Proserpine a pomegranate before her release. When Jupiter learned of the devil's deceit, he was enraged. But Pluto consoled Jupiter by offering him and Ceres a fair agreement – he'd allow Proserpine to roam the Earth's surface, where she could create new life for six months of the year (spring and summer) as long as she returns to Hades during the fall and winter. And that's why the Earth turns cold and flowers die during the long nights of fall and winter, only to be reborn in the spring and thrive every summer.

The story of Proserpine also mirrors the Sun's annual passage from the north celestial sphere to the south celestial sphere, as it journeys eastward along the ecliptic in the fall. In fact, on the first day of autumn (the autumnal equinox), the Sun is in Virgo, at a point where the ecliptic intersects the celestial equator. At that point, the Sun is passing from a time of light (the long days of summer) to a time of darkness (the long nights of winter). On the equinox, however, the hours of day and night are equal: "equi" (Latin for "equal") and "nox" (Latin for "night"). We call the autumn season "fall" because the chariot of the Sun is now plunging into the underworld of the celestial sphere. The plunge occurs because Earth's axis is tilted $23\frac{1}{2}°$ with respect to the plane of the Solar System. As the Earth orbits the Sun, its axis remains in the same relative orientation, so that during Northern Hemisphere summer, the Sun's rays are directed north of the equator; during winter, the Sun's rays are directed south of the equator. During the spring and fall equinoxes, the Earth is at a position 90° from these two points.

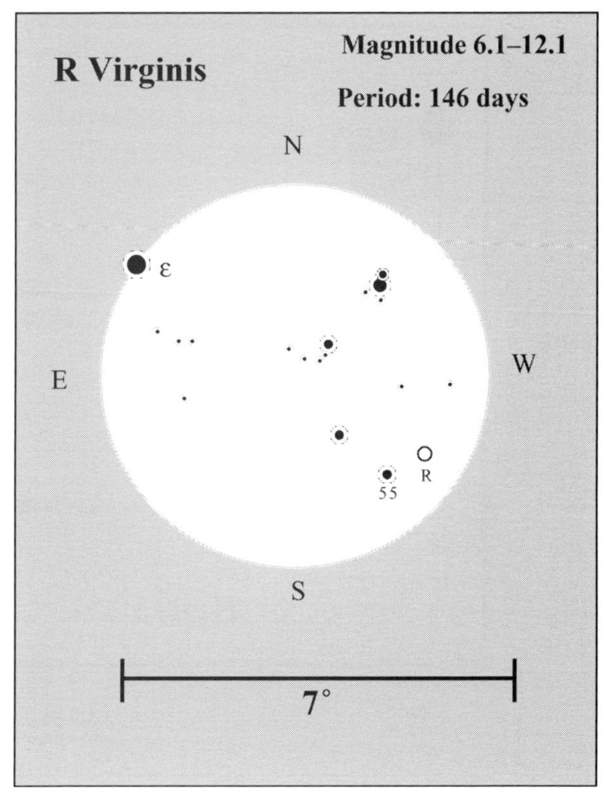

Proserpine is but one of the many identities of Virgo, all of which symbolize purity and fertility. In Babylonia, Virgo was Ishtar, Queen of the Stars. Medieval Christians saw her as the Madonna. And in ancient Egypt, Virgo was the God Mother, Isis, who among other things, possessed the powers of a water goddess, an earth goddess, a corn goddess, and a star goddess; more importantly, she was also Queen of the Underworld.

Now turn your binoculars to radiant Spica. At a distance of 260 light years, Spica is a B-type dwarf that's actually a very close double star. Through binoculars, the star shines with a spectacular ice-blue-diamond radiance. Compare the blue of Spica with the golden sheen of

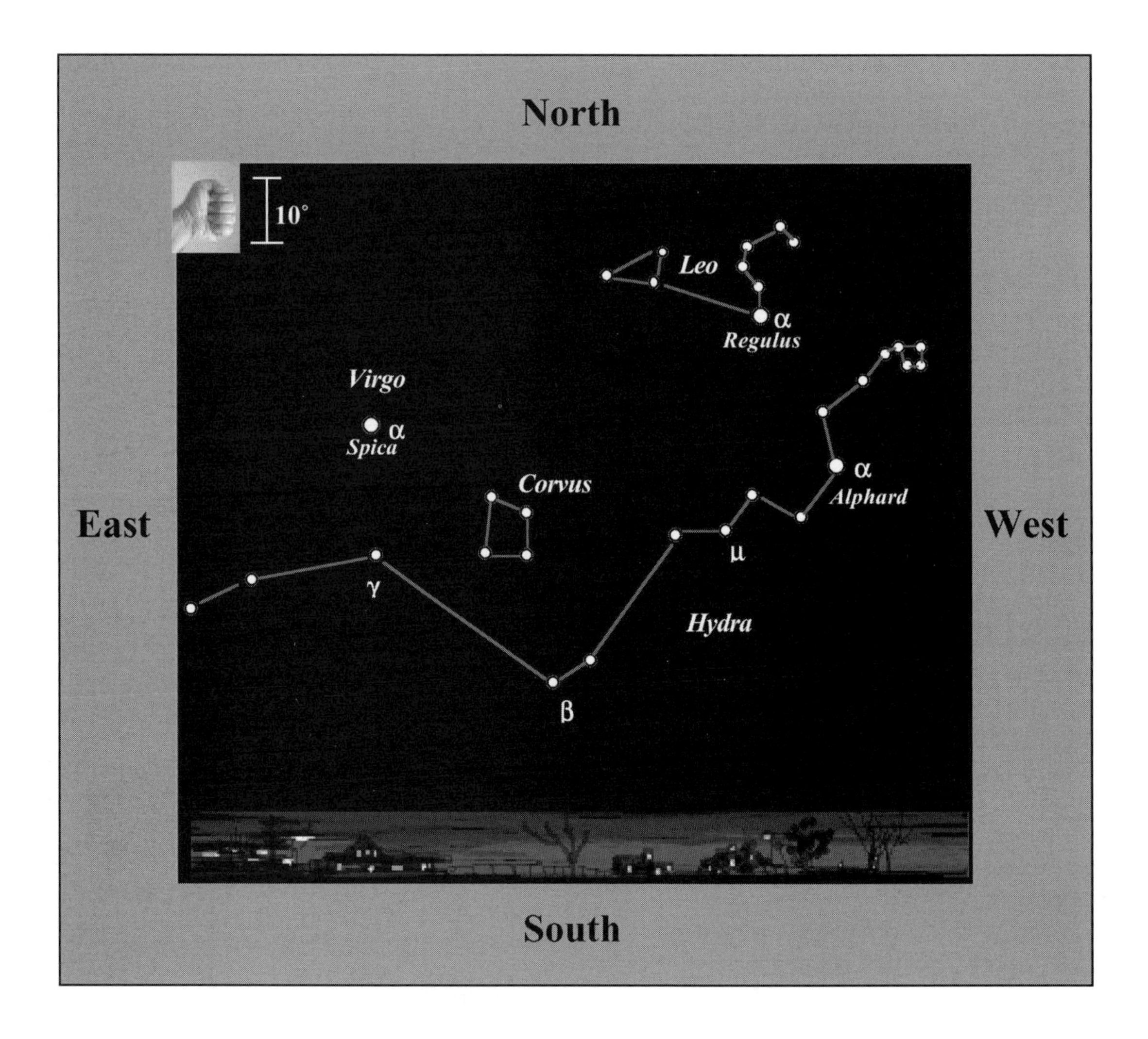
North
10°
Leo
α
Regulus
Virgo
α
Spica
Corvus
α
Alphard
East
West
γ
μ
Hydra
β
South

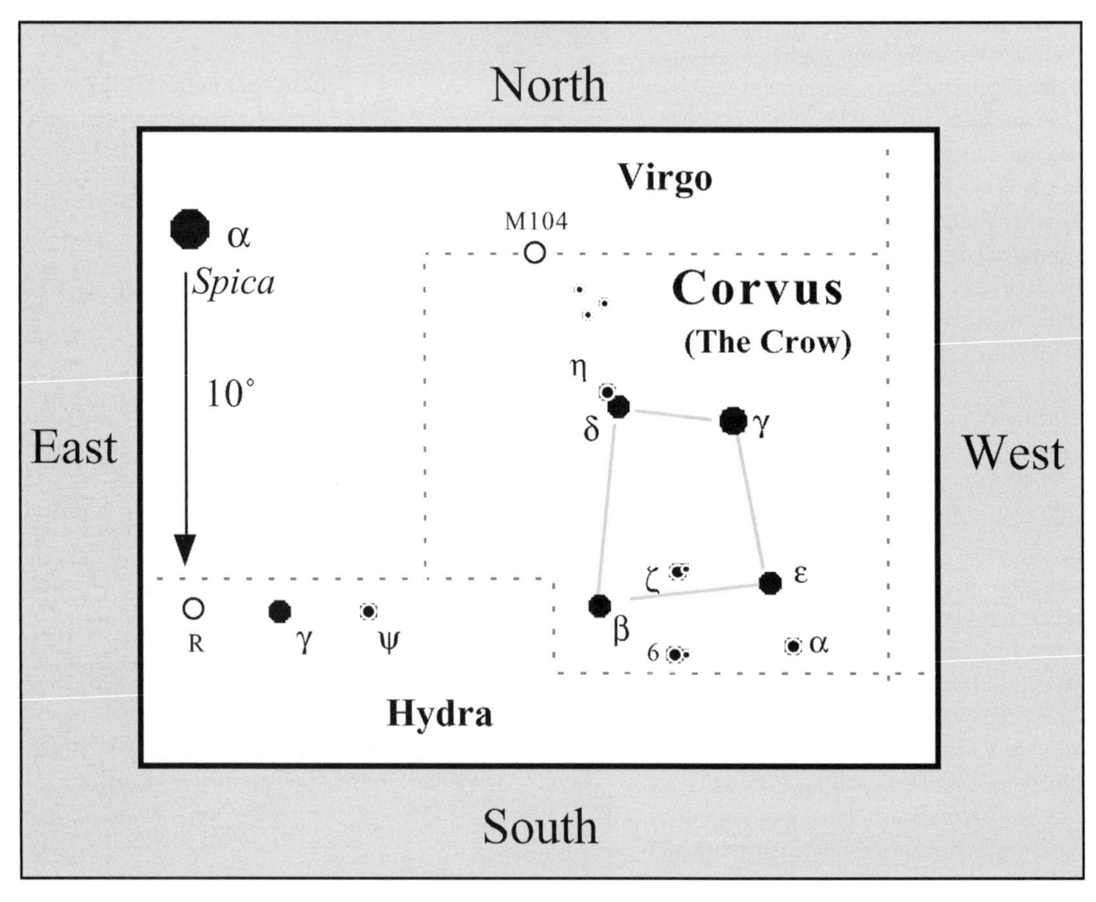
North
Virgo
α
Spica
M104
Corvus
(The Crow)
10°
η
δ
γ
East
West
ζ
ε
R
γ
ψ
β
6
α
Hydra
South

3rd-magnitude Delta (δ) Virginis – one of the few M-type red giants visible to the unaided eyes. While Delta Virginis is nearly 60 light years closer than Spica, it is only about 5 percent as luminous, so it appears dimmer to our eyes. Fourth-magnitude Epsilon (ε) Virginis (Vindemiatrix) to the north has a cooler lemon-peel hue. Indeed, the star is a yellow giant 102 light years distant.

One binocular field to the southwest you'll find the Mira-type variable star R Virginis, which rises to a maximum of 6.1 magnitude, making it within reach of the unaided eye under a dark sky. Around maximum light, the star should be slightly fainter than a 5.5-magnitude star immediately to its southeast.

If you raise your binoculars one field to the north, you will be looking 55 million light years distant into the heart of the great Virgo Cluster of Galaxies. This massive collection of some 2,500 galaxies completely dominates our corner of the visible universe. The united gravitational front of this system is so great that our Milky Way, and the entire Local Group of Galaxies, is being drawn toward it. Unfortunately your binoculars are too feeble to see these distant island universes, each with billions of suns. For now, you'll have to rely on your imagination to ponder its magnificence.

When you have finished pondering your insignificance, sweep your binoculars eastward, back to Epsilon Virginis, then move one field northeast to 5th-magnitude 70 Virginis. Once again, we have entered the realm of the imagination, for here is a G-type sun with a planet nearly seven times more massive than Jupiter. The planet is about 40 million miles from its parent star, placing it at about the the distance of Mercury from our Sun. Since 70 Virginis is only 59 light years distant, anyone on that giant planet spying on Earth through a very powerful telescope in the year this book is printed might catch a glimpse of the B-50 Superfortress Lucky Lady II making its historic, nonstop flight around the world, or see the last six survivors of the American Civil War meet in Indianapolis, Indiana (both events occurred in 1949).

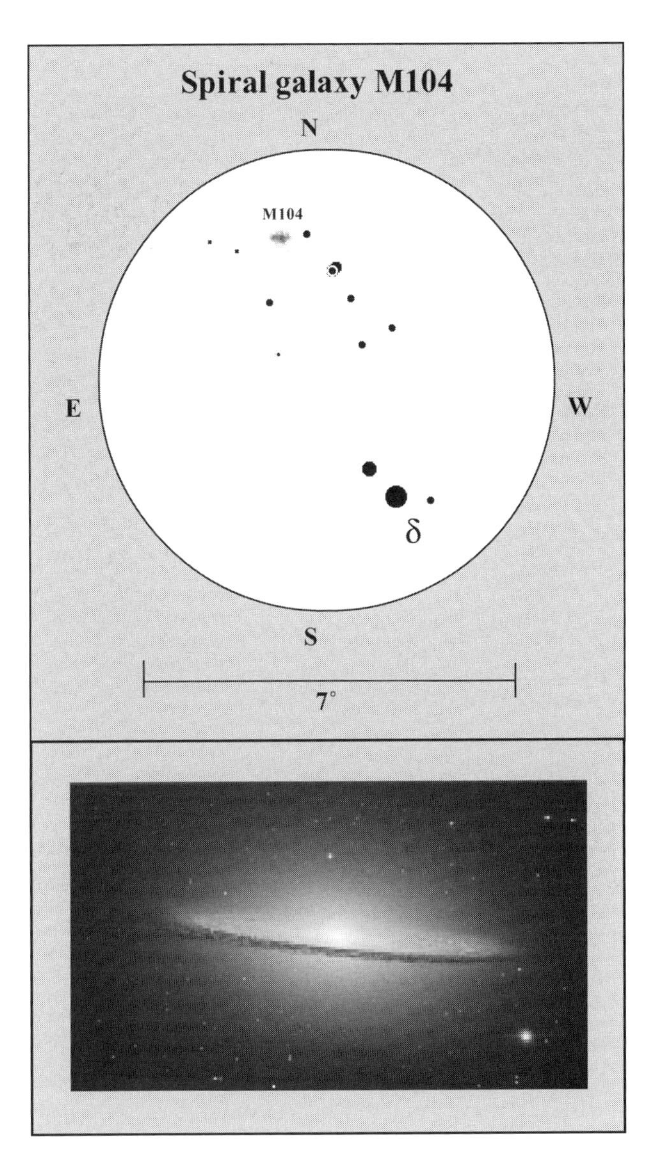

Corvis, the Crow

Look at the chart on page 38 top, which shows the trapezoidal-shaped constellation of Corvus, the Crow, which is about 15° southwest of Spica. Your fist should fully cover the formation, which is also called the Sail. Now sweep your binoculars across the Sail and scrutinize the colors of its brightest stars. Delta (δ) and Gamma (γ) Corvi should appear blue, while Epsilon (ε), Beta (β), and Eta (η) will all be varying shades of yellow. I see Epsilon having the warmest (almost golden) color and Eta having the palest hue. Zeta (ζ) Corvi has a greenish tint which contrasts well with its pale-yellow companion. Now drop your gaze to the south – to the wide double star 6 Corvi, whose primary burns with a reddish hue. You can also check the region 10° (a fist) south of Spica periodically for the appearance of the Mira-type variable R Hydrae, which ranges in brightness from a bright magnitude 4.2 to a faint magnitude 9.5 every 389 days, though the period has shown signs of decreasing over recent years.

Finally, if you're up for a challenge, try to find M104, the wonderful Sombrero Galaxy in Virgo. My wife, Donna, noticed it in a sweep of the sky with 7 × 50 binoculars. The 8th-magnitude galaxy is a nearly edge-on system some 65 million light years distant.

3 June

And stars, ringed glittering in whorls and bells,
Or bent along the sky in looped star-sprays,
Or vine-wound, with bright grapes in panicles,
Or bramble-tangled in a sweetest maze,

Sidney Lanier, *June Dreams, in January* (1869)

Since we began our journey in April, the Earth has traveled one-sixth of its way around the Sun. As a result, our perspective of the night sky has changed accordingly. It seems only a short time ago that the bright stars of Leo prowled the heavens so high and majestically in the south.

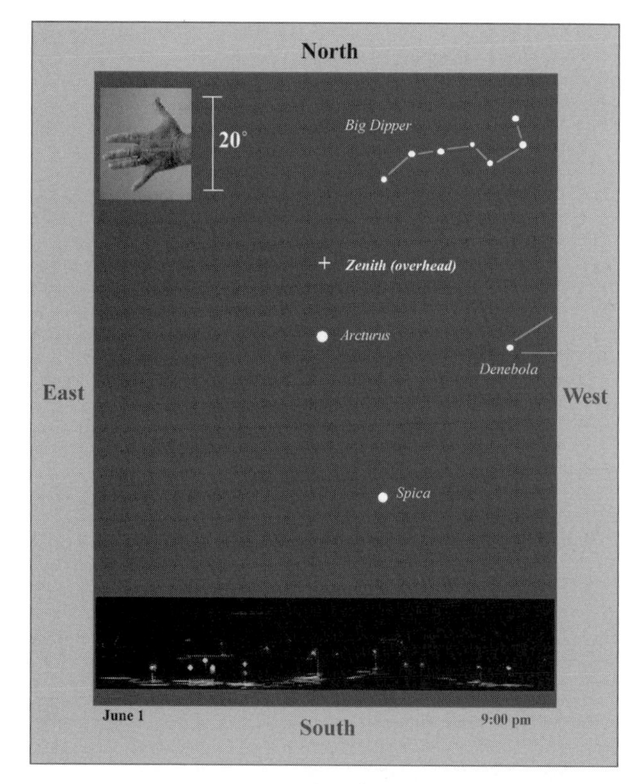

Now the Lion is leaping nose first toward its twilight den as summer approaches. Face north and you'll see that Ursa Major, the Great Bear, has summersaulted counterclockwise over the pole and landed on the tip of its nose. Spica still gleams brilliantly in the south. But if you look midway between Spica and Eta (η) Ursae Majoris, the tip of the Bear's tail, and a tad east, you'll see a brilliant topaz wonder – magnitude-0 Arcturus, the brightest star north of the celestial equator and the fourth brightest star of all. You can confirm its presence by extending the curve of the handle of the Big Dipper, which "arcs to Arcturus, then speeds on to Spica." Arcturus also marks the eastern tip of an equilateral triangle made with Denebola, the tail of the Lion, and Spica. I call this the Spring Triangle.

Bootes, the Herdsman

There's no mistaking Arcturus, even with the unaided eye. This gem blazes with the light of molten gold; through binoculars, the star's fringes flicker with reddish fire. Arcturus means "the guardian of the Bear." It is the Alpha star in Bootes (pronounced Boh-OH-Tease), the Herdsman. Face north and let the Earth turn and you can watch the star and its parent constellation follow Ursa Major counterclockwise around the north celestial pole. The star's (as well as the constellation's) heritage is somewhat confused. In Chapter 1, I introduced you to the legend of Ursa Major, starring the unfortunate Callisto and her illegitimate son Arcas (Arctos), who Jupiter transformed into Ursa Minor, the Little Bear. But other versions of the story turn Arcas into Arcturus. "Arctos," after all, is Greek for "bear," and Arctos is the root for the polar word "arctic." So in this latter version of the tragic tale, Arcas, the Herdsman, protects his mother endlessly around the pole. Then again, the problem with this version of the story is that Arcturus and Bootes are not circumpolar. Unlike the bears in the associated legend, Arcturus and many of the other stars of the Herdsman dip below the western horizon during the winter months, only to tiptoe back into the sky in the early morning hours – like a bad child who broke a parent's curfew.

Like Alpha (α) Ursae Majoris (Dubhe), Arcturus is a K-type giant. But at 37 light years distant, Arcturus is nearly four times closer, making it a near neighbor to our Sun. Now compare the brightness of Arcturus with that of Spica. Although Spica is a magnitude fainter than Arcturus, it is *seven* times farther away! Remember, Spica is a hot binary star; its absolute radiance is 2,100 times that of the Sun! Arcturus is only 113 times more luminous than our star. Still, the swollen form of that bloated K-type giant could consume 25 million Earths.

Arcturus is special for another reason: it has a high proper motion, whipping through space at about 87 miles per second in the direction of Virgo. Over the last 1,600 years, in fact, Arcturus has moved against the starry backdrop by about two Moon diameters (1°). Not only is this rate of movement much faster than any of the other stars in the Sun's immediate neighborhood, but it's also traveling in a rogue direction – toward the star Spica. It is possible then, that Arcturus is an adopted star – one from a dwarf galaxy that our Milky Way cannibalized some 6 or 7 billion years ago.

Aside from all its visible wonder, Arcturus entered the history books when its light was used to open the Century of Progress Exposition as part of the 1933 World Fair in Chicago. On the evening of May 27, at 9:15 pm central time, light from Arcturus was focused onto photocells through telescopes at four observatories, including the great 40-inch refractor at Yerkes Observatory in Wisconsin; the other telescopes included those at Harvard College Observatory in Massachusetts, the Allegheny Observatory in Pittsburgh, and a 29.5-inch Cassegrain telescope just south of the University of Illinois campus. The current generated from all these instruments was sent to Chicago over telegraph lines and was great enough to trip a switch that lit up the fair grounds, "turning on" the Expo. The event was a keen publicity stunt driven by astronomical foresight. In 1933, astronomers believed that Arcturus was 40 light years distant. Therefore, it only seemed appropriate that the 1933 Expo should be opened by light that *started* its journey 40 years ago at the Columbian Exposition, Chicago's previous "World Fair," held in Jackson Park in 1893. No doubt, the 39 million attending the 1933 Expo enjoyed the show as shown in this postcard from the Dr. Andrew Wood collection, used here with permission.

Don't worry if you can't see the Herdsman in your mind's eye. Few modern observers can fathom some of the constellations created by the ancients. Time has a way

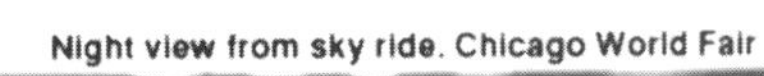
Night view from sky ride. Chicago World Fair

KAUFMANN-FABRY PHOTO

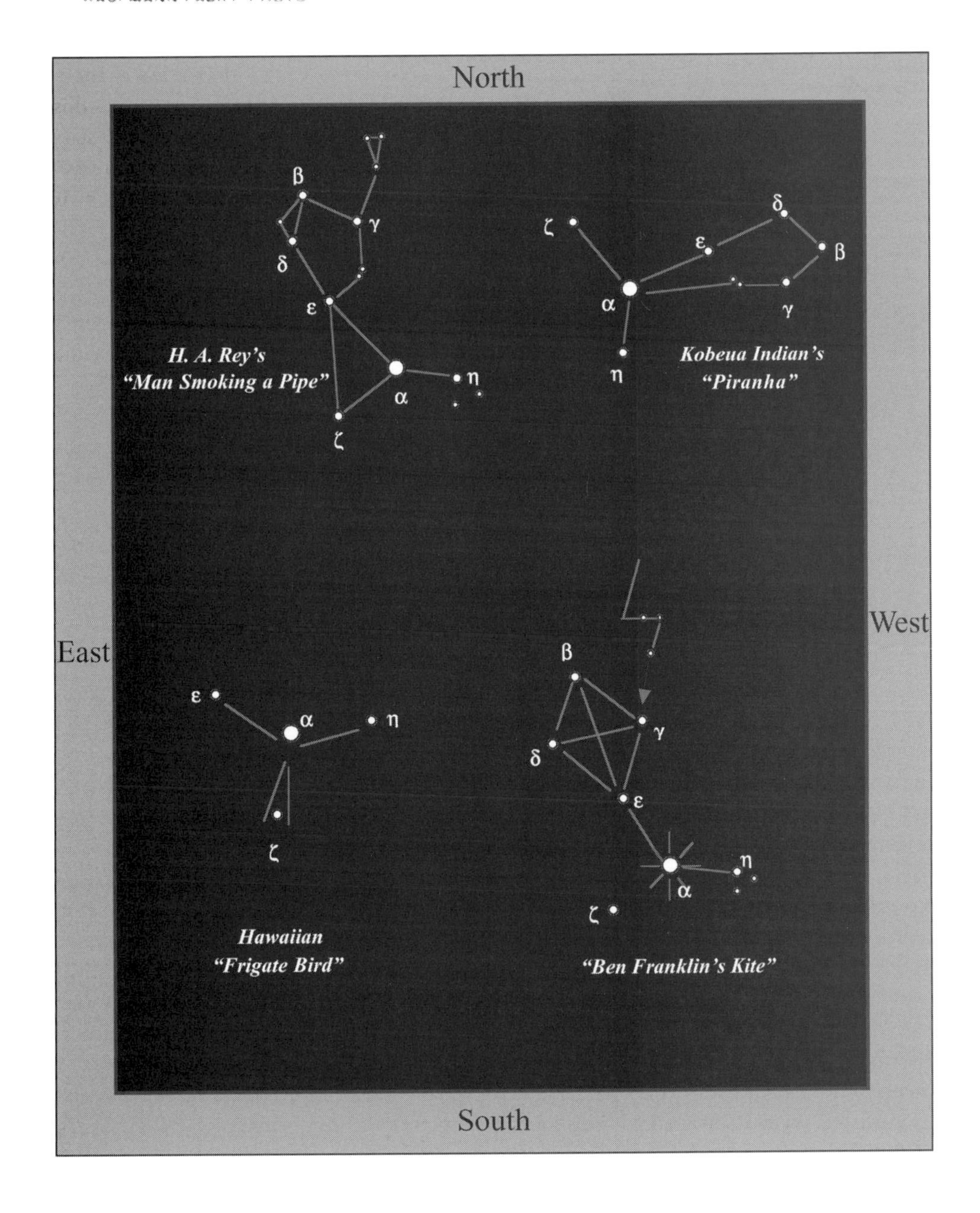
North
East
West
South
H. A. Rey's
"Man Smoking a Pipe"
Kobeua Indian's
"Piranha"
Hawaiian
"Frigate Bird"
"Ben Franklin's Kite"

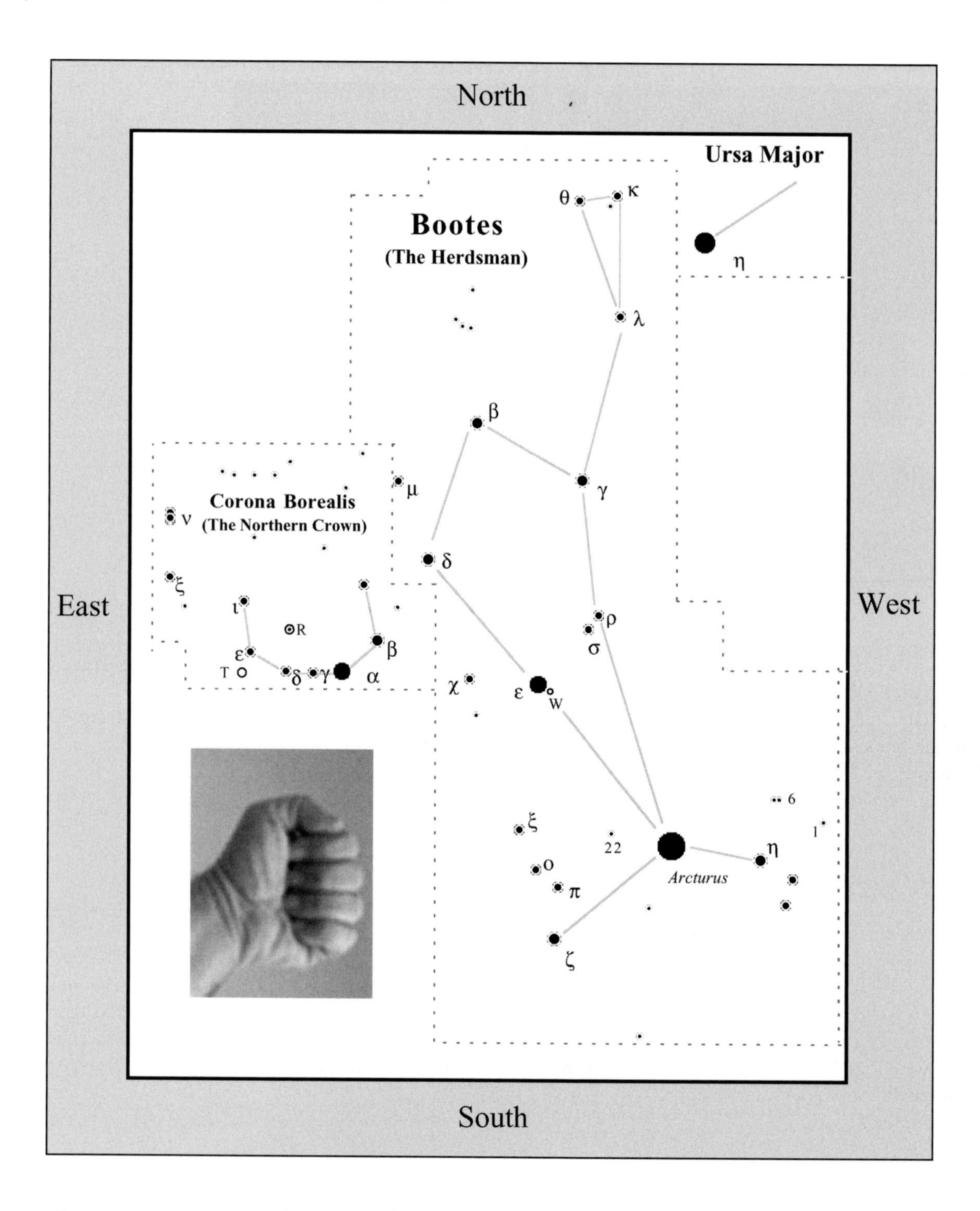

of wearing away memory (especially of things historical) and replacing old thoughts with new. This point came to me clearly one night when my wife, Donna, and I were under the stars, splayed out on a blanket with our tired papillon Daisy. When Donna asked about some of the springtime constellations, I started to show her the old patterns. After contemplating the view, she said, "I'd never join the stars that way. Someone should come up with a new way of looking at them."

Actually, humans have been doing just this for years. Take the Kobeua Indians of northern Brazil. They did not converse with the ancient Greeks or Romans, so they did not see Alpha (α) Bootis (Arcturus) as a guardian of the Bear. They had their own mythologies and star patterns. To them, the brightest stars of Bootes were a Piranha, with Beta (β) Bootis being the fish's snout and Zeta (ζ), Alpha (α), and Eta (η) Bootis, the fish's tail. In Hawaiian lore, Arcturus is the head of a frigate bird hovering over the Earth with wings extended – Epsilon (ε) and Eta (η) Bootis; Zeta (ζ) Bootis marks the bird's majestic tail. And in his classic book *The Stars: A New Way to See Them* (Boston: Houghton Miffin Company, 1952), the late H. A. Rey depicts Bootis as a man with a large head and small body, smoking a pipe [Gamma (γ), Lambda (λ), Kappa (κ), and Theta (θ) Bootis]. I see Beta, Delta (δ), Epsilon and Gamma Bootis as Ben Franklin's Kite, with Alpha and Eta Bootis as its tail; Arcturus is Franklin's key charged with electrical current (a fitting image given Arcturus's role in "charging" the lights of the 1933 Expo) and the kite's western tip is being struck with lightning – the stars Gamma, Lambda, and Kappa.

Take the time now to use the chart on page 42 and scan the brightest stars of Bootes with your binoculars. Note how Eta (η) Bootis (Muphrid), like Arcturus, shines with a golden light. Compare the two stars with your binoculars. Eta and Arcturus both lie 37 light years distant. But while Arcturus is a classic, full-fledged *K*-type orange giant, Eta is an orange *G*-type *sub*-giant only nine times brighter than the Sun; its core has just ceased hydrogen fusion and is starting the process of evolving into a giant. Now check out Epsilon (ε) Bootis (Izar). Like Arcturus, it too is a *K*-type giant but some 200 light years distant. It shines so brightly (3rd magnitude) because it radiates 400 times more light than does the Sun. So Epsilon is also like Dubhe in Ursa Major, only slightly more distant, which explains why Epsilon's light looks more lemon yellow than gold.

Gamma (γ) Bootis (Seginus) is an *A*-type giant star with a pale-green hue. At 85 light years distant, it is about as far away from our Sun as are the *A*-type dwarfs in the Big Dipper. Seginus is surrounded by a small cloud of dust that remains a mystery to astronomers.

Bootes has no bright star clusters or galaxies, but it has plenty of pretty stellar pairs and groupings. Be sure to check out the binocular fields around 1, 6, Xi (χ) and Beta (β) Bootis for a delightful array of star color and geometrical variety. For a challenge under a dark sky look at 5th-magnitude 22 Bootis, which is followed by a string of three dim pairs of stars to the east.

Two interesting variable stars hug Epsilon Bootis (see the chart at right). The closest is the semiregular variable W Bootis, which ranges from magnitude 4.7 to 5.4 with an unknown period. From 1966 to 1990, for instance, W Bootis pulsated with a period of about 25 days. From 1991 to 1994, its period was about 50 days, except in 1993 when its period was about 35 days. New periods of 25 and 33 days were found in 1996. So it is a fascinating and mysterious star. R Bootis is a more classic Mira-type giant that varies from magnitude 6.2 to 13.1 every 223 days.

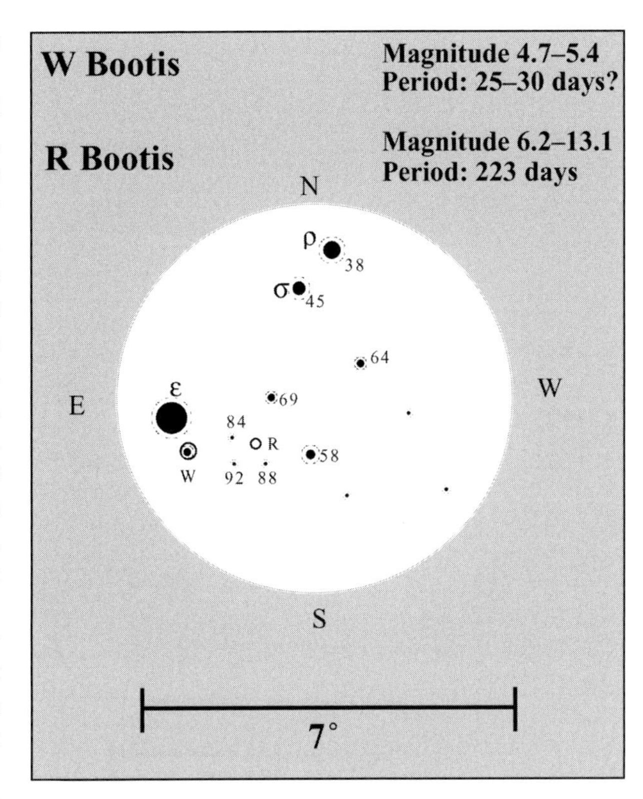

Corona Borealis, the Northern Crown

Look at the chart on page 42. Notice the beautiful C-shaped gathering of suns a little more than 10° east of Epsilon Bootis. This is Corona Borealis, the Northern Crown, whose brightest member, Alpha (α) Coronae Borealis (Alphecca), is the crown's central jewel. For this reason, the star is also known by the name Gemma, for jewel.

The story of the Crown centers around Ariadne, the daughter of King Minos. Ariadne's claim to fame was helping Theseus get out of the Minotaur's labyrinth after the mighty hero slew the beast with his bare hands; Ariadne had advised Theseus to tie some yarn at the maze's entrance, then pay out the thread on his way to the Minotaur; if the hero succeeded in killing the beast, he could then find his way out by following the string. To this day, spelunkers still use this trick whenever they enter deep underground passages. Ariadne, of course, fell in love with Theseus upon his return and the two planned to wed. But the return home to Crete was so long and arduous, that Theseus changed course for the isle of Naxos, where the maiden could get some sleep.

The next morning, however, Ariadne found herself alone. Theseus had taken the ship and abandoned her. Fortunately Bacchus, the Roman God of wine, frequented the island and heard Ariadne's cries of despair. Bacchus immediately fell in love with the helpless girl and took her for his bride. As a token of his undying love, Bacchus presented Ariadne with his golden crown (one forged by Vulcan and beset with precious jewels from India). The two lived happily but not forever. Ariadne was mortal and destined to die. When she did, Bacchus tossed her crown high into the heavens so that all may witness that symbol of a god's love for a mortal woman. As the crown ascended, the jewels flared in brightness and turned to stars.

The Crown's stars are all fairly common. But they do fit comfortably in your binocular field of view and make for a pretty sight. Most noteworthy is 2nd-magnitude Gemma, which is closely attended by three roughly 8th-magnitude stars, and stunning Nu (ν) Coronae, which is a pair of equally bright orange "eyes" in the northeast part of the constellation.

Most interesting, though, are the variable stars R and T Coronae Borealis. T Coronae Borealis, also known as "The Blaze Star," is the best representative of a special class of variable stars known as recurrent novae. T Coronae Borealis normally has a magnitude of about 10, but twice it flared – once in 1866, then again in 1946 – reaching 2nd-

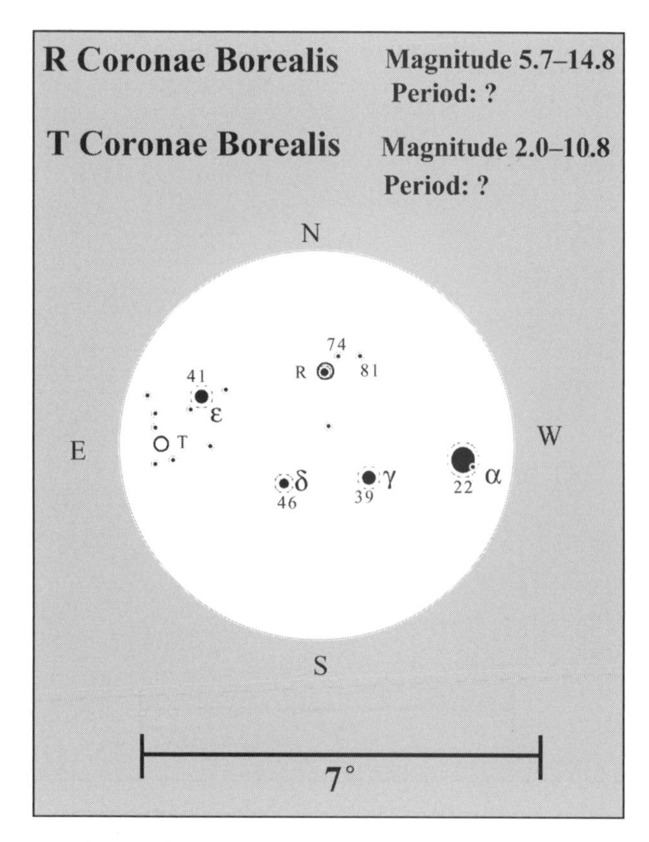

and 3rd-magnitude, respectively. It's now believed that T Coronae Borealis is a binary star system comprised of a swollen red giant star that's being orbited closely by a white dwarf. The outbursts, then, probably occur when the white dwarf star strips gases from the red giant star. The stolen matter forms a disk around the white dwarf that ultimately "overloads" and explodes in a thermonuclear reaction, which we see as a flaring of the "star." Unlike with a supernova event, the stars in a recurrent nova remain untouched in the explosion, giving them a chance to repeat the process. The illustration below shows the nova system before the blaze.

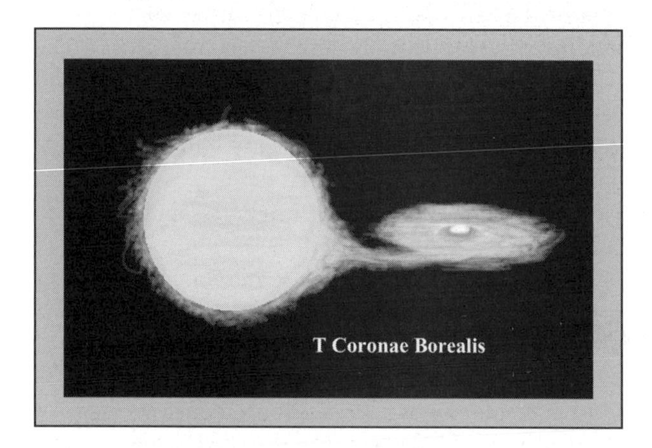

R Coronae Borealis, on the other hand, is the prototype of what I call an "upside down" variable star. Unlike a nova, which starts out faint, flares to prominence, then fades again, R Coronae Borealis – a carbon-rich supergiant – does just the opposite. It goes into "outburst" by fading, then returning to normal brightness. It's a remarkable phenomenon. Normally, R Coronae Borealis hovers around 6th magnitude, which is noticeable to the unaided eye and a cinch to see in binoculars. As the nights go by, the star might vary by a few tenths of a magnitude until one day it simply vanishes from view. It dips in brilliance by up to a thousand times within a few weeks, before it recovers its usual brightness over the course of months.

This spectacular fading may be caused by a phenomenon called the "Dust Puff" – a theory first proposed by Louisiana State University astronomer Geoffrey C. Clayton. In this theory, R Coronae Borealis burps up sooty clouds of carbon dust. The dust clouds rise above the star's surface where they cool and condense. The opacity of the cloud dims the view, causing the star's intensity to drop (imagine a bright Sun suddenly being eclipsed by smoke from a raging forest fire). In time, light pressure radiating from the star blows the carbon dust away. As the dust thins, the star's surface becomes more and more apparent, which we see as a gradual rise in brightness. What makes watching this star particularly exciting is that the period of these dimmings is highly irregular – a celestial form of peek-a-boo guaranteed to take you by surprise.

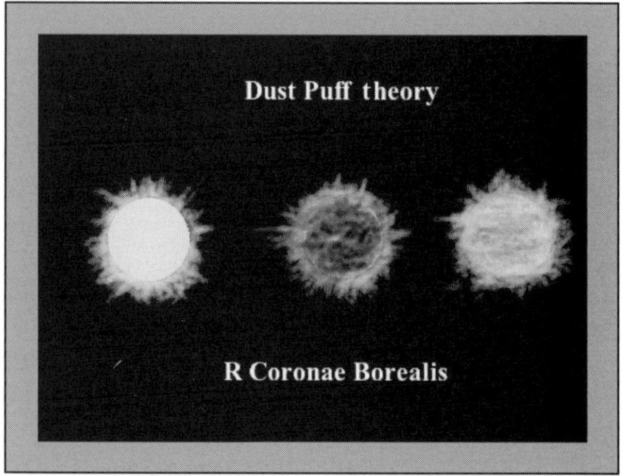

Ursa Minor, the Little Bear

The time has finally come for us to face north and squint in an attempt to see the form of the Little Dipper – an asterism in Ursa Minor, the Little Bear. I call the Little Dipper the Measuring Cup, because it looks like the one I use to scoop coffee from its tin. But be warned, unlike its popular sibling, the Big Dipper, the Little Dipper is truly a *bear* to see. Most beginners I've met have had no trouble identifying Polaris, the North Star, which marks the tip of the Little Dipper's handle or the Little Bear's tail. But that's about as far as they get.

Not to worry. June is a wonderful time for Little Bear hunting, because the star pattern has achieved its highest point above the north celestial pole. Start your search by locating the North Star (see page 2). Your next step is to look for the Little Dipper's second brightest luminary, Beta (β) Ursae Minoris (Kochab), which rivals Polaris in color and intensity. Kochab, which marks the front of the bowl

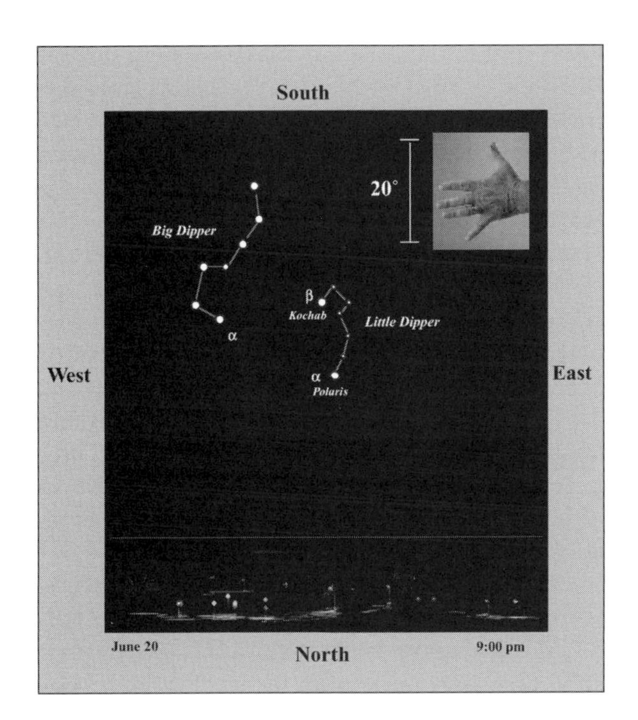
South
20°
Big Dipper
β
Kochab
Little Dipper
α
α
Polaris
West
East
June 20
North
9:00 pm

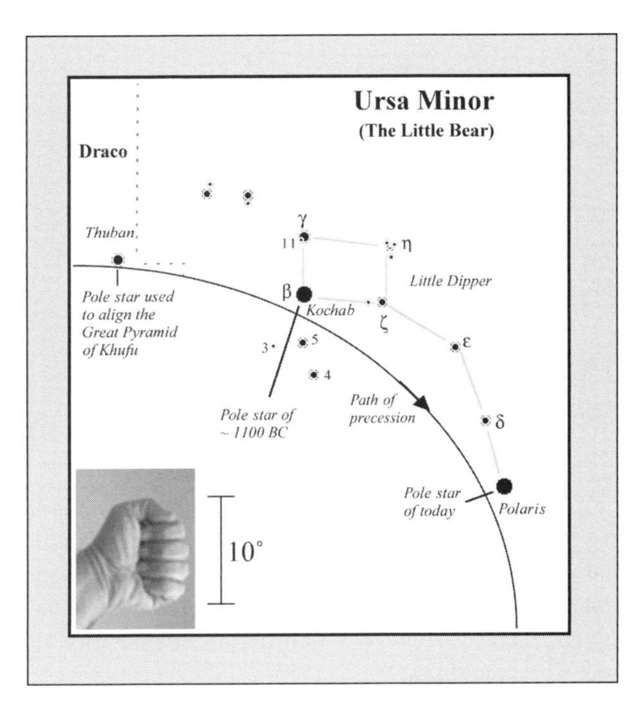
Ursa Minor
(The Little Bear)
Draco
Thuban
γ
η
Little Dipper
β
Kochab
ζ
Pole star used
to align the
Great Pyramid
of Khufu
ε
Pole star of
~ 1100 BC
Path of
precession
δ
Pole star
of today
Polaris
10°

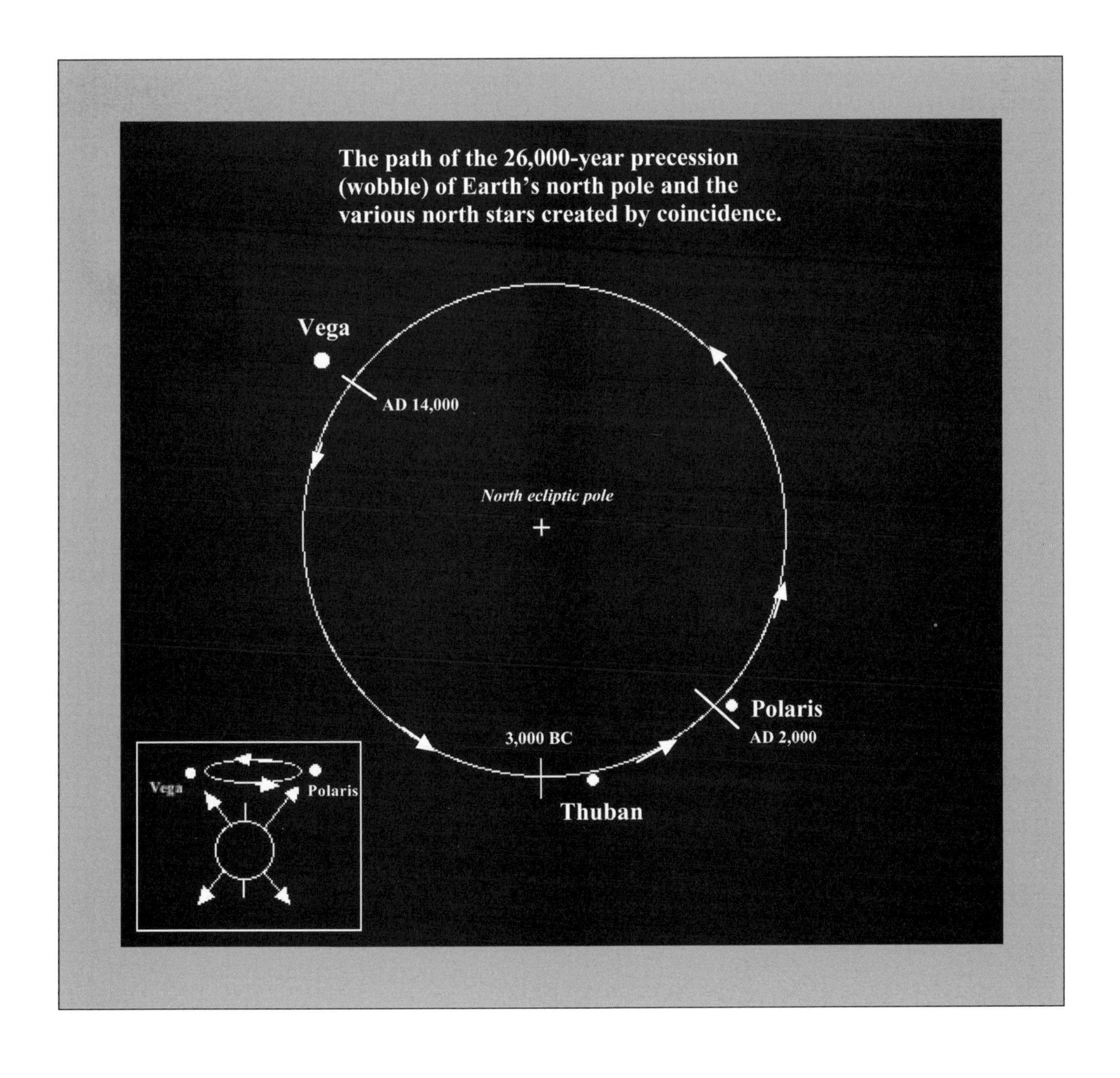
The path of the 26,000-year precession (wobble) of Earth's north pole and the various north stars created by coincidence.
Vega
AD 14,000
North ecliptic pole
Polaris
AD 2,000
3,000 BC
Thuban
Vega
Polaris

of the Little Dipper, was, for centuries, regarded as one of the guardians of the north celestial pole. In fact, in about the year 1100 BC, Kochab was closer to that imaginary point where Earth's northern axis intersects the celestial sphere, than Polaris is today. At that time, Kochab was our pole star, not Polaris.

As I described in Chapter 1, it is only by chance that Polaris is so close to the north celestial pole today. Polaris was not always our pole star and it will not always be our pole star. As the Earth spins on its axis, gravitational tugs by the Sun and Moon cause our planet to wobble like a top. Since one wobble takes about 26,000 years to complete, Earth's imaginary axis slowly precesses over the years, transcribing an invisible circular path in the northern sky. Right now, that point on the circle just happens to bring it close to Polaris, our present day North Star. In AD 14,000, the bright star Vega will be our pole star. The pole star of ancient Egypt was Thuban, in Draco (see page 71). It was to Thuban (not Polaris) that the northern air shaft of the Great Pyramid of Khufu was aligned. From this pyramid, the pharaoh Khufu spirited to the sky, where he joined the realm of immortality – the circumpolar zone where stars never set or perish.

At a distance of 126 light years, 2nd-magnitude Kochab shares traits with both Epsilon (ε) Bootis and Alpha (α) Ursae Majoris. Like them, Kochab is an evolving K-type orange giant. Through 10 × 50 binoculars, I see the star shining with a golden tulip luster. By comparison, Polaris has a more lemon or buttercup-petal tint. What's amazing is that although both Kochab and Polaris shine at 2nd magnitude, Polaris is almost $3\frac{1}{2}$ times more distant. That's because while Kochab is 500 times more luminous than the Sun, Polaris – an F7 yellow supergiant – is 2,500 times more luminous.

But don't let these numbers fuel the belief that Polaris, our North Star, is the brightest star in the night sky. Polaris ranks 48th among its fellow luminaries. The misconception most likely arises from the star's primary position in the heavens – namely that since recorded history, the pole star has remained a symbol of consistency and faithfulness; as Confucius expressed in the fifth century BC, "He who rules by moral force, is like the Pole star, which remains in its place while all the lesser stars do homage to it." In the ancient legends of Asia, the North Star is regarded as the pinnacle of the world's cosmic mountain or the axis of the universe – a reflection of the primal belief that we are the center of the universe. Virtually all the temples of India, such as those at Khajuraho (see below), are symbolic representations of that cosmic manifestation.

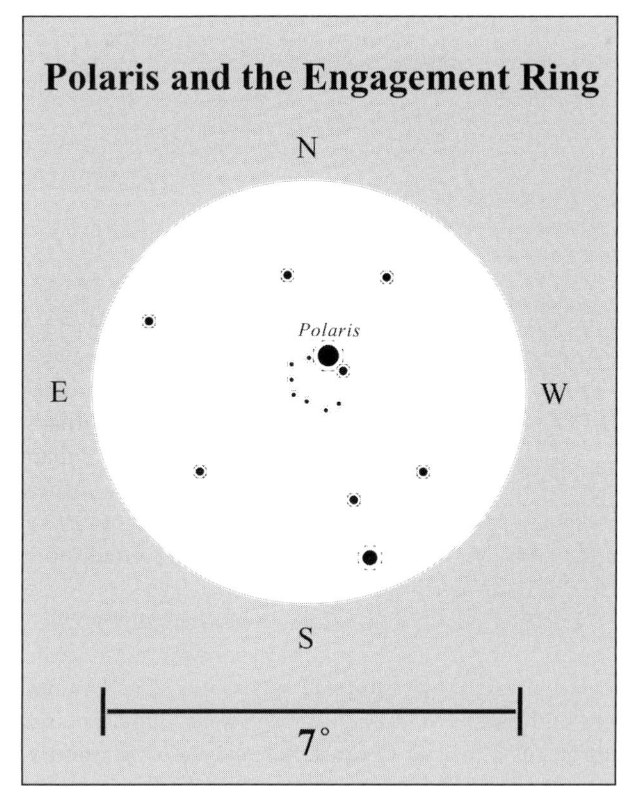

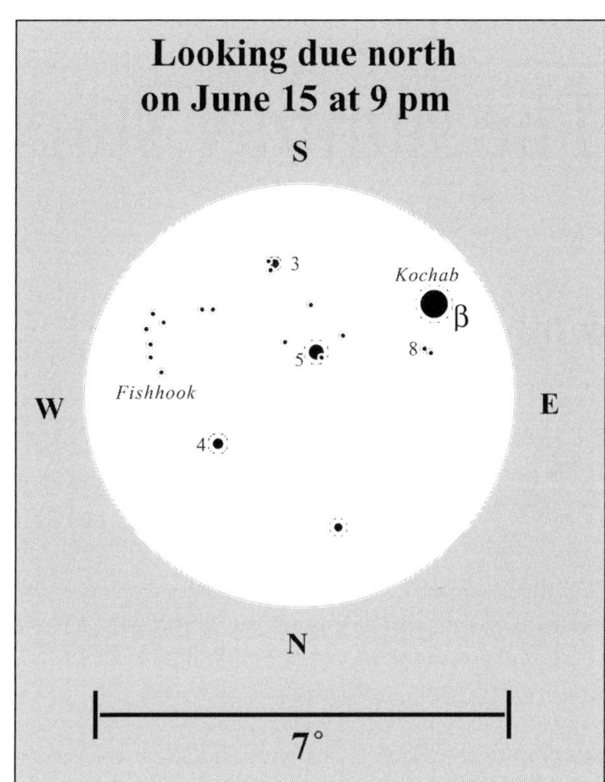

In Shakespeare's *Julius Caesar*, the Roman leader foresees his death:

> But I am as constant as the northern star,
> of whose true-fix'd and resting quality
> There is no fellow in the firmament.
> The skies are painted with unnumber'd sparks, –
> They are all fire, and every one doth shine;
> But there's one in all doth hold his place:
> . . .
> Hence! wilt though lift up Olympus?

And in his 1832 *Hymn to the North Star*, William Cullen Bryant refers to Polaris as *A beauteous type of that unchanging good, / That bright eternal beacon, by whose ray / The voyager of time should shape his heedful way*.

There is something else beautiful about the North Star that goes unnoticed by casual stargazers. If you raise your binoculars to that yellow beacon this month around 9 pm, and look closely beneath its radiant form, you should see a subtle semicircle of eight stars, forming an asterism known as the Engagement Ring, with beautiful Polaris as the radiant topaz jewel.

Owing to the ring's faded appearance (and its association with a symbol of eternity) I see it as the faded funerary garland in Walt Whitman's poem, "The Pallid Wreath": *For but last night I woke, and in that spectral ring saw thee, / thy smile, eyes, face, calm, silent, loving as ever: / So let the wreath hang still while within my eye-reach, / It is not yet dead to me, nor even pallid*.

The significance of this association became magnified one day when I discovered that the Mount Auburn Cemetery – an historical landmark near my childhood home in Cambridge, Massachusetts – has a finely polished gravestone on a hill facing north with the word "North" etched wisely on the south side of its face; someone had a grave sense of humor.

Now turn your attention to the other Guardian of the Pole – 3rd-magnitude Gamma (γ) Ursae Minoris (Pherkad). This *A*-type giant, 480 light years distant, has a delicate aqua hue, which may be a contrast illusion owing to the pumpkin color of its 5th-magnitude *K*-type attendant 11 Ursae Majoris. In Arabic, Pherkad means "dim one of the two calves," clearly a reference to it and Kochab.

If you go outside on or around June 15 at 9:00 pm and place Kochab in the top of your binocular field, you should see a nice assortment of stellar groupings dancing around the field. Below (nearly due north) of Kochab is 8 Ursae Minoris – a tight pair of 7th-magnitude suns. You'll need averted vision to spy the dim companion to golden 5 Ursae Minoris. Then look above and slightly to the right for 6th-magnitude 3 Ursae Minoris, which is part of a tight triangle of stars. Finally, use your averted vision to look for the 1°-long fishhook of 6.5- to 8.5-magnitude suns.

CHAPTER 3

The summer stars

4 July

They are immortal, all those stars both silvery and
golden shall shine out again,
The great stars and the little ones shall shine out again
they endure,

Walt Whitman, *Leaves of Grass* (1871)

Summer has arrived. The nights, though shorter, are warmer, making the time spent under the stars a more comfortable experience. In my youth, summer brought family and friends together under the stars, even if for a little while. This was especially true whenever I went camping in the early 1970s with friends in the White Mountains of New Hampshire. To us, summer was the rich smell of the north pines mixed with the smoke of campfires, the cold sound of the White River as it rushed against granite boulders after a spring thaw, and the Milky Way swirling over the ancient silhouettes of worn Appalachian peaks. I never tired of answering the many questions my friends asked me as we lay supine in the grass, looking up in wonder at the stars, and trying to fathom the possibility of life on other worlds . . . or what we'd do if a UFO descended from the heavens and landed in the open field that surrounded us.

When I married Donna in 1987, summer under the starlit skies took on new significance. I had a new friend to share them with. And in the sleepy hours of the waning night, huddled in a blanket that we laid out on a lounge chair outside her parents' home, deep in the heart of the Connecticut wilderness, we watched the Milky Way stretch across the sky like a misty morning bridge and gasp as meteors poured out from it like tears. Today, Donna and I continue that tradition in Hawaii, having a blanket stored neatly in the trunk, firewood ready and available. Donna packs a cooler full of food and I pack the books and binoculars, both of which we share on special evenings with one another, so we can ponder the stellar riches above. Summer nights under the stars tend to bring out that relaxed state of mind, which opens our eyes to these distant celestial horizons and fuels our thoughts with endless possibilities.

Hercules, the Strongman

Since we began observing the stars of spring, the Earth has advanced one quarter of the way around the Sun. Leo, the Lion, has finally arrived at the western horizon. Arcturus and Spica have slipped away from the south meridian. We have journeyed past the date of the summer solstice. Now the majestic summer Milky Way is making its presence known east of the meridian, seemingly drawn by the "red horse" Antares, low in the southeast. While high above it, sailing east of the zenith is brilliant white Vega. But we'll study that star later. For now, we'll use it as a guidepost to our next constellation, Hercules, the Strongman.

Your first task is to find the "Keystone" of Hercules. Return your gaze to Arcturus. Now look for Vega high in the northeast about two hand spans away. The Keystone lies midway between Arcturus and Vega, about one fist northeast of Iota (ι) Coronae Borealis, the northeasternmost star in the Northern Crown. The Keystone is comprised of four 3rd-magnitude stars (all about as bright as Delta (δ) Ursae Majoris in the Big Dipper), which are virtually directly overhead, so you might want to lie down to enjoy this one. Your fist held at arm's length will cover the Keystone, which is a little less than 10° in extent.

Confirm the Keystone immediately by identifying the 5th-magnitude globular star cluster M13. Just look about one-third of the way along the line from Eta (η) to Zeta (ζ) Herculis – the two stars marking the western corners of the Keystone. The globular cluster will appear as an obvious swollen orb of diffuse light with a bright core. Many skywatchers have spotted it with their unaided eyes under a dark sky. It should be a cinch to see in binoculars, even from a bright suburban sky.

Messier 13 is arguably the finest cluster of its type in the northern skies and is popularly known as the Great Cluster in Hercules. It is the 13th object in Charles Messier's famous catalog of comet-like objects, though it was discovered by Edmond Halley (of Comet Halley fame) in 1714; Halley also noted that "it shews it self to the naked Eye, when the Sky is serene and the Moon absent [*sic*]". We see this misty globe of a half-million suns, which spans 140 light years, at a distance of 25,000 light years.

Any thoughts I had about extraterrestrial life while camping under dark New Hampshire skies swelled to magnificence on November 16, 1974, my 18th birthday. That day, astronomers used the enormous radio telescope at Arecibo Observatory in Puerto Rico to send their first and only deliberate radio message to potential listeners in space. Their target was the great globular cluster M13, whose multitude of suns offered maximum potential of reaching a civilized world in a small area of sky. The perceived hope was that an ET would receive the pictorial message on page 50, decipher it, and phone home. But the

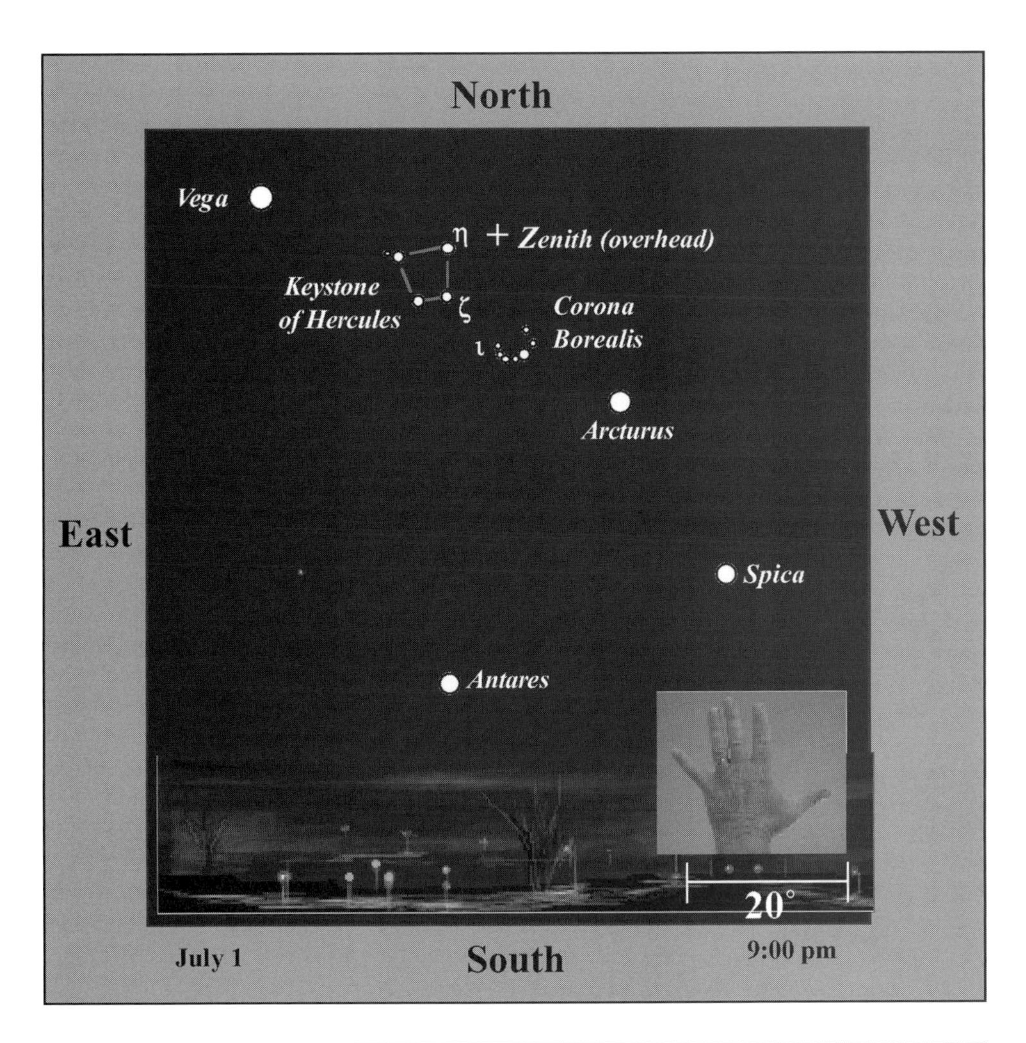

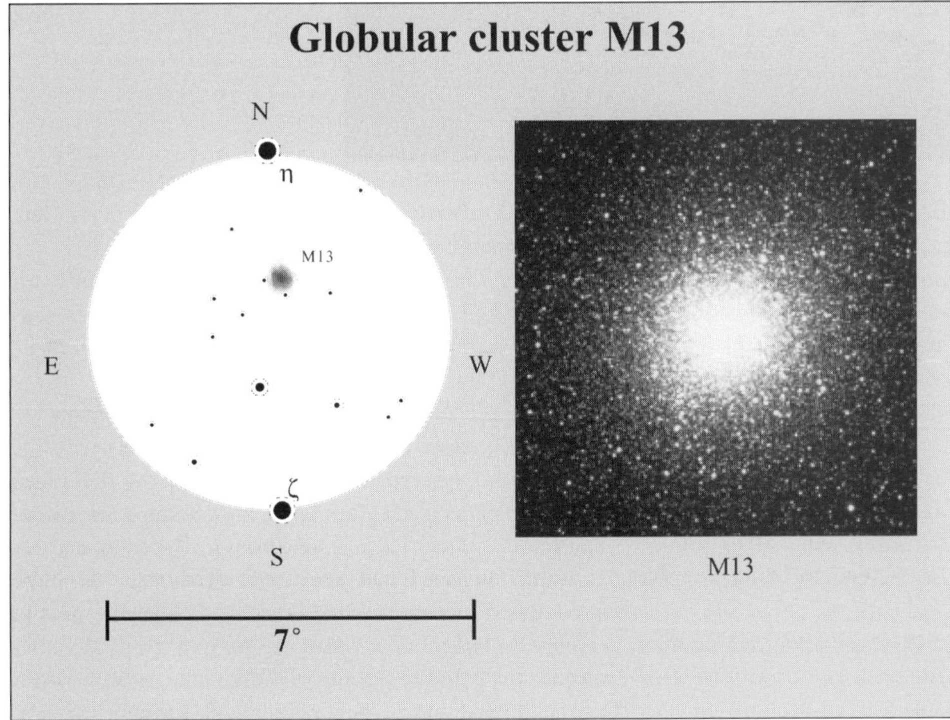

gesture was largely symbolic – aimed at getting us to think about the difficulties involved in extraterrestrial communication. Besides, by the time the signal arrived at its destination 25,000 years later, the cluster would have moved in its elliptical path around the galaxy and out of the way of the beam. But the three-minute-long transmission did get "out there" some vital information about the Earth and its inhabitants. *Anyone* who might intercept the beam, would see the radio brightness of our star increase by 10 million times for a period of three minutes. Of course,

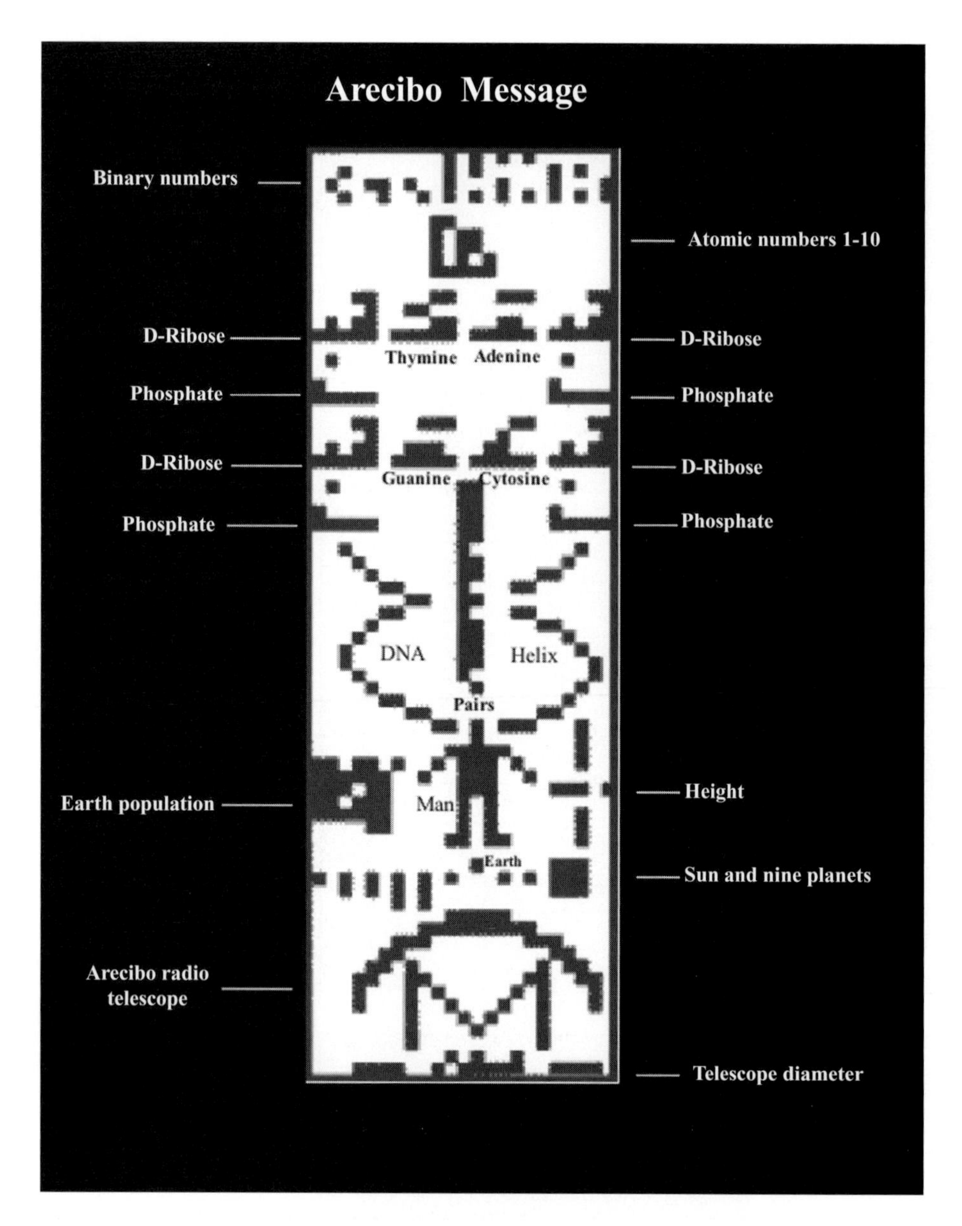

a hypothetical return signal from M13 would take yet another 25,000 years to reach the Earth, during which time the Sun would have moved in its orbit around the galaxy. So you see the futility in expecting results using targets at such great distances. Besides, you know how short human attention spans are today – imagine who'd remember anything about the experiment that far in the future. Can you recall a specific event that happened 50,000 years ago? Still, the promise of that attempt fueled the imagination of my friend John Ehlers and I, who won our high-school's science fair that same year with a project on the possibility of life in the universe and the search for extraterrestrial intelligence.

The mythology of Hercules is more about brawn than brains. The Strongman was the great grandson of Perseus and performed near impossible feats of strength, including those associated with the legend of his *Twelve Labors* – a curse, of sorts, placed upon him by Juno for being the son of a mortal mother impregnated by her unfaithful husband. We see in the sky, the antagonists of several of these restless labors, including his fight with the Nemean Lion (Leo) and the slaughter of the Hydra. His last exploit was bringing Cerberus, the three-headed hound from Hell, to the upper air. After Hercules completes this task, the Strongman, perhaps stressed by his workload, kills his friend Iphitus in an unfortunate fit of madness. As punishment, he is condemned to be the slave of Queen Omphale for three years.

While serving time, something odd happens to Hercules; he dons a dress and lives effeminately with the Queen, who, in turn, wears his lion's skin (so the queen wears the pants in this family – symbolizing a reversal of manhood). Upon his release, though, Hercules marries Dejanira. He lives happily for three more years – until he is poisoned by a bewitched dress that, when he tries to remove it, strips off his skin. In the end, poor Hercules ascends the volcanic summit of Mount Etna in Sicily, builds a funeral pyre, and commands a friend to lay a torch to it. But the Roman gods take pity on Hercules and intervene (as they usually do in the case of heroes). They make sure that only half of Hercules (his mortal half) is consumed in fire. Dead to the mortal world, Jupiter then makes sure the immortal half of Hercules is carried to the star-filled heavens in a chariot drawn by four ashen horses (another

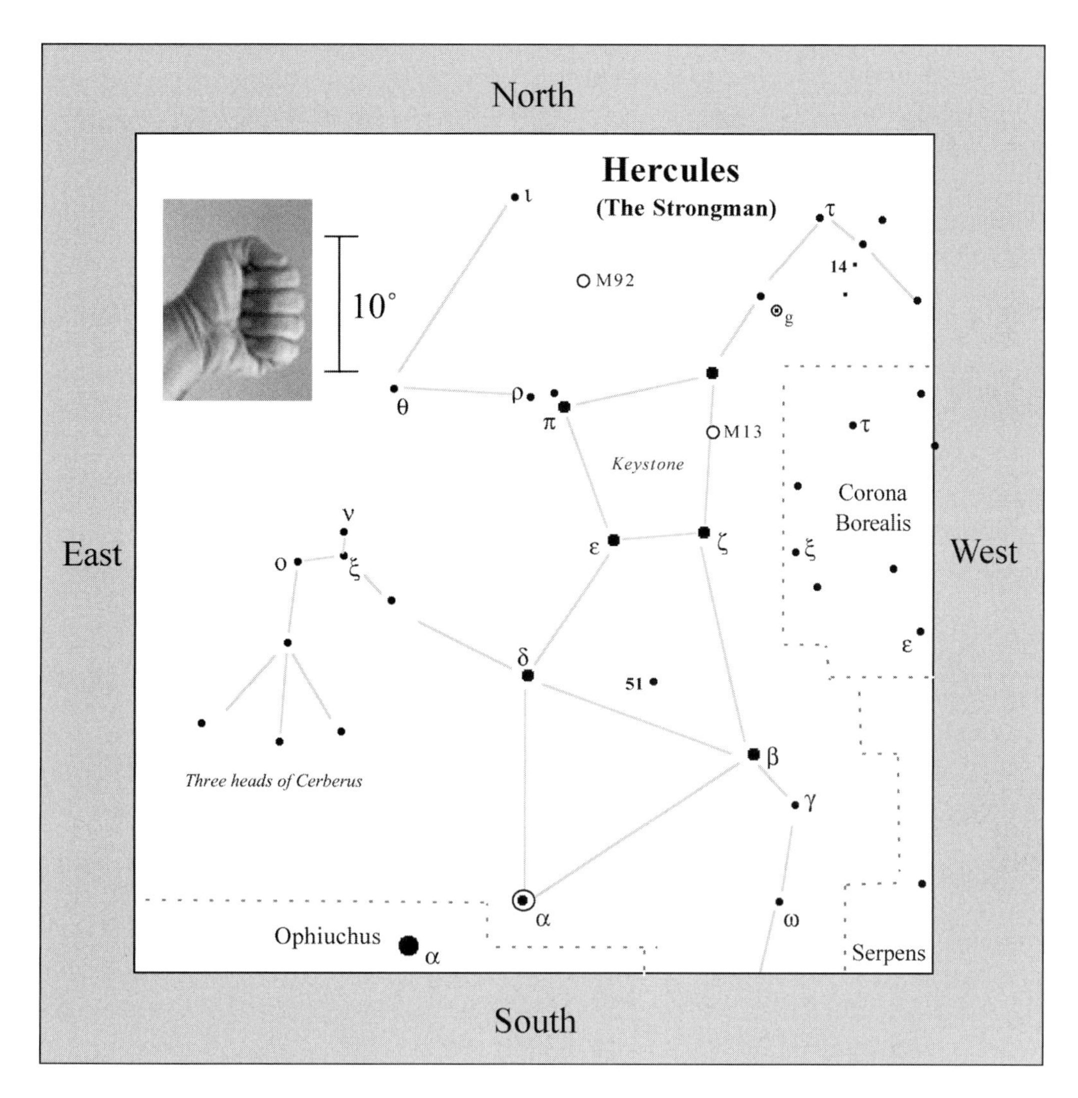

reference to volcanic activity). And so it is that every time we see the immortal flames of Etna gallop skyward with blackened clouds of ash, we are reminded of the spiritual ascension of Hercules.

As for our visual journey in the night sky, each summer we see the Strongman reigning high overhead with dignity. We see him, not in a restive state, but kneeling with one foot on the Head of Draco, the Dragon (another one of his labors), a hand grasping a club, and his other hand clutching the neck of Cerberus, which is now a defunct constellation, meaning it no longer is recognized as one of the official 88.

The chart above portrays the stick figure of Hercules, who is seen upside down: Third-magnitude Alpha (α) Herculis (Rasalgethi) marks the Strongman's head; Delta (δ) and Beta (β) are his shoulders; Gamma (γ) and Omega (ω) form his club; the Keystone is his lower torso; Tau (τ) represents the kneeling leg, and Iota (ι) is the foot holding down the Head of Draco (not seen). Cerberus is a bouquet of dim stars south of his hand – Omicron (ο), Nu (ν), and Xi (ξ), Herculis. If you forget the club, Hercules looks like he's on bended knee proposing. But few observers like to see Hercules in this way. One problem is that Alpha Herculis is fainter than a nearby star Alpha (α) Ophiuchi (Rasalhague). Besides, the Keystone is the pattern of stars that really stands out.

The ancient Greeks did not see the constellation as Hercules. The Greek poet Aratus called it "the Phantom whose name none can tell." Aratus's American-Indian-tribal-sounding name is actually quite appropriate for the constellation, whose stars (except the four making up the Keystone) are for the most part relatively dim, spread out, and inconspicuous. But, like Bootes, the classical stellar arrangement of Hercules as a Strongman can be seen in other ways, four of which are depicted on the next page.

Now use the chart above and the one on page 52 to focus on the point midway between Pi (π) and Iota (ι) Herculis and a bit west, where you will see yet another distinct globular star cluster, M92, which is only about one magnitude fainter than M13. This tiny cocoon of light lies at about the same distance of M13 (perhaps a tad more distant) and is a bit smaller in true physical extent. When you peer at this fuzzy disk of light, imagine that you're seeing more than 300,000 suns packed together in a sphere about 110 light years across.

Hercules has several attractive stars; use the chart above and the ones on page 53 to find them. Alpha (α) Herculis, for instance, is a semiregular variable attended by three dimmer suns. The primary shines with an imperial yellow glow that changes hue as the star pulsates between its two extremes – from a high of magnitude 2.7 to a low of magnitude 4.0; though the star usually varies from

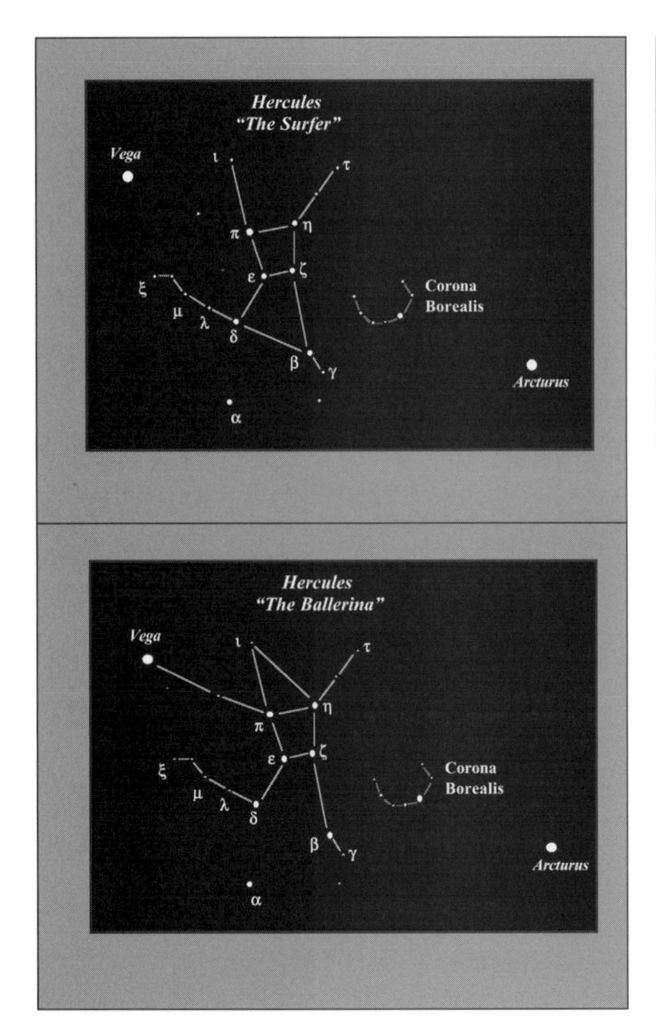

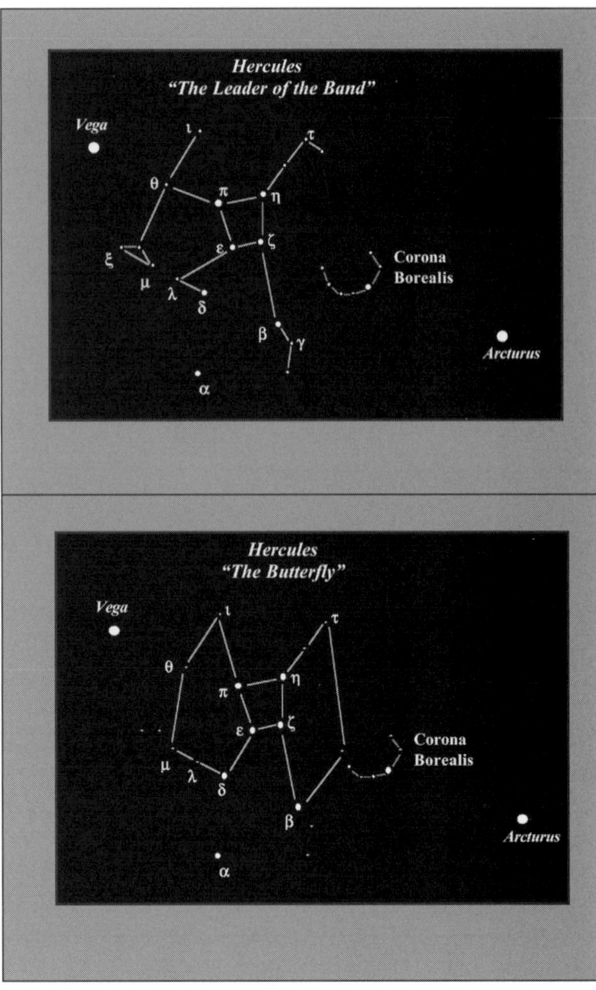

magnitude 3.1 to 3.7 about every 128 days. Another beautiful semiregular variable is pumpkin-orange g Herculis (the first variable star I observed as a youngster), which has a nontraditional lower-case designation; the star varies semiregularly between magnitude 4.3 and 6.3. Mu (μ) has two asterisms close to the northwest that spell the word "No" when seen with west up.

Now locate the 6th-magnitude, K-type star 14 Herculis, which is being orbited by a planet 4.5 times more massive than Jupiter. The giant planet orbits its parent star every

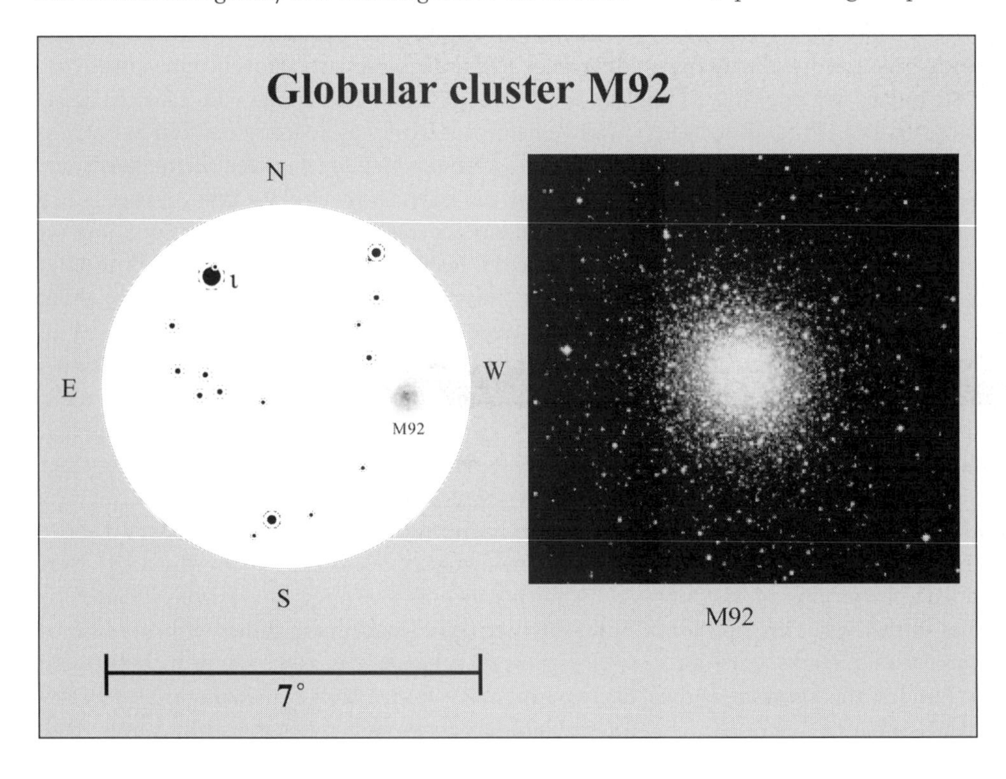

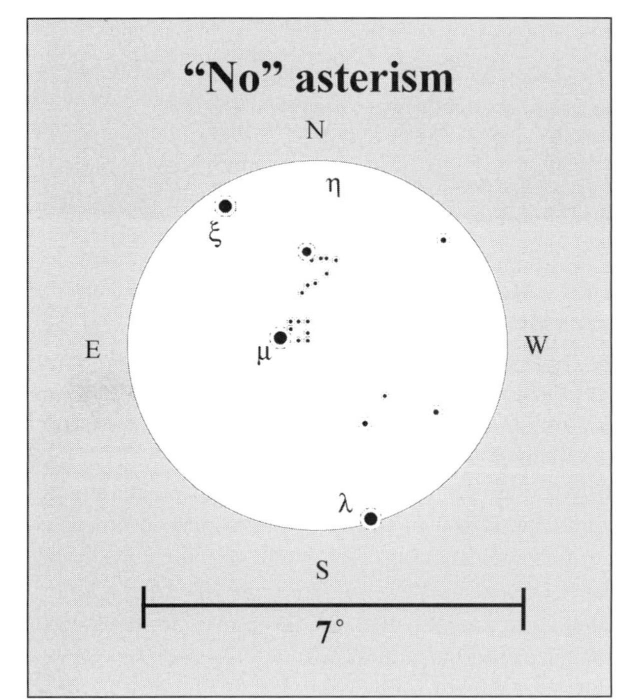

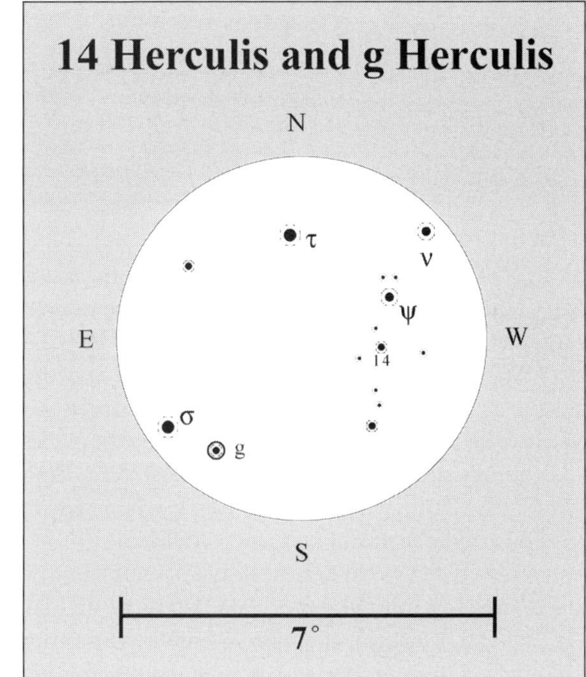

five years at a distance of 670 million miles – a tad farther than Jupiter's distance from the Sun. A planet roughly half its size may lie further out from 14 Herculis, but that planet's existence is, at the moment, unconfirmed.

Scorpius, the Scorpion

Let's turn our attention to the south. If you go out around July 15 at 9:00 pm and look about two and a half fists above a clear, flat horizon, you should see the brilliant red star Antares punctuating the meridian. Shining at 1st magnitude, Antares is the Alpha (α) star of Scorpius, the legendary Scorpion that stung Orion, the Hunter, to death (see page 120). Of course, the gods felt sorry for Orion and placed the Scorpion opposite him in the sky, so that it may never sting the Hunter again; thus whenever the Scorpion rises, Orion sets, and vice versa.

Scorpius is the 8th sign of the zodiac, and the ecliptic cuts across the sky very close to Antares. Therefore, on occasion, the Moon and planets can congregate near – or, in the case of the Moon, occult – Antares. Because of its stunning reddish glow, Antares (from the Greek *ant-Ares*,

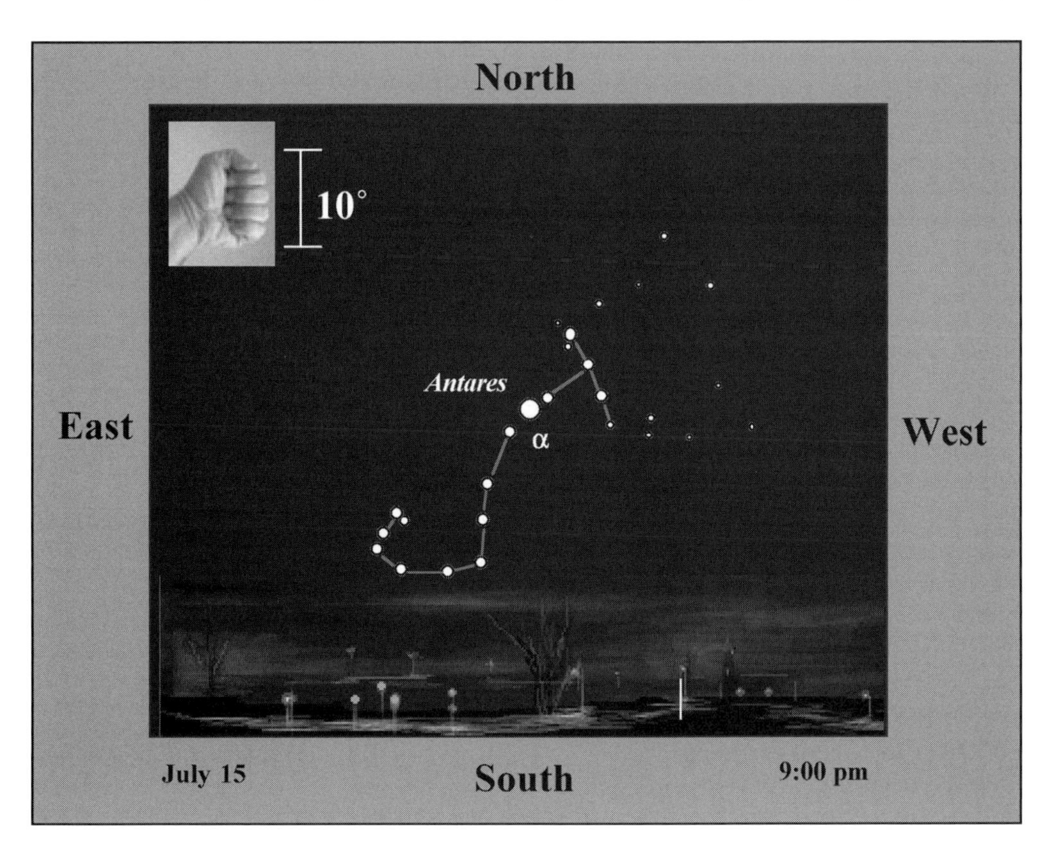

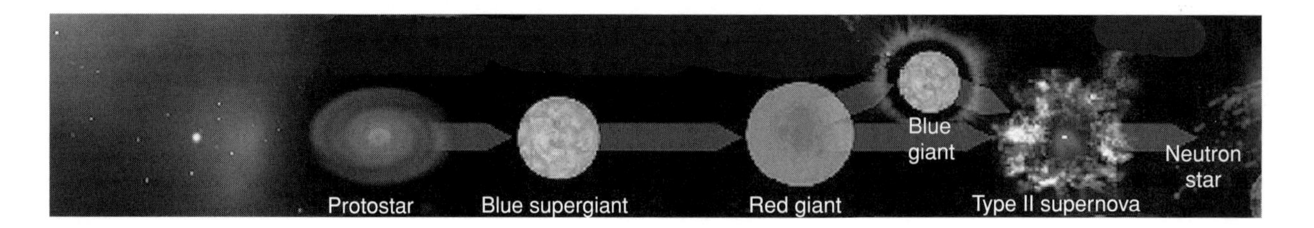

meaning "like Mars") is popularly known as the Rival of Mars. The star's color appears more hot-sauce orange through 10 × 50 binoculars. Actually, Antares is only the rival of Mars when the planet appears most red to the unaided eye – which happens either early on, or late, in its *apparition* (defined by the period of time when the planet first becomes visible before sunrise in the east to the time it departs the sky after sunset in the west); when Mars is closest to Earth its reddish hue cools to yellow.

Antares is a cool (3,600 Kelvin) M-type supergiant that deserves attention through binoculars. Although the star is 600 light years distant, it appears 10,000 times more luminous than the Sun. Being so bright and cool means that the star is also extremely large. In fact, Antares is so large (nearly 400 million miles wide) that its body would stretch from our Sun out to the orbit of Jupiter; Earth, in other words, would be enveloped in the star's tenuous atmosphere.

Like most red giants, Antares is a semiregular variable that can change by several tenths of a magnitude over a period of years. With a mass of some 15 suns, Antares is now nearing the end of its life and may soon (astronomically speaking) go supernova. Unlike a nova, a supernova obliterates the parent star. Antares is now forging iron in its core and cannot create any heavier elements. At some critical point in the near future, the star will stop releasing energy and succumb to the force of gravity. When it does, the star's core will collapse within seconds into a sphere merely six miles across (the size of Manhattan island); the rest of the star will rush inward and crash into the core before rebounding a quarter second later as a cataclysmic explosion that will shine with the light of a billion suns! The blast will hurl several sun masses of material outward into space at speeds of several thousand miles per second. This spewed ejecta will ultimately collide with nearby interstellar material, heating it up and causing it to glow. Sprinkled among the debris will be newly forged heavy elements, like gold, silver, and calcium. Here on Earth, an Olympian's gold medal, your grandmother's silver brooch, and a child's molar – are all made of atoms dispersed by ancient supernova explosions whose ejecta collided with the material from which our Solar System was born. So keep vigil on Antares, because, although the big boom may not happen until a million years from now, there's no reason why it also couldn't happen tomorrow night . . . *Kaboom!*

Scorpius is one of the few constellations that needs little imagination to see as its mythological namesake – a Scorpion. As Garret Serviss writes in his 1888 book *Astronomy with an Opera-Glass* (New York: D. Appleton and Company), "It does not require a very violent exercise of the imagination to see in this long, winding trail of stars a gigantic scorpion, with its head to the west, and flourishing its upraised sting that glitters with a pair of twin stars, as if ready to strike." The Scorpion's claws are two long Y-shaped extensions to the west. The ancient Hawaiians saw the Scorpion's tail stars as the Fishhook of their demigod Maui, who used it in a valiant attempt to join the Hawaiian islands – until a mishap prevented him from achieving the task; that's why we have to take planes or boats today to get from one Hawaiian island to the next. In ancient Egypt, the Scorpion was the personal symbol of the pharaoh known as "the Scorpion King," who ruled Egypt shortly before the "Two Lands" unified around 3100 BC.

From mid-northern latitudes, the Scorpion may be difficult to see in full unless you have a clear, unobstructed horizon and the constellation is at its maximum height above the horizon. Still, there are many binocular wonders to behold in the constellation's upper reaches. Perhaps in your binocular view of Antares, you noticed its two third-magnitude companions, Sigma (σ) and Tau (τ) Scorpii, which the ancients saw as the Scorpion's arteries. And if the night was particularly clear and free of pollutants, you might have also noticed a bright fuzzy glow midway between, and a tad south of, Antares and Sigma Scorpii. This is the 5th-magnitude, loose globular star cluster M4 – the 4th object in Messier's catalog of nebulae and clusters. Under a dark sky, the cluster is visible to the unaided eye as a tiny breath of light. But in 10 × 50 binoculars, it is a compelling glow that intensifies toward the center to an elliptical nucleus – especially when seen with averted vision. The cluster, which lies 6,800 light years distant, is a senior member of the Milky Way, having an estimated age between 10 and 13 billion years!

Now look north of Antares with your binoculars for three roughly 5th-magnitude stars, which form a "roof"

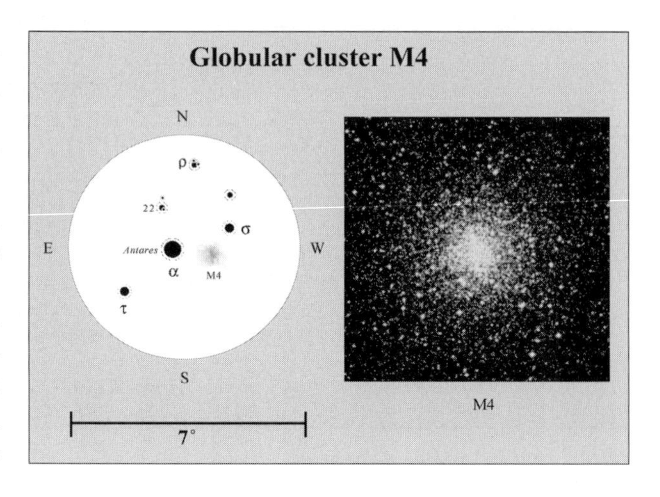

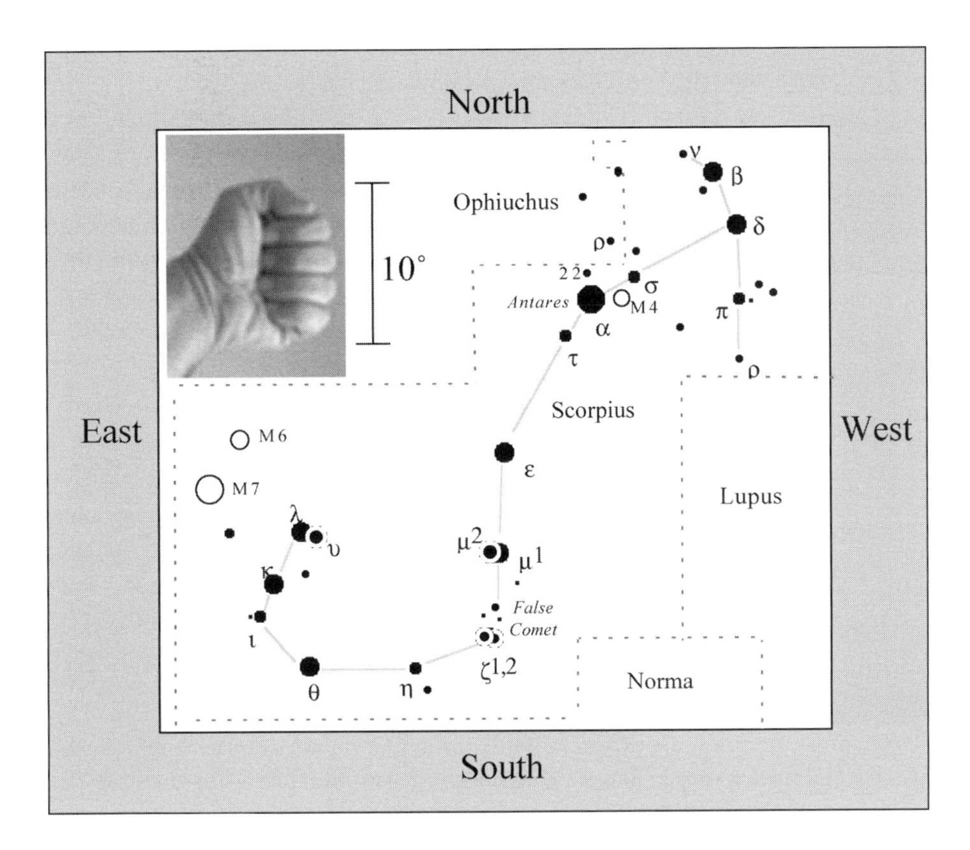

over Alpha and Sigma Scorpii. When all five stars are seen together, they form a pentagram. The star 22 Scorpii has a dim companion "under its wing" to the southwest, and a similarly bright companion a little farther out to the north. Now look at the star labeled Rho (ρ) just over the border in Ophiuchus. It is a stunning binocular triple, though you may need to use averted vision to see all the stars well. If you happen to be under dark skies, sweep your binoculars around Antares. I've noticed that the blanket of stars to the west are mostly blue-white, while those to the east have warmer hues. Now sweep them across the head of Scorpius, from the double star Nu (ν) to Pi (π), because there are many pretty starfields around them. Icy crisp Delta (δ) Scorpii is of particular interest. Beginning in July 2000, this remarkable 2nd-magnitude B-type star began increasing in brightness until it achieved 1st magnitude in 2002. Many skywatchers around the world independently noticed the gradual change. Apparently, Delta Scorpii is a rapid rotator and may be periodically hurling mass from its equator every 10 years or so, which we see from our distance of 400 light years as a flare up.

If you are ever in the southern United States under a dark sky that's very transparent (no dust or mist) all the way to the horizon, look at Zeta ($\zeta^{1,2}$) Scorpii at the western bend in the Scorpion's tail. If you use averted vision, you should see what appears to be a comet with a 2°-long tail fanning to the north. This visual phenomenon, known as the False Comet, is created by the close association of two large and bright open star clusters. Small and condensed NGC 6231 forms the comet's head, while the tail is the large and elongated open cluster Trumpler 24 to the north. The two clusters are truly related, belonging to the same group of luminous stars known as the Scorpius OB1 Association – a widely scattered family of young, hot stars that all formed from the same parent cloud many years ago.

Begin your binocular survey by first inspecting the colors of Zeta^{1,2} ($\zeta^{1,2}$) Scorpii; Zeta^{1}, the westernmost star of the pair shines with a saffron glow, while Zeta^{2} is more marigold. Then swing up to the two glorious clusters mentioned above. Shining at magnitude 2.5, NGC 6231 is so condensed and bright that it can be seen clearly in binoculars even in bright twilight. As the sky darkens, the cluster gradually energizes into a ball of shimmering starlight. Shortly thereafter, the spire of stars forming Trumpler 24 will sparkle into view. The two clusters are joined by parallel rows of stars that make it appear like

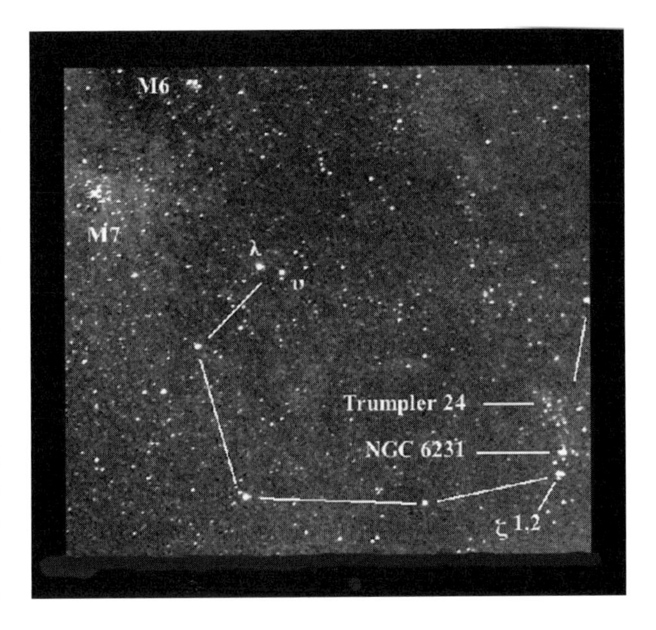

strands of mozzarella cheese being stretched from a slice of pizza.

Now focus your gaze at the tail's two stinger stars, Lambda (λ) and Upsilon (υ) Scorpii, then sweep northeastward to the large and bright masses of stars known as M6 and M7 – two of the sky's most striking examples of open star clusters; 3rd-magnitude M7 is 780 light years distant and 100 million years young, while 4th-magnitude M6, also known as the Butterfly Cluster, is twice as distant and twice as old. Unlike the many distant globular clusters we have been enjoying, these clusters display some 20 stars each that can be glimpsed clearly through 10 × 50 binoculars. Finally, look at the wide-field photograph of the region above. We see M6 and M7 projected against the very hub of our Milky Way galaxy, which is rising to prominence in the south. Messier 7 is also one of my seven wonders of the naked-eye Milky Way, the others you will be introduced to shortly. The photograph also reveals a wealth of dark nebulosity, which becomes amazingly apparent through binoculars under a dark sky. These cold patches of obscuring dust appear to drape the stars like thick strands of black webbing, which seems to sag under the immense weight of starlight in its delicate grasp.

Libra, the Scales of Justice

We end our tour of Scorpius by taking a look at its claws, or what used to be its claws. The area occupied by them was, in fact, known to the ancient Greeks as Chele, which means "claws." But apparently, ancient Roman astrologers of the first century BC did not like working with 11 zodiacal constellations, so they clipped the Scorpion's claws and created Libra, the Scales of Justice, with them. This explains why Libra is the only inanimate object in the now 12 zodiacal constellations. Then again, the scales are symbolic of Astraea (the goddess of justice, often identified as Virgo), who sits by Jupiter on her throne to give him counsel.

No matter, the Romans professed another reason for their actions. As Ian Ridpath writes in his delightful 1989 book *Star Tales* (Cambridge: Lutterworth Press), the Roman writer Manilius saw the creation of Libra a just move, because the Moon was said to have been in the location of Libra when Rome was founded. "Italy," Manilius said, "belongs to the Balance, her rightful sign. Beneath it Rome and her sovereignty of the world were founded." Still, being born under the zodiacal sign of "Scorpio," I cannot help but feel the injustice of such an action. Not only

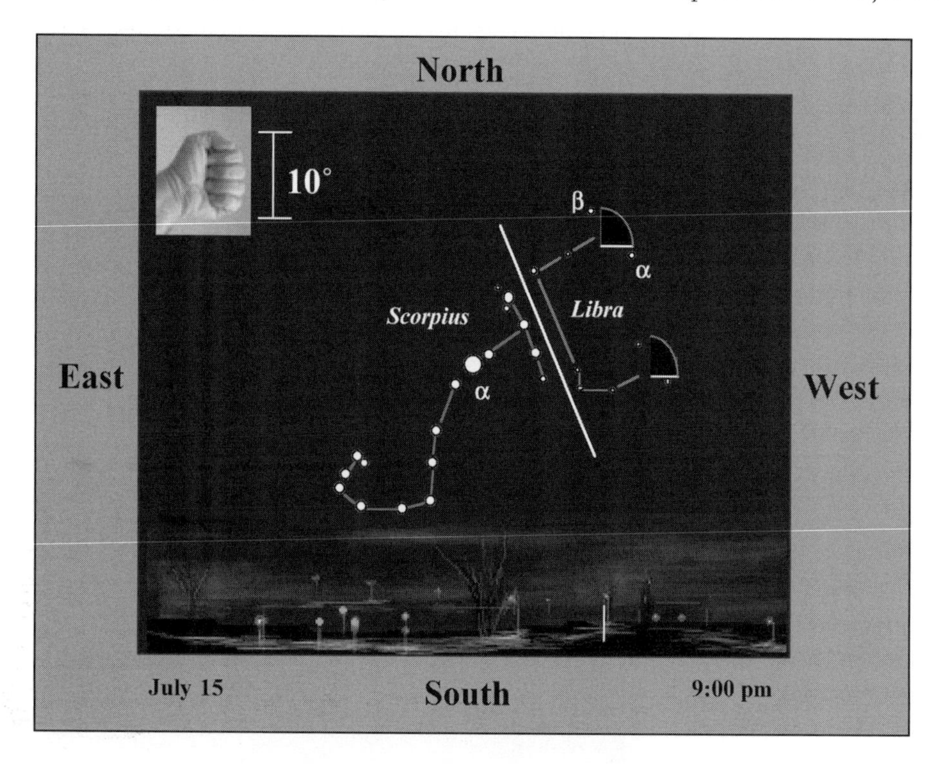

was the Scorpion declawed, but the new manifestation of the Scales of Justice were handed over to a neighboring constellation (Virgo).

It's also possible that the Scales represent the moment of the autumnal equinox. When the Sun entered the sign of the Scales around the year 300 BC, day and night were equal or, "in balance." Milton, in his "Paradise Lost," suggests a different reason for creating the constellation. Here's his account of Gabriel's discovery of Satan in Paradise:

> . . . Now dreadful deeds
> Might have ensued, nor only paradise
> In this commotion, but the starry cope
> Of heaven, perhaps, or all the elements
> At least had gone to wrack, disturbed and torn
> With violence of this conflict, had not soon
> The Eternal, to prevent such horrid fray,
> Hung forth in heaven his golden scales, yet seen
> Betwixt Astrea and the Scorpion sign.

Well, at least Libra's two brightest stars remind us of their association with the Scorpion. Alpha (α) Librae (Zubenelgenubi – pronounced zoo-ben-el-jah-new-be) is Arabic for "the southern claw," while Beta (β) Librae (Zubeneschamali – pronounced zoo-ben-ah-sha-mah-li) means "the northern claw." If you can identify these two third-magnitude stars, try showing them to your friends while saying their names fast and in one breath; they might chuckle with delight. Also make sure they see Zubeneschamali through your binoculars. Many observers see this star as green or pale emerald. In fact, the star holds the distinction of being the only green star whose color is easily detectable with the unaided eye. Remember, though, color is in the eye of the beholder; I see the star smoldering with a pasty-blue light. Compare this star in your binoculars with Alpha Librae – a wonderful double with an aqua-colored primary and a salmon-colored secondary.

Ophiuchus, the Serpent Bearer and Serpens, the Serpent

Standing on the back of the Scorpion and butting heads with Hercules is another huge and muscular man of old – Ophiuchus, the Serpent Bearer (see the chart on page 59). We see Ophiuchus from behind, holding an enormous serpent, whose head (Serpens Caput), "snakes off" to the west, while its tail (Serpens Cauda) wiggles off to the east. The rest of the snake's body lies between Ophiuchus's legs.

No one's absolutely certain why Ophiuchus is depicted wrestling a snake. But we do know that the man is identified with Aesculapius, the son of Apollo and the god of healing or medicine. Apparently his association with snakes has to do with Aesculapius seeing two snakes and killing one of them; he then watched with surprise as the companion snake revived its partner by feeding it a special herb. Through the action of the snakes, Aesculapius learns the medicinal powers of plants and how to revive the dead. Aesculapius's newfound power angered Pluto, because they took away what was rightfully his. So Pluto complained bitterly to his brother Jupiter and persuaded him to kill Aesculapius with a thunderbolt and to command that all mortals must one day die. (So, if you fear death, now you know why.)

Anyway, Jupiter complies and zaps Aesculapius. As his body descended into the Underworld, Jupiter kept his spirit alive in the sky; as a perk, Jupiter also tossed up his friend the snake. To this day, the symbol of the medical profession is the caduceus – two serpents coiled around a staff flanked by Mercury's wings. It is also said that Hippocrates, the famous Greek physician and the father of medicine, was Aesculapius's (Asclepius to the Greeks) fifteenth grandson. Anyone entering the medical profession must take the Hippocratic Oath, which is supposed to prevent doctors from becoming immoral "snakes" – like

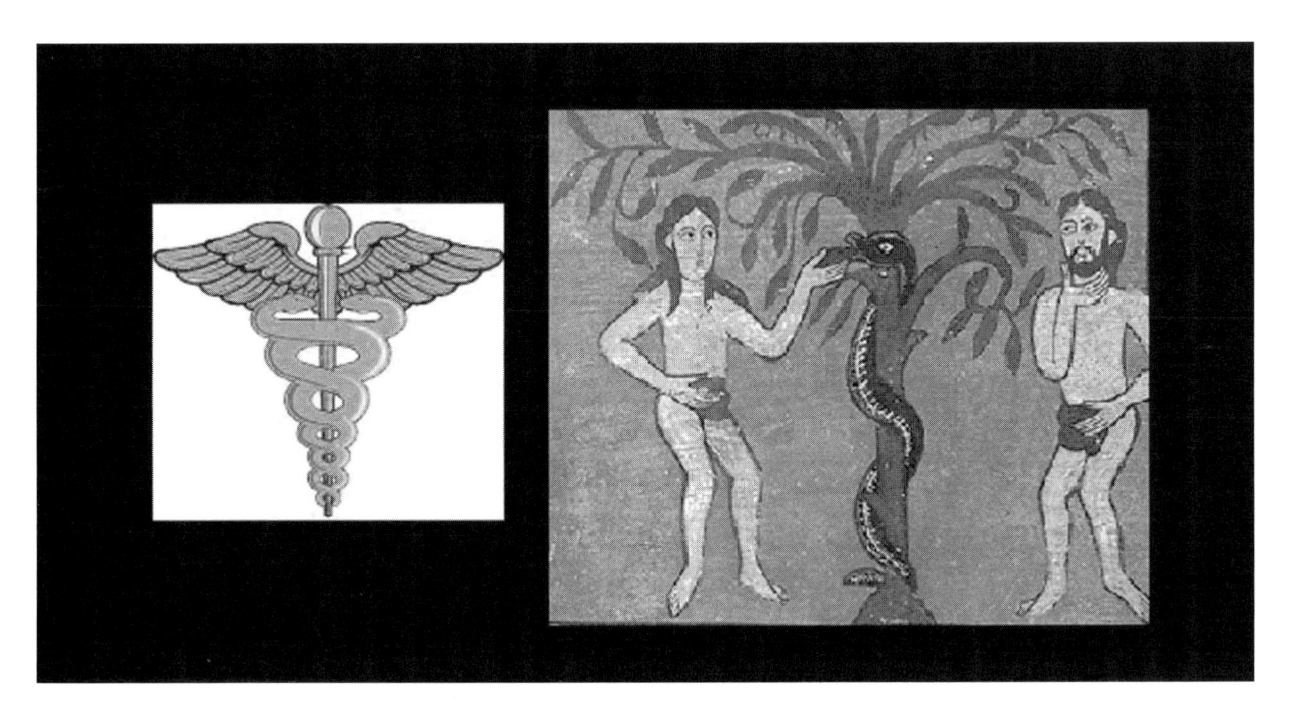

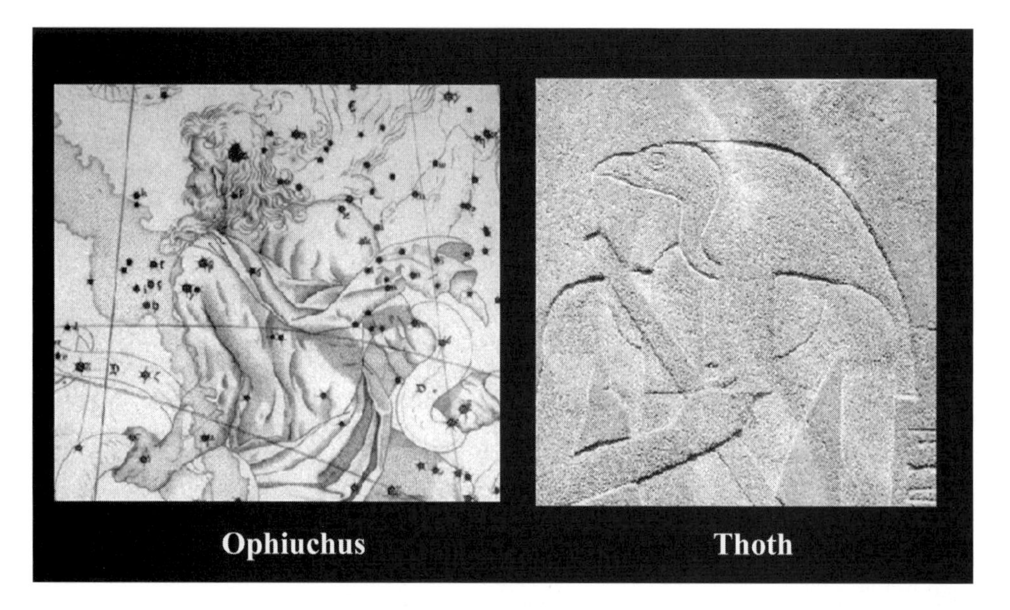

the snake that tempted Eve in the Garden of Eden; that serpent is often depicted coiled in a tree illustrating the dark side of the caduceus.

The caduceus, however, is much older than any medical profession symbol. For instance, in Egyptian hieroglyphs, Thoth – the god of wisdom and the originator of all medical knowledge – is often portrayed with caduceus in hand.

As you can see, we cannot escape the influence of ancient Egypt on modern thought – a fact that did not go unnoticed by Hollywood. In the classic 1932 version of *The Mummy*, we see Boris Karloff as the "mummy reincarnate" using the Sacred Scroll of Thoth – which, we learn, contains the ancient spell by which the goddess Isis brought the dead Osiris back to life – to raise his lover from the dead. In the many sequels, a high priest uses extinct tanna leaves to return the mummy to life. All reflections of the myth of Aesculapius.

When the Sun is deepest in the Underworld (south of the celestial equator), just prior to the time of the *winter solstice*, it passes briefly through Ophiuchus before ascending toward life in the spring. As American astronomer Royal Hill noted, "Out of the twenty-five days,

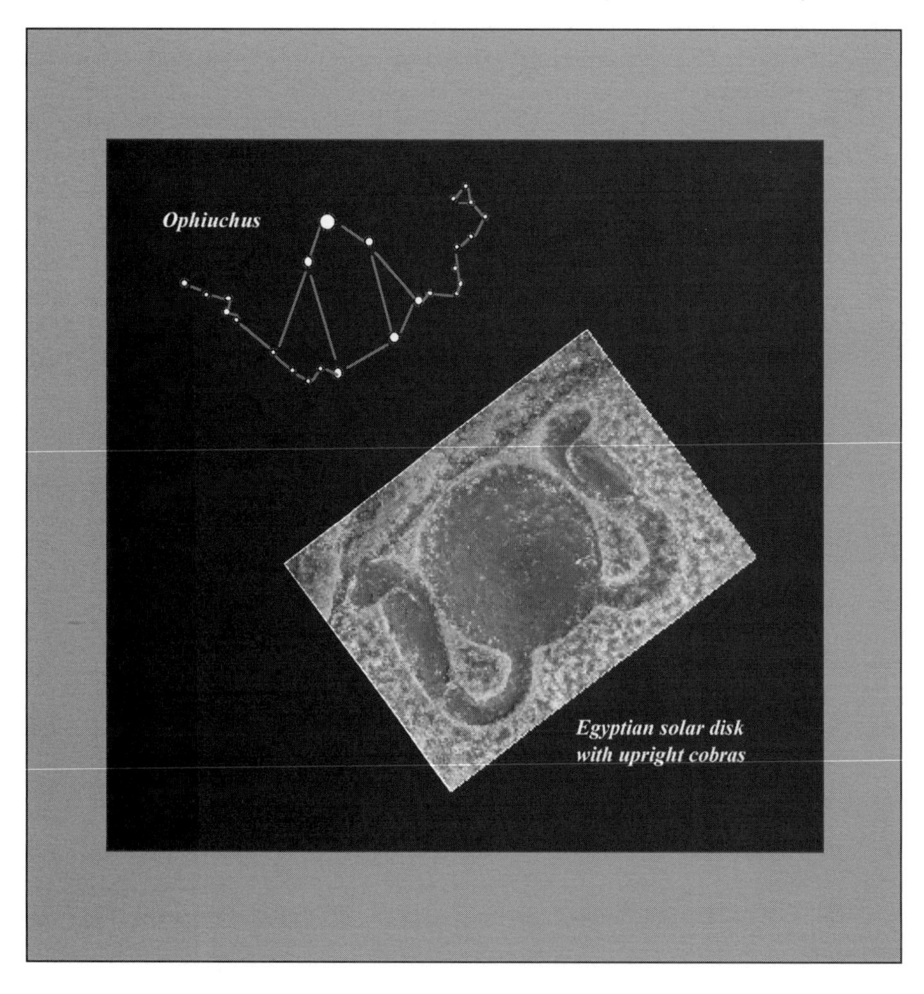

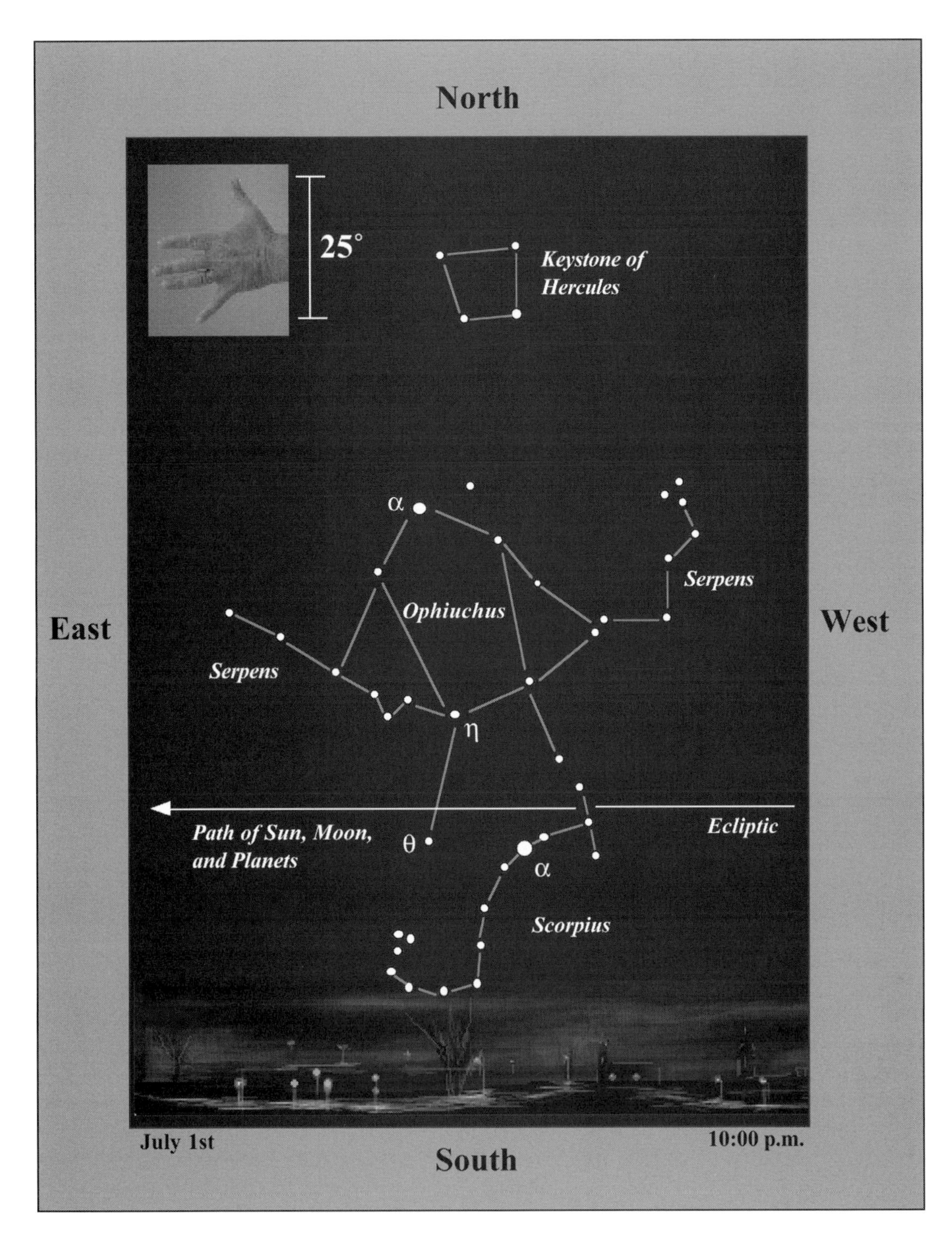

from the 21st of November to the 16th of December, which the sun spends passing from *Libra* to *Sagittarius*, only nine are spent in the *Scorpion*, the other sixteen being occupied in passing through Ophiuchus." For that reason some astrologers have added a thirteenth zodiacal constellation – Ophiuchus. So the ancient tale of Ophiuchus as Aesculapius (or Thoth) is a metaphor for the Sun's passage from death to life. Seen through the eyes of an ancient Egyptian, Ophiuchus, flanked on either side by a serpent, just may have been viewed as the solar disk flanked on either side by an upright cobra (the uraeus). The rearing uraeus connotes protection and is often incorporated into scenes from the *Book of the Dead*. Thus the western snake protects the Sun on its initial descent into the underworld, while the eastern serpent protects the Sun on its ascent after the winter solstice, which is represented by the "solar disk" of Ophiuchus.

Of course, if the Sun passes through Ophiuchus, so too do the Moon and planets. (See the chart above and the one on the next page.)

Ophiuchus has six Messier globular star clusters, but two of them, the 6th-magnitude pair M10 and M12, are by far the easiest to see in binoculars. Find them using the chart on page 60. First locate 2nd-magnitude Alpha (α) Ophiuchi (Rasalhague) near Alpha Herculis. Then look about 10° (one fist) to the southwest for 3.5-magnitude Kappa (κ) Ophiuchi; Kappa is easy to identify because it is flanked to the northwest by 4th-magnitude Iota (ι) Ophiuchi. Next look another fist-width southwest for 4th-magnitude Lambda (λ) Ophiuchi. Raise your

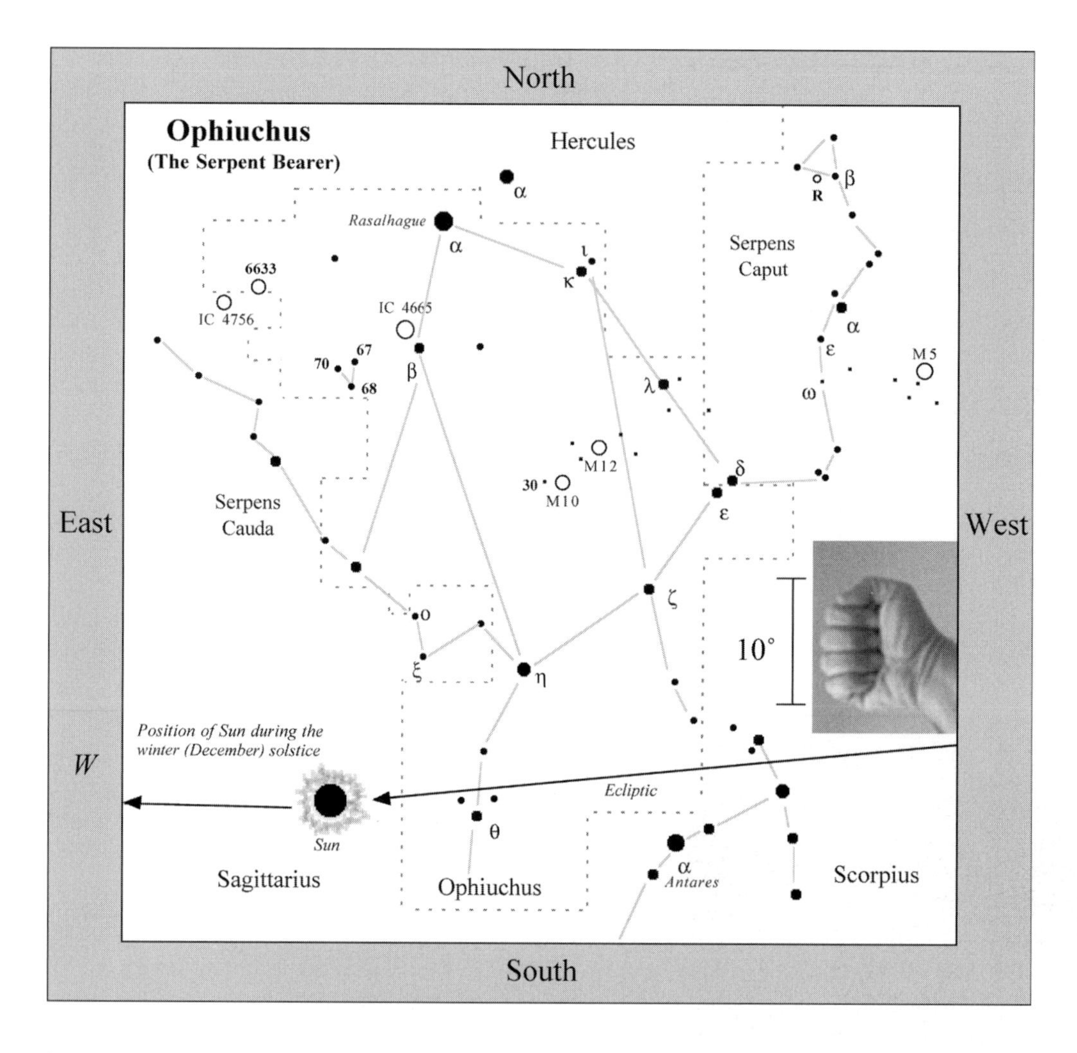

binoculars to this pure white star and place it in the field of view at the 2 o'clock position angle. Messier 12 should be about three-quarters of the way across the field to the southeast. If you center M12, M10 will suddenly pop into view toward the lower left edge of the field. Each globular cluster should appear as a tight wad of diffuse light. Messier 10 is 14,000 light years distant and M12 is about 4,000 light years farther off. Actually, the view is a triple treat, because M10 is a little west and slightly north of 5th-magnitude 30 Ophiuchi – a beautiful orange K-type giant.

Now return your attention back to Lambda Ophiuchi, then look a little more than one fist-width to the west–northwest for 2.5-magnitude Alpha (α) Serpentis, whose common name is a real throat gurgler – Unukalhai (oo-nook-ul-high). Alpha Serpentis is another K-type giant 73 light years distant. You should be able to fit Alpha and its attendants – Epsilon (ε), and Omega (ω) Serpentis in a single binocular field. Messier 5 lies a little more than one binocular field west of Omega. There's no mistaking this fine 6th-magnitude globular, which appears as a swollen orb of light half the size of the full Moon.

For a special treat look for the stunning swarm of glittering suns known as IC 4665 – the 4,665th entry in the *Index Catalogue* of clusters and nebulae. Train your binoculars $1\frac{1}{2}°$ northeast of 3rd-magnitude Beta (β) Ophiuchi in the shoulder of the Serpent Holder. The 4th-magnitude cluster spans nearly two full Moon diameters and resolves into some 15 or so binocular stars. The brightest form an irregular circle with vast double overlapping wings attached to the west and east. I call it the Black Swallowtail Butterfly Cluster, because the beautiful insect's black body and wings are lined with bright white spots, like the stars in IC 4665. It's also known as the "Little Beehive."

If you are under a dark sky and can see IC 4665 with the unaided eye, take a moment to reflect on the fact that while IC 4665 is only about a magnitude fainter than Beta Ophiuchi, the star appears much more apparent. Why? Because the size of a deep-sky object like IC 4665 affects its visibility. Beta concentrates all its light into a tiny stellar point, while the light of IC 4665 is spread across some 70′ of sky. Again, imagine a flashlight with an adjustable beam. Think about how dim the beam appears at its widest setting compared to at its narrowest setting. The same effect is seen in the sky with concentrated and diffuse objects that have the same intrinsic brightness.

While most of the Milky Way's open clusters are found in the plane of its spiral arms, IC 4665 is at relatively high galactic latitude, almost 16° above the galactic plane. It lies 1,100 light years distant and has an age of only about 35 million years.

A little more than one binocular field due east are the magnificent open star clusters NGC 6633 in Ophiuchus and IC 4756 in Serpens Cauda. I call them Tweedledum and Tweedledee after the "fat little," pugilistic twins in

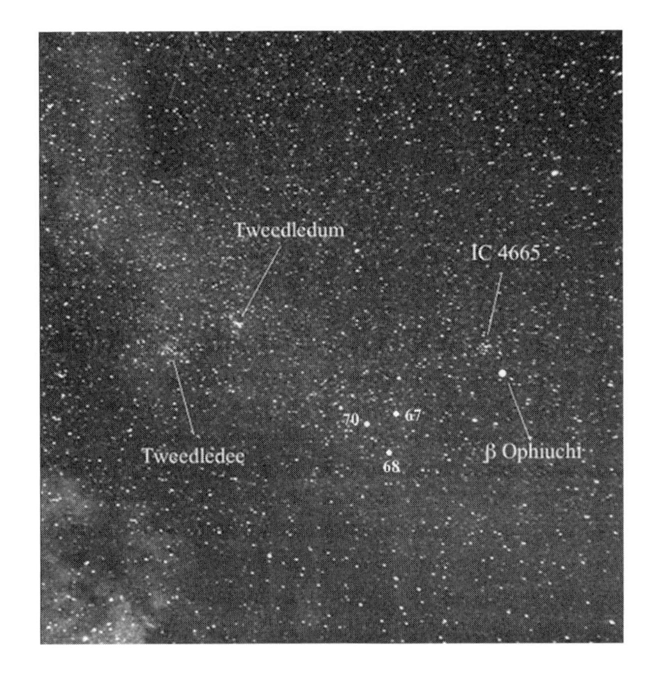

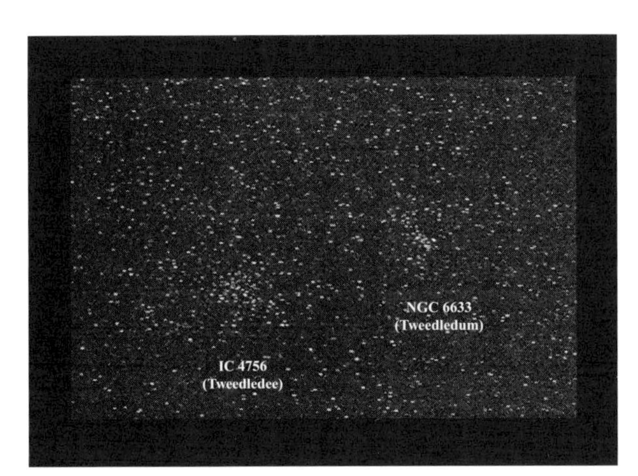

Lewis Carroll's classic children's book *Through the Looking Glass and What Alice Found There*. Both shine at magnitude 4.3 and can be seen without optical aid. They also lie at comparable distances from the Sun. NGC 6633 is 1,000 light years away and spans 20′ in apparent diameter; IC 4756 is 300 light years farther away and twice as large. In true physical extent, NGC 6633 spans 5.8 light years of space, while IC 4756 spans 15.2 light years. Both have stars that clump together to make the clusters appear patchy under the darkest of skies. They are a cinch to resolve in binoculars.

By the way, the nearby stars 67, 68, and 70 Ophiuchi form the V-shaped face of the now defunct constellation Taurus Poniatovii, the Polish Bull. Finally, take a look at the stick-figure configuration of Ophiuchus on page 60. Try to forget how the ancients patterened the stars and use a more modern eye. What do you see? I'll give you a hint: Think of a famous photograph of Marilyn Monroe from the 1955 film "The Seven Year Itch." Do you see her standing over a subway grate, holding down her dress from an updraft?

5 August

I HEARD *the trailing garments of the Night*
Sweep through her marble halls!
I saw her sable skirts all fringed with light
From the celestial walls!

Henry Wadsworth Longfellow, *Hymn to the Night* (1839)

The most glorious section of the Milky Way is sweeping its sable skirt across the meridian. On August 1 at 9:00 pm Scorpius, the Scorpion, has moved a bit to

the west, making room for another southern zodiacal constellation – mighty Sagittarius, the Archer, home to the galactic center and all its wonder. We are now peering 30,000 light years distant at the most magnificent of my seven wonders of the naked-eye Milky Way. This is the month you need to get out under a dark sky, far away from city lights. Lie down on a blanket, raise your binoculars to the Milky Way, and just sweep and enjoy. The milky-white expanse resolves into a breathtaking multitude of suns that can, with contemplation, overwhelm the mind.

Although the Milky Way splinters into starlight through your binoculars, there is no final resolution. Look carefully and you'll see that the visible stars burn before a more distant curtain of smoky light. Toss your binoculars aside and replace them with a 5-inch telescope, and you'll see more stars in front of a more distant curtain of Milky Way. Attend a star party and look at this region through a 24-inch telescope, and there'll be even more starlight and more Milky Way smoldering in the background. The heart of the Milky Way is like a Russian nesting doll.

There is too much beauty here to describe in detail, so I will be selective and tell you about some of the wonders that have amazed me over the years. Sharing is such an important part of amateur astronomy. But remember that "discovery" is a personal adventure, and I encourage you to take note of your own binocular finds and pursue them to the fullest.

Sagittarius, the Archer

We'll start by learning a new constellation, actually, an asterism – the Teapot of Sagittarius. Look just one fist to the upper left of the stinger star Lambda (λ) Scorpii and you'll come right to the Teapot's spout – Gamma (γ), Delta (δ), and Epsilon (ε) Sagittarii. The open star cluster M7 lies midway between these points. Your fist held at arm's length should also roughly cover the Teapot from spout to handle: the handle stars are Sigma (σ), Tau (τ), Phi (Φ), and Zeta (ζ) Sagittarii; Lambda (λ) Sagittarii marks the tip of the Teapot's lid. The Milky Way is curling vapor rising from the Teapot's spout.

In mythology, Sagittarius was a centaur – a "monster" with the torso of a man (from head to loins) and the legs and body of a horse. I put "monster" in quotes because despite their beastly appearance, centaurs were generally wise and kind. Of course, what would mythology be without tragedy? So, yes, there were bad centaurs. One rowdy bunch attended the wedding of Pirithous and Hippodamia. After consuming much wine, some of the centaurs set out to violate the bride; soon others followed suit. This led to a dreadful conflict in which several centaurs were slain. But Sagittarius was not among them. He was a friend of Apollo and Diana, and renowned for, among other things, his skill in hunting. That's why in the sky we see Sagittarius facing westward drawing a bow. The

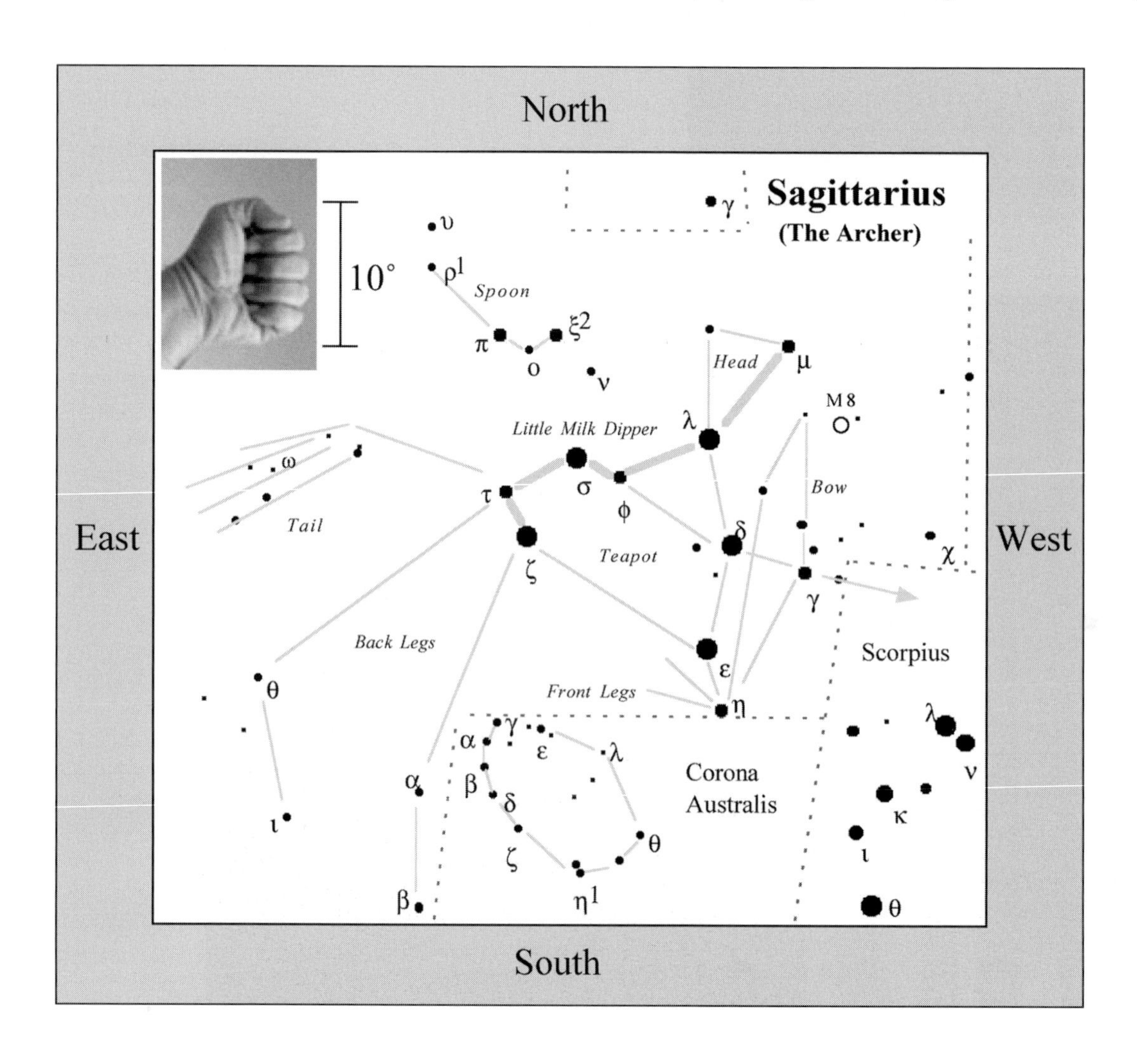

arrow is pointed to the Scorpion. (Again, being someone born under the sign of this unfortunate arachnid, I cannot help but feel that our ancient ancestors had an unhealthy phobia about this creature; good thing the Scorpion has a stinger, to protect itself from centaurs and other gigantic hunters.)

Looking at the Teapot and its surroundings, we can truly appreciate the illusion of the two-dimensional sky. Although the Teapot's stars are roughly of the same magnitude – between 2nd and 3rd – they are all disproportionately distant. Start by looking at the Teapot's tip. At a distance of 77 light years, Lambda (λ), Sagittarii is a 3rd-magnitude, *K*-type orange giant that shines with an emperor's gold light as seen through 10 × 50 binoculars. Lambda is part of the Little Milk Dipper, an asterism within an asterism, formed by the stars Zeta (ζ), Tau (τ), Sigma (σ), Phi (ϕ), Lambda (λ), and Mu (μ) Sagittarii. See how it resembles a smaller version of the Big Dipper? There's also a Spoon asterism to its northeast, comprised of the stars Rho^1 (ρ^1), Pi (π), Omicron (o), and Xi^2 (ξ^2) Sagittarii.

Lambda is one of four *K*-type orange giants in the Teapot. The other three are Delta (δ) (305 light years distant), Tau (τ) (120 light years), and Gamma (γ) Sagittarii (96 light years). I see Delta as marigold in color, Tau as liquid gold, and Gamma Sagittarii with a golden-tulip hue. Now look to the southwest at the rich monarch color of Eta (η) (149 light years), which contrasts splendidly with the blue blaze of Epsilon (ε), which is nearly at the same distance. The difference in the intensity and color of the two stars is due to Epsilon being a hot, *A*-type blue giant compared to Eta Sagittarii, which is a cool M-class, red giant.

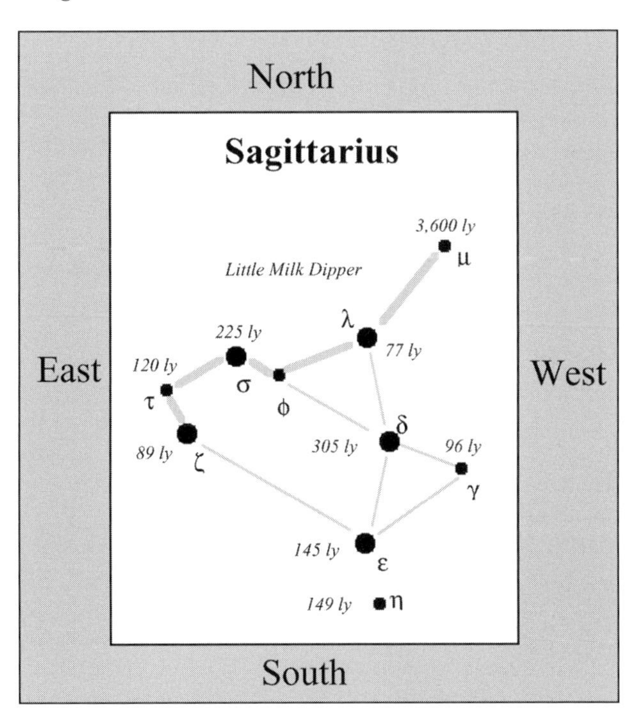

As a stellar finale, turn your binoculars to the tip of the Little Milk Dipper's handle – to cool blue Mu (μ) Sagittarii. Although Mu shines modestly at 4th magnitude, the star is inconceivably distant – perhaps 3,600 light years! Mu is a true-blue supergiant 180,000 times more luminous than the Sun. Only eight million years old, the star is on its way to becoming an enormous red supergiant before it terminates its existence in a cataclysmic supernova explosion.

Return your gaze to Lambda Sagittarii and use the chart on page 64 to see three star clusters in the same field of view but at vastly different distances. First look at 5th-magnitude M22. This bundle of diffuse light – a full Moon's width in apparent diameter – is actually a globular star cluster 10,000 light years distant, making it one of the closest clusters of its kind to our Sun. Under a dark sky, it is visible to the unaided eye as a tight bead of light $2\frac{1}{2}^\circ$ northeast of Lambda. Through 10 × 50 binoculars, its half million suns appear as a large and diffuse cometary glow with a bright core and fluffy halo.

Now compare the light of M22 with that of tiny M28, another globular star cluster about 1° northwest of Lambda. While M28 appears about one-third the size of M22 and is two magnitudes fainter, it also lies twice as far away. Still, M28 holds its own through 10 × 50 binoculars, looking like a small condensed glow of diffuse light. A stone's throw away is the 5th-magnitude open star cluster M25; it appears just as large as M22 in apparent diameter but it is 10 times closer, so some of its stars can be resolved in binoculars. Overall M25 is a wide and loose aggregation of dim suns with a small and tight core whose halo stars are scattered against a scrim of unresolved starlight. But now for a mind-bending thought. Messier 25's 600-odd suns shine together as a single 4.6-magnitude star as seen with the unaided eye. That's nearly a full magnitude fainter than the glow from the solitary supergiant Mu (μ) Sagittarii, which is 1,300 light years more distant!

From Mu Sagittarii, look about $2\frac{1}{2}^\circ$ north for M24 – a most impressive star cloud and another one of my seven wonders of the naked-eye Milky Way. Messier 24, also known as the Small Sagittarius Star Cloud, is a vast citadel of starlight nearly 10,000 light years distant. To the unaided eye, the $1\frac{1}{7}^\circ$ rectangular glow looks like a flying carpet fleeing the fringes of the galactic center.

Messier 24 lies in the plane of the Milky Way and helps to define one of its spiral arms. But unlike an open star cluster, which also lies in the galactic plane, M24 is not an isolated "patch" of starlight but the effect of an optical illusion – one caused by the chance framing of dark dust in the line of sight. The dust forms the dark walls of a three-dimensional tunnel through which we happen to glimpse a section of one of the Milky Way's inner spiral arms at the position of M24. Sweep these dusty cobwebs away and you'd see the Large and Small Sagittarius star clouds blend and bloom into a garden of delight. While the Large Sagittarius Star Cloud shines with a distinct golden hue, M24 has a subtle emerald polish.

To see my next naked-eye wonder, look 4° southwest of Mu for a large curdle of galactic vapor off the western edge of the Milky Way. This is M8, the Lagoon Nebula

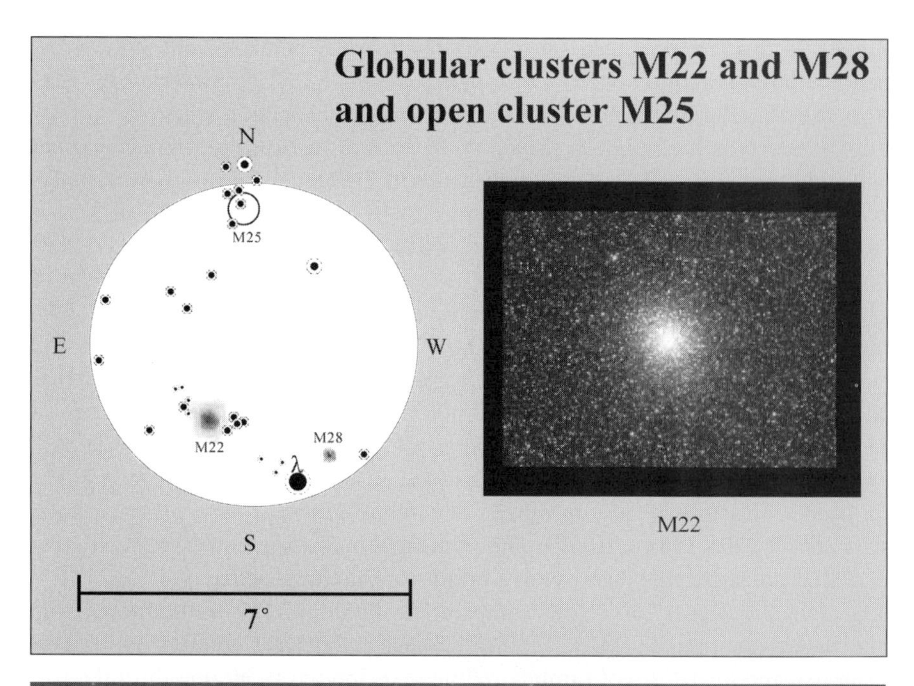
Globular clusters M22 and M28
and open cluster M25
N
M25
E
W
M22
M28
λ
S
7°
M22

Small Sagittarius
Star Cloud
μ
M8
Jupiter
Large Sagittarius
Star Cloud
Teapot
M7
Tail of Scorpius
View from Hawaii
Latitude +19°

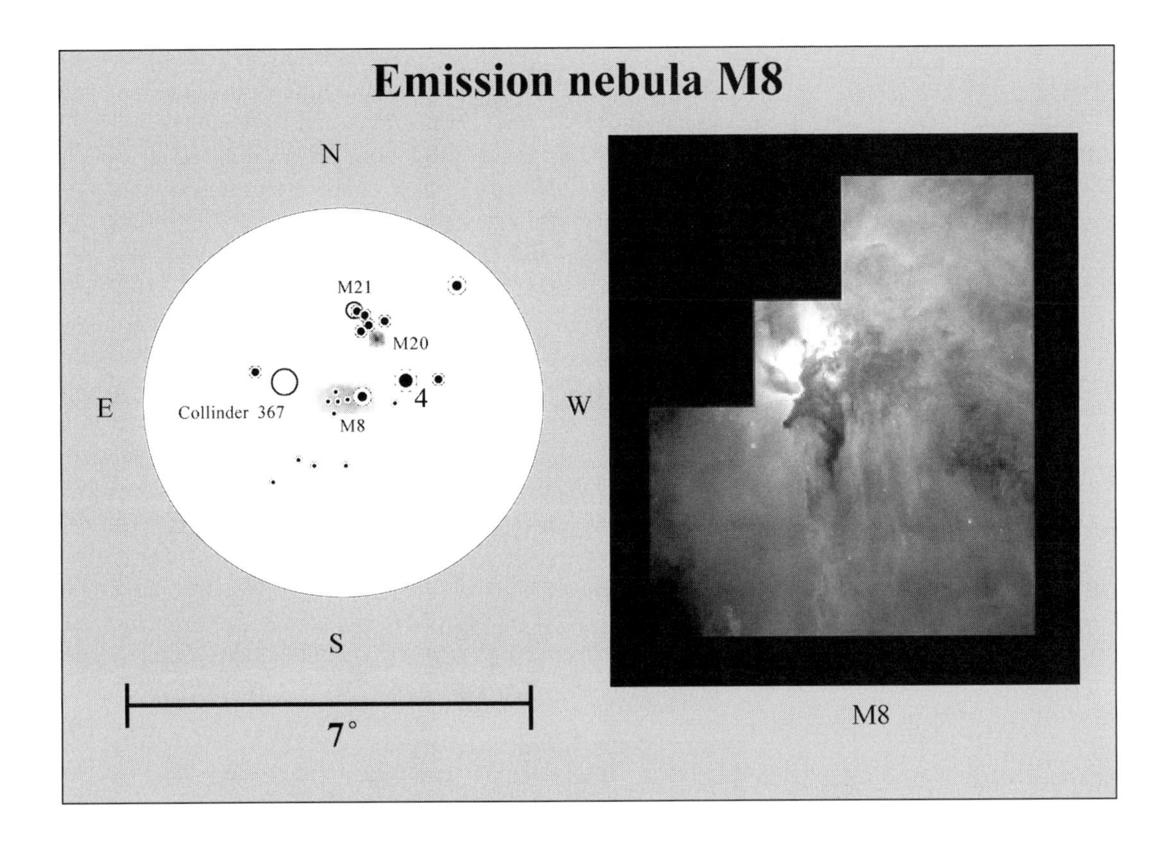

and cluster (see the chart above). You can confirm its position by looking for Lambda (λ) Sagittarii, which should be about 5° to the east and slightly south. Look at M8 with your binoculars. The radiative energy from a very hot 6th-magnitude star and some other fainter ones immersed in the nebula power the light emitted from this gaseous glow, which surrounds a cross-shaped asterism lying on its side. Through binoculars, the cross-shaped asterism stands out prominently and dwarfs the fainter nebulosity, which is best seen under a dark sky using averted vision. Messier 8, which lies 5,000 light years distant, is called the Lagoon Nebula because telescopic images show a large mass of obscuring matter separating the nebula's brightest region from its fainter shores like water in a black lagoon. The Lagoon Nebula is a place of celestial nativity, a site where new stars are being born from dusty molecular clouds.

The Hubble Space Telescope image above shows M8's bright central region. The image reveals a cacophony of small-scale structures in the interstellar medium, including giant twisters. And like tornados on Earth, M8's twisters appear to form by a large difference in temperature between the hot surface and cold interior of the clouds. The pressure of radiating starlight and strong horizontal shear forces twist the clouds into their tornado-like appearances.

Messier 8 lies in a rich region of the Milky Way and has some dimmer deep-sky companions: M20 (the Trifid Nebula) and M21 (a tiny open star cluster). Both lie just above and to the west of the Lagoon; they are part of a warped Christmas-tree-shaped asterism of 6th- and 7th-magnitude stars, which is more formally known as Webb's Cross. Messier 20 forms the Christmas tree's base, and M21 is the "star" topping the tree. Messier 20 lies at the same distance as M8, so the two nebulae could be part of a vast star and nebula complex. Messier 21 is only 1,000 light years more distant. A beautiful star cloud lies immediately east and a bit north of M8. Here you'll find a glowing patch of milky starlight that includes the sparse open star cluster Collinder 367 superimposed on yet another dimly glowing nebula that matches its extent. The entire region is awash in dim patches of glowing gas, starlight, and dark dust clouds. Under dark skies, this binocular field is a phantasm of ill-defined shapes of light and dark that will waft in and out of view as your eyes drift across this foggy starscape.

Using the chart on page 66 as a guide, place M24, the Small Sagittarius Star Cloud, at the bottom of your field of view and two more, albeit small, nebulae should pop into view: M17, the Swan, or Omega, Nebula (just 2° north and slightly east of M24), and M16, the Eagle Nebula (2° north–northwest of M17); M16 is in the constellation Serpens Cauda, but M24 in Sagittarius is a better guide to it than any star in Serpens – a good demonstration of the arbitrary nature of constellation boundaries.

The Hubble Space Telescope image of M17 on the next page shows the Swan Nebula's interior as a wonderland of nebulous splendor – a crib of newborn stars swaddled in nascent blankets of glowing gas. Intense ultraviolet radiation fleeing from these hot, young stars not only causes the nebula to fluoresce but also eats away at the dark mountains of cold gas from which these stars formed. The region of the nebula shown in this photograph is about 3,500 times wider than our Solar System.

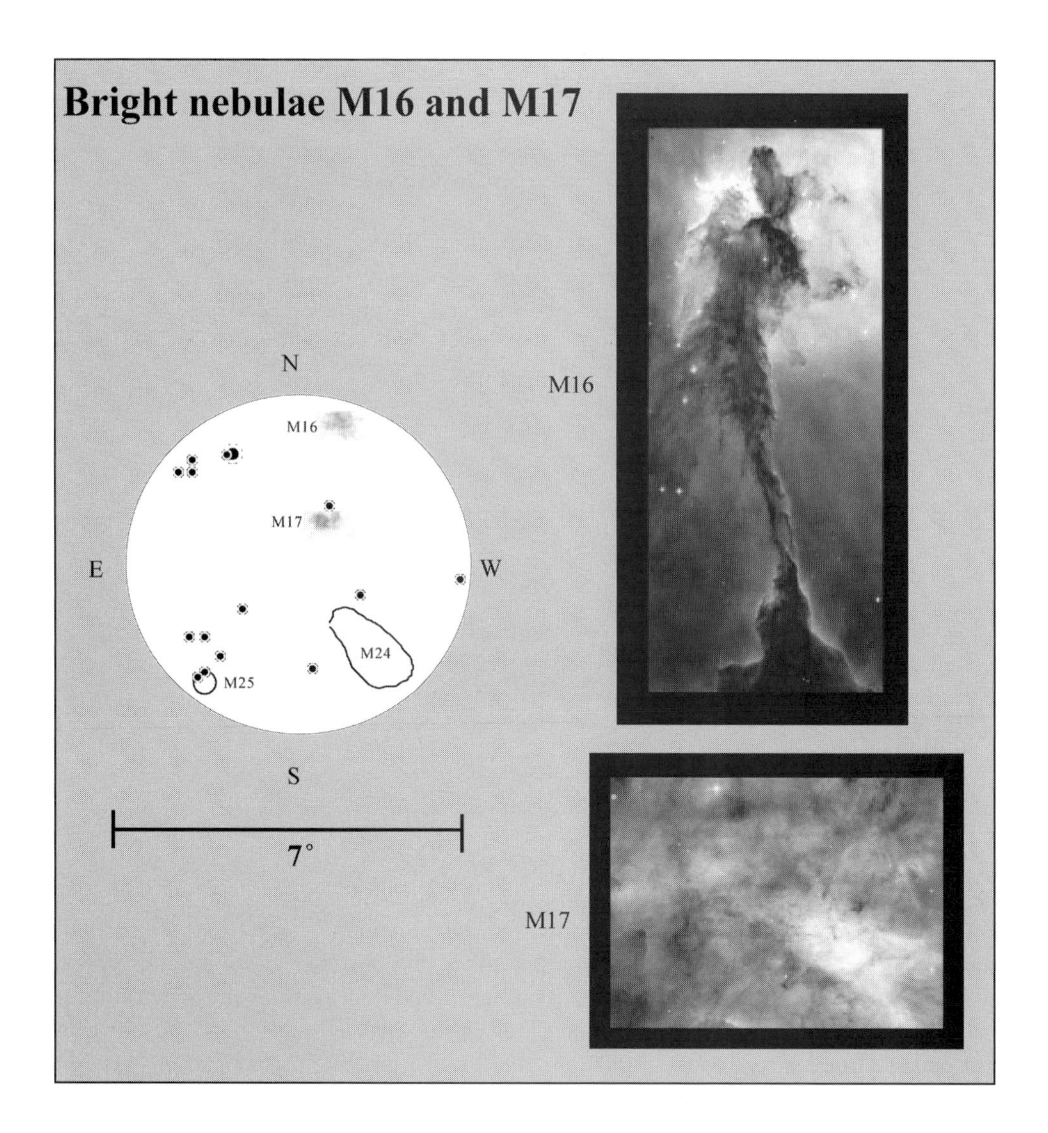

Like the dark mountains in M17, the dark sea horse of M16 (see the Hubble Space Telescope image above) is but one of several such dramatic spires in that nebula. These ebony pillars, each tens of trillions of miles long, are being boiled away by the ultraviolet radiation from new, nearby stars. The dark sea horse may be a giant incubator for those newborn stars. Ghostly streamers of gas boil off the pillar's surface, creating a haze that highlights the spire's three-dimensional shape. Astronomers liken the erosion process to that of a brush fire that quickly consumes grass but slows down when it encounters dense brush. Likewise, the ultraviolet radiation in M16 quickly consumes the thin outer region of the pillar but slows when it encounters the column's dense core.

Scutum, the Shield

We continue our journey northward to the dim and inconspicuous constellation of Scutum, the Shield. A relatively new constellation formed by Johannes Hevelius in 1684, it honors King John III Sobiesci of Poland for his defense of Vienna in 1683. While the brightest star of Scutum burns dimly at 4th magnitude, the region is blessed with yet another naked-eye wonder of the Milky Way – the Great Scutum Star Cloud, the Gem of the Milky Way. This grand ghost of sepulchral light burns like St Elmo's fire clinging to a broad sail. The cloud's brightest segment is a 3° circle of light between 4th-magnitude Beta (β) Scuti and 5th-magnitude Epsilon (ε) Scuti, between the southwestern edge of what I call the Forgotten Crown – 14, 15, Lambda (λ), and 12 Aquilae and Eta (η) and Beta Scuti – and the Scutum Trapezoid, which includes the stars Alpha (α), Delta (δ), and Epsilon (ε) Scuti. The cloud is bordered to the north by a dense arc of dust, which, with imagination, looks like an ink stain on a blotter. With averted vision, the Scutum Star Cloud becomes the bright northwestern segment of a larger crescent of milky wonder that arcs 3° south via the east.

Now place 4th-magnitude Beta (β) Scuti in your binoculars as shown in the chart at the bottom of page 67. Just 1° south–southeast is a tight trapezoid of 5th- to 7th-magnitude stars, the northwestern corner of which is the semiregular variable star R Scuti, a pulsating luminous yellow supergiant that shines with a distinct amber light when at maximum. It is a magnificent object to monitor each season in binoculars. The star's somewhat irregular pulsations cause it to rise to 4.5 magnitude, making it visible to the unaided eye, then dip to 9th magnitude, about every 146 days.

Delta (δ) Scuti is a relatively bright short-period variable, the prototype star of its class. At a distance of 187 light years, Delta has stopped fusing hydrogen in its core and will shortly be on its way to becoming a red giant. For now, it exhibits brightness fluctuations from 4.6 magnitude to 4.8 magnitude with a period of 4.65 hours.

If you return to R Scuti, then look about 1° to the southeast, you'll see a fuzzy pellet of light that has a bright core surrounded by a large halo of mottled light. This is M11, the Wild Duck. When seen through a telescope, this open star cluster has a V-shape that mimics a flock of ducks in flight.

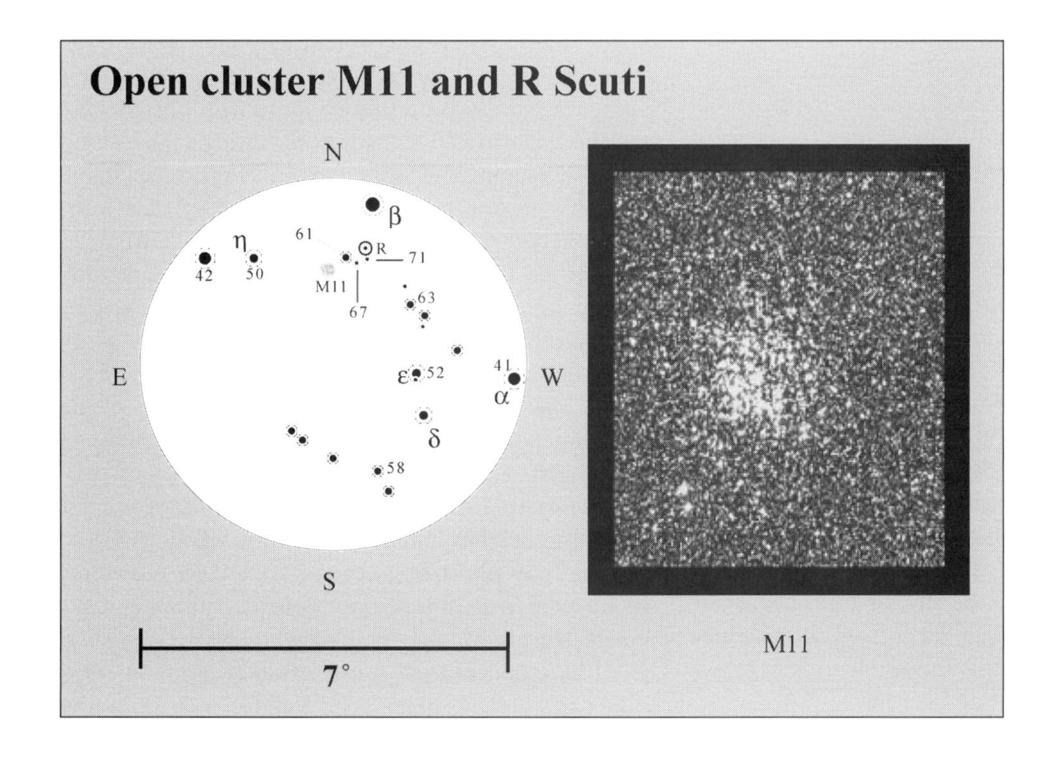

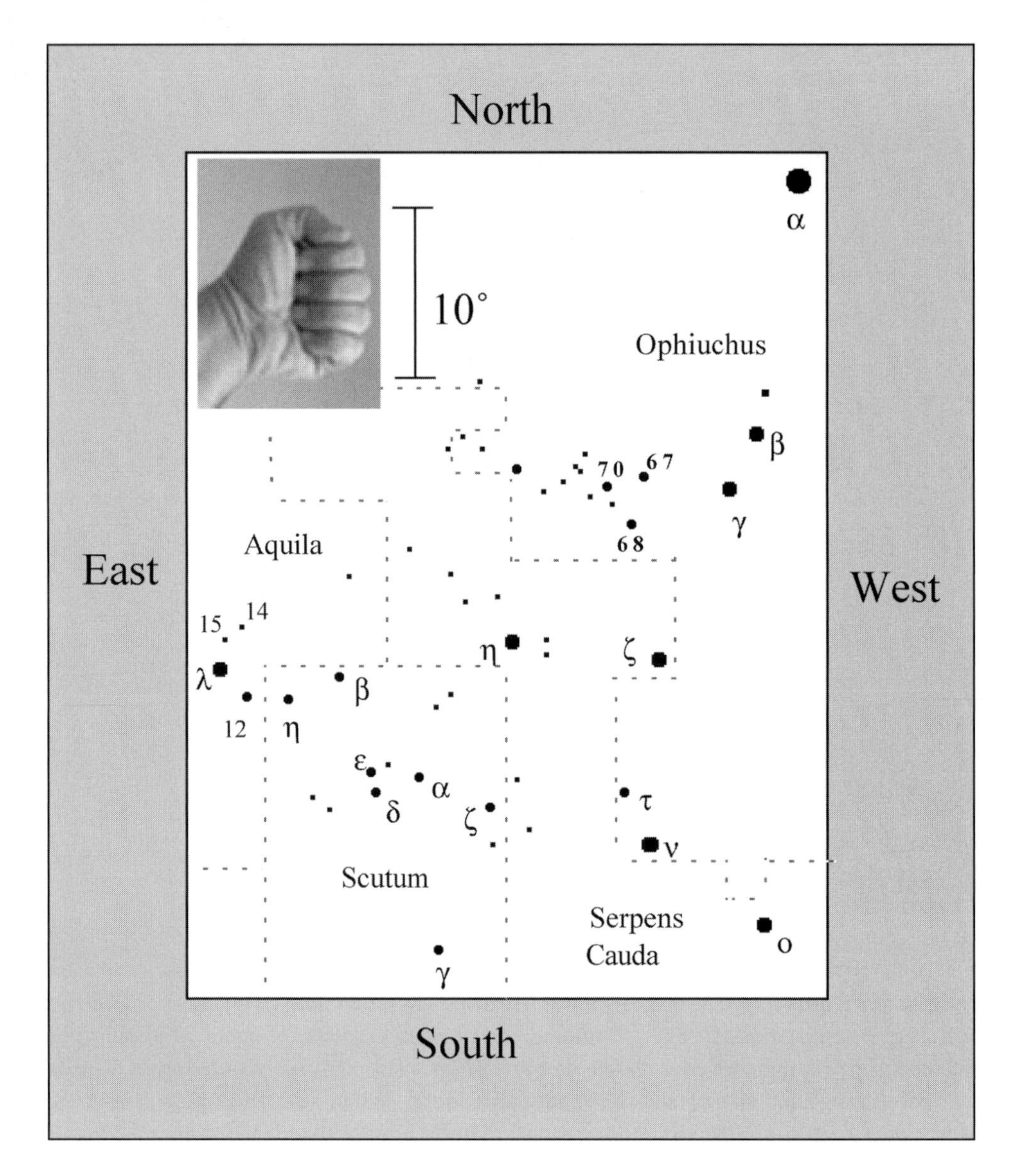

Before leaving this beautiful area, compare the colors of Alpha (α) and Beta (β) Scuti. Alpha is a *K*-type giant 175 light years distant that shines with a yellow-tulip hue. Beta is a *G*-type sun with a curious butterscotch "flavor." If you sweep your binoculars across the small but rich expanse of Scutum and let your mind relax see if the star clouds and dark nebulae don't work together like the careful brushstrokes in fine art. The sinewy shapes of the dark nebulae also cause the eye to follow various strings and loops of starlight – like the strings and bows tossed aside in wild abandon as a child eagerly rips into a wrapped present. When the great nineteenth-century Italian astronomer Angelo Secchi saw these forms, he imagined they were superimposed beds of stars, like a disorderly stack of dollies. As for the neat stellar arrangements, they were, he said, great spiral arcs whose stars curved together like beads on a string. "Sometimes they form rays," he added, "which seem to diverge from a common focus, and, what is very singular, one usually finds, either at the center of the rays, or at the beginning of the curve, a more brilliant star of red color, which seems to lead the march. It is impossible to believe that such an arrangement can be accidental."

Lyra, the Harp

In late August, go outside around 10:00 pm, put a blanket on the ground, and lie down so that your feet point due south. Now look directly overhead. Three bright blue stars should form a 30°-long isosceles triangle whose long axis points south. This is the famous Summer Triangle – summer's greatest *asterism*. Three Alpha (α) stars from three different constellations form the Summer Triangle: 0-magnitude *Vega* in Lyra, the Lyre or Harp; 1st-magnitude *Altair* in Aquila, the Eagle; and 1st-magnitude *Deneb* in Cygnus, the Swan. The chart on p. 69 shows the position of the Summer Triangle relative to Sagittarius and the Zenith. The photo at bottom shows the position of the three stars relative to the Milky Way.

Let's start our tour of these three constellations with Lyra, the Harp. Its Alpha star, Vega, is the fifth brightest star in the heavens and the Alpha Dog of the summer triangle. I say Alpha Dog because while Lyra represents the classical harp invented by Mercury, I think it looks remarkably like an alert dog wagging its tail, which is how I have outlined the constellation on page 70: Vega is the dog's nose; Epsilon[1,2] ($\varepsilon^{1,2}$) and Zeta (ζ) mark the

North
Deneb
Zenith (overhead)
Vega
20°
Summer Triangle
Altair
East
West
Sagittarius
August 20
South
10:00 pm

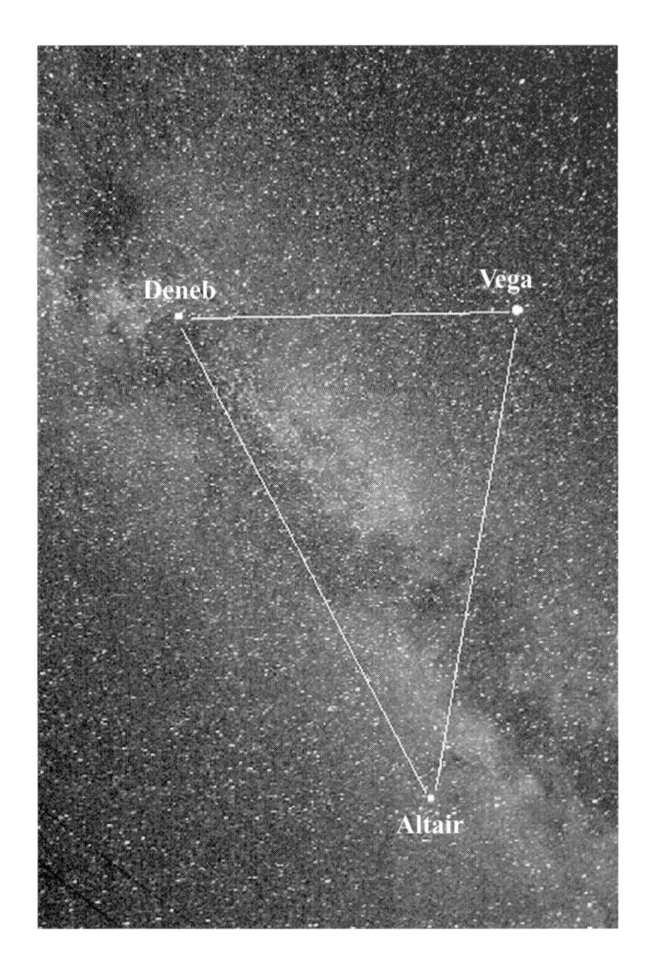

animal's head; Beta (β), Gamma (γ), and Lambda (λ) are its feet; and Eta (η) and Theta (θ) represent the tail in motion. Of course, the figure can also be seen as a leaping cat or a spraying skunk.

As for the Harp – the first ever made – Mercury fashioned it from a beautiful tortoise shell. He poked holes in the opposite edges then strung nine cords of linen through them; one string for each of the nine muses. Mercury then offered the Lyre to Apollo in exchange for his caduceus. Apollo in turn gave the Harp to his son Orpheus, who learned to play it with such perfection that nothing could withstand its lilting tone, not even the rocks and trees, which crowded around him whenever he played. Lyra was the very harp that Orpheus took with him into the Underworld. Heartbroken over the loss of his wife, Eurydice, who came to an untimely end by the bite of a viper (literally, in this case, a snake in the grass), Orpheus sought help from Pluto and Proserpine. While the rulers of the Underworld leaned back in their thrones, Orpheus strummed his Harp and sang his woeful tale. In his song, Orpheus pleaded that the dark lords renew the thread of Eurydice's life, so that she could finish out her brief existence with him – before they return to Pluto's realm of ultimate silence. The music was so pure, and the song so heartfelt, that even the lifeless ghosts wept in sorrow. So powerful was the performance that Pluto and Proserpine gave Orpheus two thumbs up.

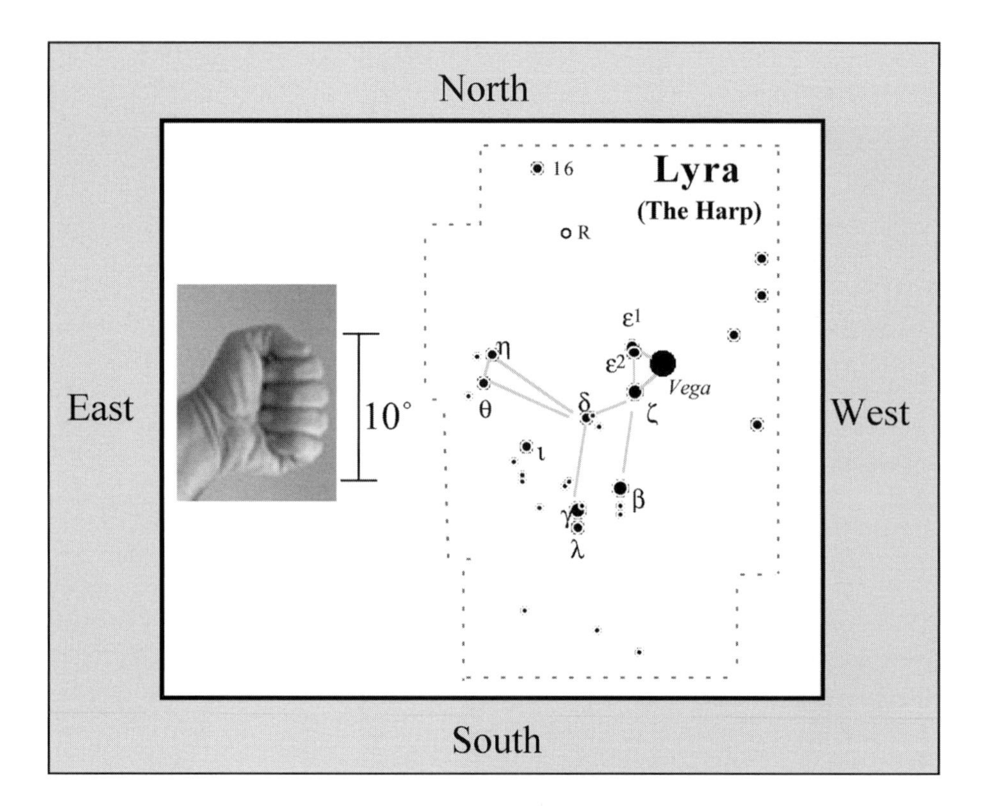

As if in a dream, Eurydice emerged radiant from the crowd of ghosts, though still limping from her fresh wound. Pluto allowed her to leave with Orpheus under one condition: that Orpheus not look at her until they reach the upper air. After Orpheus had led Eurydice through many dark and steep passages, he spied the light of day. Alas, in his excitement, he turned around to make sure his lover had kept pace with him. Suddenly he realized his mistake. But it was too late. His eyes had met hers. In an instant Eurydice was sucked violently back into the cavernous Underworld. Orpheus had only enough time to hear her fading voice: "Farewell, my love."

All attempts Orpheus made to reenter Pluto's domain failed. Orpheus returned to the surface and mourned Eurydice's second death. For the rest of his days, Orpheus failed to recognize the beauty of any other mortal woman. One day, during a raucous celebration, several maidens tried to seduce Orpheus but to no avail. Angered at his neglect, the maidens threw javelins and stones at the sad man. But the stones and spears only fell to Orpheus's feet when they encountered the Lyre's soothing voice.

Finally, in frustration, one maiden screamed so loud that it drowned out the Lyre's music long enough for a spear to hit its mark. In a maniacal frenzy, the vengeful maidens ripped Orpheus apart, limb by limb. They threw his head and Harp into the river Hebrus, where they continued to murmur their sad music. The shores began to sob and the willows to weep at the mournful song. Fortunately Jupiter had witnessed the fate of Orpheus and ordered the nine muses to gather up his body parts and bury them at Liberthra. To this day the nightingale sings over his grave with an unparalleled melody. As for the harp, Jupiter had it placed among the stars as a reminder of its beauty and power. Orpheus entered the Underworld and became forever united with his one true love. Like our eyes upon the stars, Orpheus can now glance upon Eurydice whenever he wants.

Some of the details in this tragic Greek tale were most likely borrowed from the legend of Osiris, one of the most ancient and important myths of ancient Egypt. (Indeed, versions of the Osiris myth have cropped up in every culture throughout history, including the works of Shakespeare.) A great pharaoh, Osiris married his sister Isis, and, like Orpheus and Eurydice, had an inseparable love. But Osiris had a jealous brother, Seth, who was constantly trying to kill him; but Isis, like a lyre's voice, thwarted all attempts by Seth to harm her husband. During a celebration, though, Seth lured Osiris into a beautifully jeweled box, which he thought was a present. But Seth quickly nailed a lid on the coffin, sealed it with lead, and tossed it (and Osiris) into the Nile. Learning of her husband's death, Isis traveled the Earth in search of the casket, which she discovered had washed ashore at Byblos. With the help of the land's ruling king, Isis claimed the body of Osiris and headed home. But one night, as she slept with the body by a river bank, Seth ambushed her, took hold of Osiris and ripped his body into 14 pieces, which he scattered haphazardly across the land. Isis, this time with the help of her sister, Nephthys, set off to collect the body parts. Once retrieved, Isis pieced together Osiris, mummified the remains, then transformed herself into a hawklike Kite. In that incarnation, Isis uttered spells so that Osiris was able to impregnate her. Afterwards, she pleaded to Anubis – the great jackal-headed god of cemeteries and embalming – to bring her lover back to life. Anubis replied that while he could not return Osiris to the land of the living, he could assure her that he would live as King of the Underworld and be the Judge of the Dead. Can you now see Lyra as Isis in the form of a Kite? Vega is the bird's eye; Epsilon and Zeta are its head, the parallelogram marks the great bird's wings (which are both flapping down), and Eta and Theta are the Kite's upright tail.

Now turn your binoculars to Vega, Lyra's luminary. Vega, is an *A*-type main-sequence dwarf that has a convenient apparent magnitude of 0.0 and a surface temperature of about 9,500 Kelvin. So the star shines with a bright white hue. The purity of Vega's light and the exactness of its apparent brightness makes it a standard candle against which astronomers can compare the apparent magnitudes of all other stars. Vega is a rapid rotator, spinning on its axis once every 12.5 hours, which is 90 percent of its maximum speed of rotation. (By comparison, our Sun takes 27 days to complete one rotation.) If, for some reason, Vega's spin exceeds its critical rotation speed, gravity will not be able to hold the star together and it will begin to fly apart. Although you cannot see it, Vega is a bloated star, one that's 23 percent wider than it is tall.

Through binoculars, Vega is a blue-white spark, like a glint of sunlight seen through crystal. This brilliant beacon was the ancient Babylonian Messenger of Light and, some 13,000 years ago, the Gem of the North, meaning it marked the position of Earth's north celestial pole; Vega will hold that position again around the year AD 14,000 (see page 45).

A near neighbor to our Sun, Vega lies only 25 light years distant and is a good star to show anyone celebrating their twenty-fifth birthday. By the way, on page 40, I mention that golden Arcturus is speeding through space in a rogue direction (toward the direction of the star Spica in the south) compared to other stars in the Sun's neighborhood. Just where is the Sun heading then? Toward the star Vega, in Lyra, which marks the *solar apex*, or the *apex of the Sun's way*.

The binocular view of Lyra is most interesting, because its brightest stars all but fit in the field of view. Just northeast of Vega is one of the sky's most celebrated double stars, Epsilon (ε) Lyrae. It is commonly known as the Double-Double, because, while binoculars will split the single star to the unaided eye into two stars (Epsilon^1 and Epsilon^2 Lyrae), the smallest of telescopes will split those two stars again, like quicksilver. Actually, those with the keenest of eyesight can split Epsilon into two components without optical aid. The stars shine around 4th magnitude and are separated by 3.5′, which is near the limit of human detection by someone with 20/15 (or better) vision. At a distance of 25 light years, that translates into a true physical separation of 10,000 astronomical units. Epsilon^1 and Epsilon^2 Lyrae take at least half a million years to complete an orbit.

Look directly below Epsilon for equally bright Zeta (ζ) Lyrae. It marks the southern apex of a triangle with Vega and Epsilon Lyrae; it's also the northwestern corner of a pretty parallelogram with Delta (δ), Gamma (γ), and Beta (β) Lyrae. Zeta is a much overlooked binocular double star, probably because its subtlety is eclipsed by Epsilon's grandeur. Zeta is a "doable double" in 10 × 50 binoculars, especially in deep twilight. The *A*-type primary has a fainter 6th-magnitude F-type companion 44″ distant. Seeing color in these stars might require a telescope but give it a try.

Now slip your gaze to Delta Lyrae, which is a very colorful and wide optical double (even to the unaided eye) superimposed on a dim and sparse open star cluster. So Delta's environs are fairly rich and appealing, especially with averted vision and some concentration. Actually, bright Delta is known as Delta^2 Lyrae, while its fainter companion to the west is known as Delta^1 Lyrae. The pair is striking in that Delta^2 has a rich red color, while Delta^1 is decidedly cool blue. Delta^2 is an M-type giant 900 light years distant; Delta^1 is a B-type dwarf just 200 light years further away.

R Lyrae is a perfect variable star to monitor through binoculars. In fact, its full range of fluctuations can be traced with the unaided eye, because the star never dips below 5th magnitude. Like other M-type red giants R Lyrae is a semiregular variable and lies 350 light years distant. Its brightness varies between magnitude 4.0 and 5.0 every 50 days.

Our last object is a challenge, but perhaps worth every binocular viewer's time. Roughly midway between Gamma (γ) and Beta (β) Lyrae is one of the sky's most celebrated planetary nebulae – M57, the famed Ring Nebula. Lying at a distance of some 2,000 light years, M57 is a torus of gas expelled from a red giant star like R Lyrae. We see the gaseous wreath as it was blown off the dying star some 20,000 years ago. Skilled observers have seen its dim glow through binoculars, so give it a try, especially since the chart below shows you exactly where to look. The planetary nebula will not look like the Hubble Space Telescope photo accompanying the chart. Instead, it will shine like a small star but one just ever-so-slightly swollen. Like Osiris, here is the "ghost" of a once living entity roaming the dark Underworld of space.

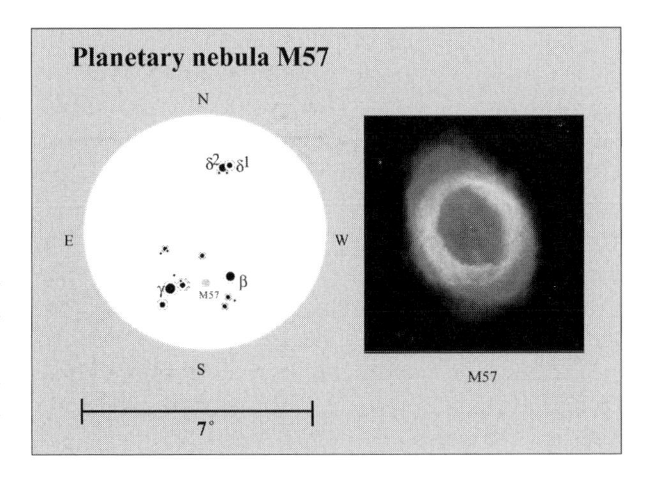

Draco, the Dragon

Keep Vega in view, but turn to face north. Now look a little more than 10° (a fist) to the lower right

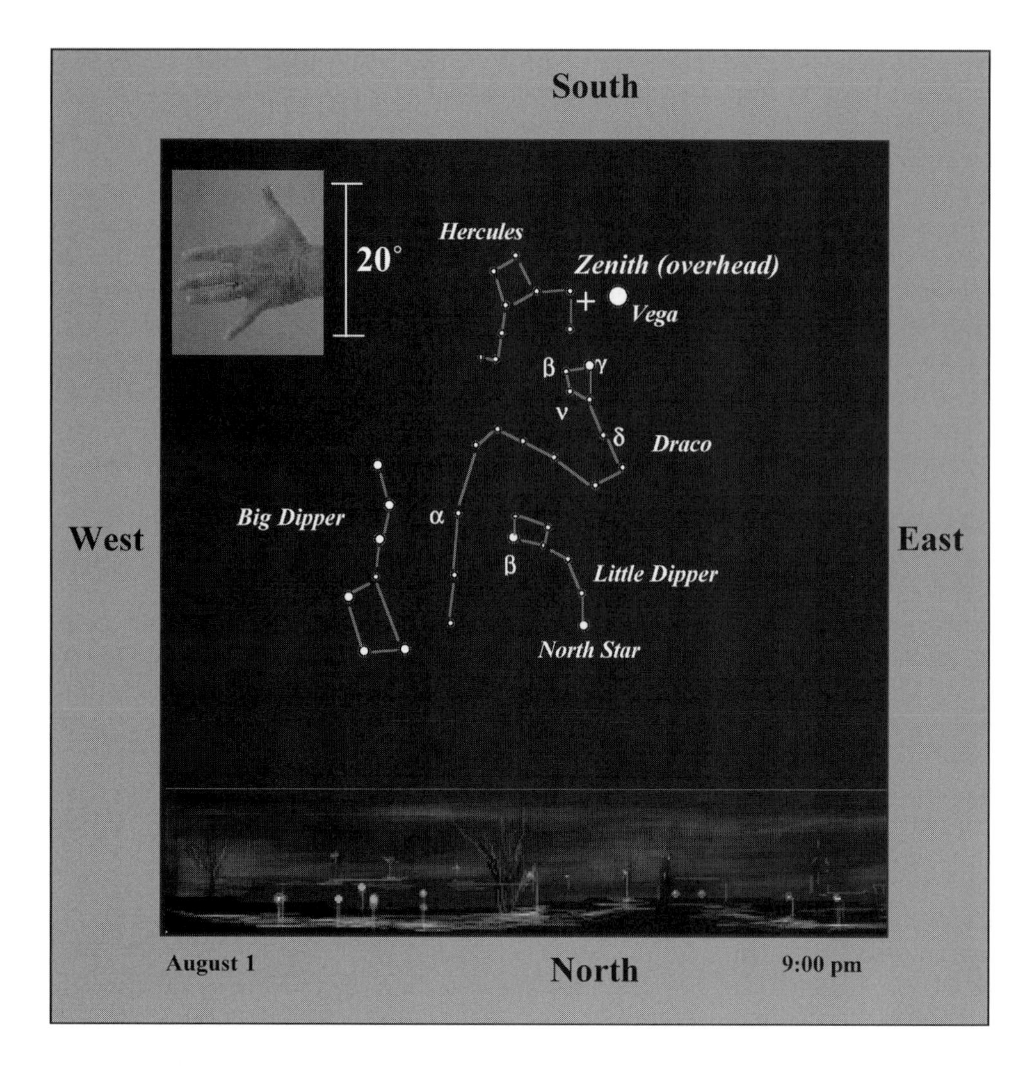

(northwest) of Vega. You should see a distinct 5°-wide trapezoid of four suns, the brightest of which is 2nd-magnitude Gamma (γ) Draconis. The trapezoid forms the head of Draco, the Dragon – the beast that guarded the tree of golden apples (a wedding gift to Juno from Jupiter).

As one of his labors, Hercules was challenged to get past Ladon the Dragon and steal some of those delicious golden apples. No problem. When Hercules arrived in the Garden of Hesperides, he saw Ladon resting under the shade of the tree. Shrugging his shoulders at the lack of a contest, the Strongman shot the helpless beast with one of his poisoned arrows, stepped over the twitching body, and plucked the fruit from the tree. Saddened by Ladon's death, Juno placed the tree's guardian in the sky, where you can still see its long, serpentine body skirting the north celestial pole.

Draco is, in fact, one of the circumpolar constellations visible from mid-northern latitudes. But most of its coiled body is lined with inconsequential stars. Still, it's fun to trace the Dragon's body across the heavens, from its head near Vega, to the tip of its tail, which lies between the tip of the Big Dipper's bowl and the Little Dipper's handle. In fact, with imagination, one can easily transform the Little Dipper into an apple tree around which the Dragon curls his body as a form of protection: Beta (β) Ursae Minoris makes a wonderful golden apple that still clings to the tree's plucked branches (the Little Dipper's bowl), while the North Star looks like a golden apple which has fallen to the base of the tree (the Little Dipper's handle).

Besides Alpha (α) Draconis, whose importance as a pole star some 4,000 years ago is discussed on page 45, most of the binocular stars of interest in Draco are found in the immediate vicinity of the Dragon's head. Ironically, in keeping with the mythical golden apples, the Head of Draco contains three golden stars. Shining at 2nd magnitude, Gamma (γ) Draconis (Eltanin) is the brightest star in the constellation. This K-type orange giant 148 light years distant is making a beeline for our Sun. About 1.5 million years from now, Eltanin will sail within 28 light years of Earth and appear as the brightest star in our nighttime sky. Now compare Eltanin's marigold hue with that of slightly fainter Beta (β) Draconis (Rastaban). With a surface temperature of 5,100 Kelvin and a brilliance 950 times that of the Sun, Rastaban is a yellow supergiant 362 light years distant that shines with the color of a cool lemon spritzer. Then, there's Nodus, Delta (δ) Draconis. This borderline K-type giant is conveniently situated exactly 100 light years distant and glows like golden tulip.

Now for the *piece de resistance*. Shining at 4th magnitude, Nu (ν) Draconis, the dimmest star in the Dragon's head is a remarkable . . . no, *stunning*, binocular double. Even the smallest pair will cleanly split this 4th-magnitude pip of

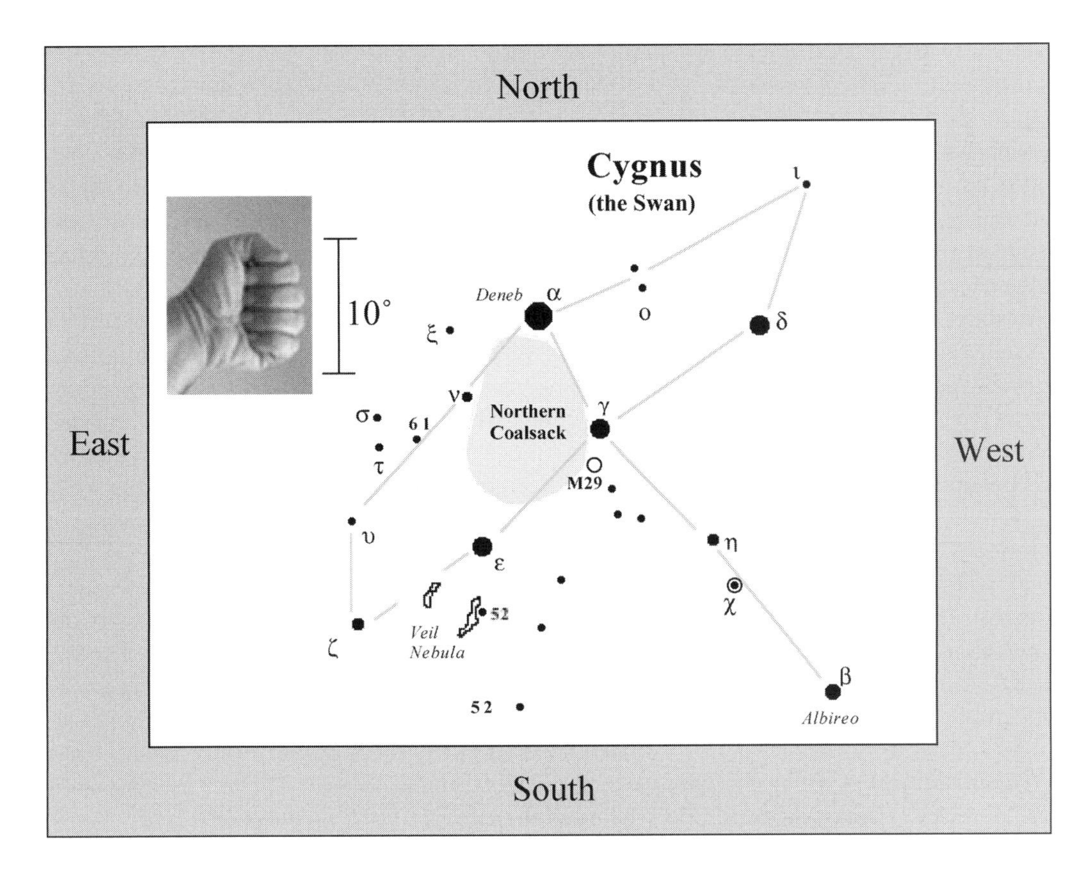

light into two pearls of near-equal intensity. But you have to look closely and carefully, because the components are separated by only a little more than 1′. It's the kind of sight, that once captured, can make the heart skip a beat.

Cygnus, the Swan

Cygnus is one of summer's two high-flying birds. According to legend, the Swan is simply one of many disguises Jupiter took in order to fool some fair maiden so that he could have his way with her. Jupiter was the original wolf in sheep's clothing. Anyway, after failing to win the affection of Nemesis (the avenging goddess), Jupiter turned himself into the Swan and had Venus (the goddess of love and beauty) chase him as Aquila, the Eagle. Fearing the Eagle might strike the Swan, Nemesis plucked the Swan from the air and wrapped her arms around the frightened bird. When the Eagle departed, Nemesis found herself in the warm embrace of Jupiter. To commemorate his victory, Jupiter placed the Swan and the Eagle in the sky as symbols of his brilliance – at least in the arena of trickery.

Let's look first at Cygnus. Alpha (α) Cygni (Deneb) is the Swan's tail, while Beta (β) Cygni (Albireo) marks the bird's beak. We see the Swan flying nose first down the Milky Way. The bright stars Epsilon (ε) and Delta (δ) Cygni mark the tips of the bird's outstretched wings. These four stars also mark the four "stations" of the Northern Cross, which I've outlined in the photo illustration at right. The Northern Cross is one of the most prominent asterisms in the night sky; it's also a landmark constellation in the summer Milky Way, which it helps to define. The Cross stands upright on the western horizon just after sunset in December, the month in which modern Christians celebrate Christ's birth.

The photo also shows two more wonders of the Milky Way: the Northern Coalsack (see the chart above) and

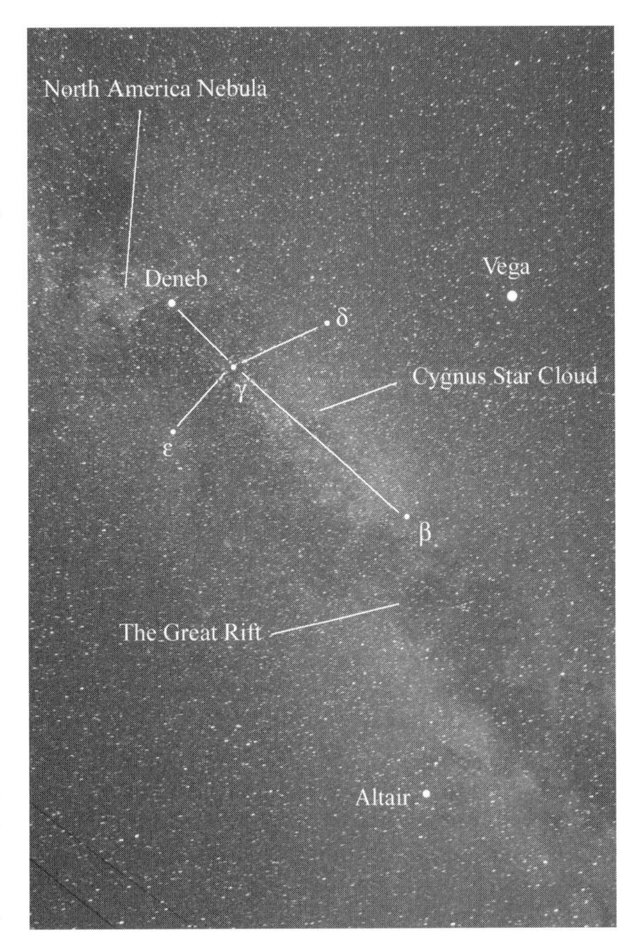

the Great Rift – two deliciously dark treats that meld into one – as well as some bright star clouds and nebulae.

To see the Coalsack, simply focus your attention on the magnificent gap in the Milky Way between Deneb, Gamma (γ), and Epsilon (ε) Cygni. This gap – a giant (8° × 5°) ellipse of obscuring dust – is bordered to the northeast by a bright and beautiful star cloud that contains the famed North America Nebula. Under dark skies, many skywatchers have spied this enhancement in the Milky Way with their unaided eyes, looking like a box that has been kicked by a mischievous child. Through binoculars, however, the shape of the North American continent is much more apparent. The North America Nebula lies some 1,800 light years distant and spans nearly 50 light years of space.

Now look around Gamma Cygni where you'll find the Great Cygnus Star Cloud, which rams into the Northern Coalsack from the southwest. That shoal of light officially separates the Northern Coalsack from the headwaters of the Great Rift – a fantastic belt of dark dust that splits the Milky Way into two streams that run parallel with the Swan's neck.

Years ago, any dark patch seen projected against the Milky Way was called a "coal sack." The courageous seafarers of the high southern seas coined the name centuries ago after they saw a black pit in heaven next to Crux, the Southern Cross. They believed it possessed some occult connection with the Cross, so whenever they chanced upon it, they would "cross themselves" and shudder. Upon the sailors' return to the Northern Hemisphere, word about the Southern Coalsack spread. It is no wonder then that astronomers suddenly took notice of the enormous vacancy in the Milky Way near the Northern Cross and applied the "coalsack" name to it.

In his 1882 novel, *Two on a Tower*, Thomas Hardy describes the Northern Coalsack and its influence on man: "You see that dark opening in [the galaxy] near the Swan?" Swithin St. Cleeve says to his lover Lady Constantine. "There is a still more remarkable one south of the equator, called the Coal Sack, as a sort of nickname that has a farcical force from its very inadequacy. Those are deep wells, for the human mind to let itself down into. . . ."

The Northern Coalsack appears, albeit indirectly, in the ancient myths associated with Cygnus the Swan. In one tale, Cycnus [*sic*] – a brother of Phaeton (the mortal son of Helios, the Sun god) – let Phaeton test-drive the Sun chariot across the sky. It was a mistake. Phaeton, the proverbial impetuous youth, lost control of the steeds driving the chariot and became the world's first reckless driver. As the steeds roamed wild, the Sun threatened to scorch the Earth and its inhabitants. Jupiter put an end to Phaeton's recklessness by striking him down with a lightning bolt. Although the Sun chariot righted itself, poor Phaeton was burned to a crisp; his charred remains fell from the sky like a meteor and plunged with a hiss into the river Eridanus. Heartbroken over the loss of his brother, Cycnus dove into the river, retrieved Phaeton's remains, and gave him a proper burial. Moved by such devotion, Jupiter turned Cycnus into a Swan and changed his name to Cygnus. He positioned Cygnus high among the stars of the Milky Way, so he could look down upon the river where Phaeton fell. Today, when we look up at the Northern Coalsack, we see the very point at which Phaeton lost control of the Sun chariot, for it marks the northern end of the scorched path of Phaeton's disastrous ride – the Great Rift.

If the Large Sagittarius Star Cloud is the poetic soul of the Milky Way, then the Northern Coalsack and the Great Rift represent the depths of human fear and imagination. Throughout history, these dusty coalsacks were regarded as mysterious voids, deep caverns that dared the human mind to descend. Today, we know these features are only cold clouds of hydrogen gas that cloak or dim the light of whatever lurks beyond them.

Still, seeing these "deep," black pits silhouetted against the brilliant Milky Way mirrors our immortal fears. As Garret Serviss writes in his 1919 book, *Curiosities of the Sky*, "Infinity seems to acquire a new meaning in the presence of these black openings in the sky, for as one continues to gaze it loses its purely metaphysical quality and becomes a kind of entity, like the ocean. . . . [T]hey produce upon the mind the effect of blank windows in a lonely house on a pitch-dark night, which, when looked at from the brilliant interior, become appalling in their rayless murk."

To see another mind-altering effect of dark nebulosity, turn your binoculars to 2nd-magnitude Gamma (γ) Cygni. This pale-yellow, F-type supergiant lies 1,500 light years away. Despite its great distance, Gamma shines brightly in our sky because it is 65,000 times more luminous than our Sun. And yet its light is being dimmed further by intervening dust. If you stare at Gamma and the bright Milky Way immediately surrounding it (which has a greenish tinge under the darkest skies) through your binoculars for a minute or more, your eyes will start to relax. When that happens, a black cobweb of dark nebulosity begins to emerge to prominence. Do you see a pattern to the darkness? When I look, I see a dark counterclockwise spiral out to the edges of the field.

Now use your binoculars and the chart on the next page to find open star cluster M29 about 2° south–southwest of Gamma. Look for a beautiful tight knot of fuzzy stars, whose total light equals that of a 6.5-magnitude star. The cluster lies at a distance of 4,400 light years and packs some 80 suns into a tiny disk only 10′ across. It is one of many clusters you will encounter in a sweep of the Cygnus Milky Way.

Using the same chart, drop your gaze to the opposite edge of the field of view, where you will find 3rd-magnitude Eta (η) Cygni. Near it is Cygnus X-1 – one of the sky's best black-hole candidates some 7,000 light years distant. Although we have to use our imagination to see the black hole (if it exists), we can at least see the 9th-magnitude O-type supergiant star it orbits. That star, with the beastly Henry Draper Extension (HDE) catalog name of HDE 226868 is just visible with binoculars if you take the time to look; it is the lower of the two close

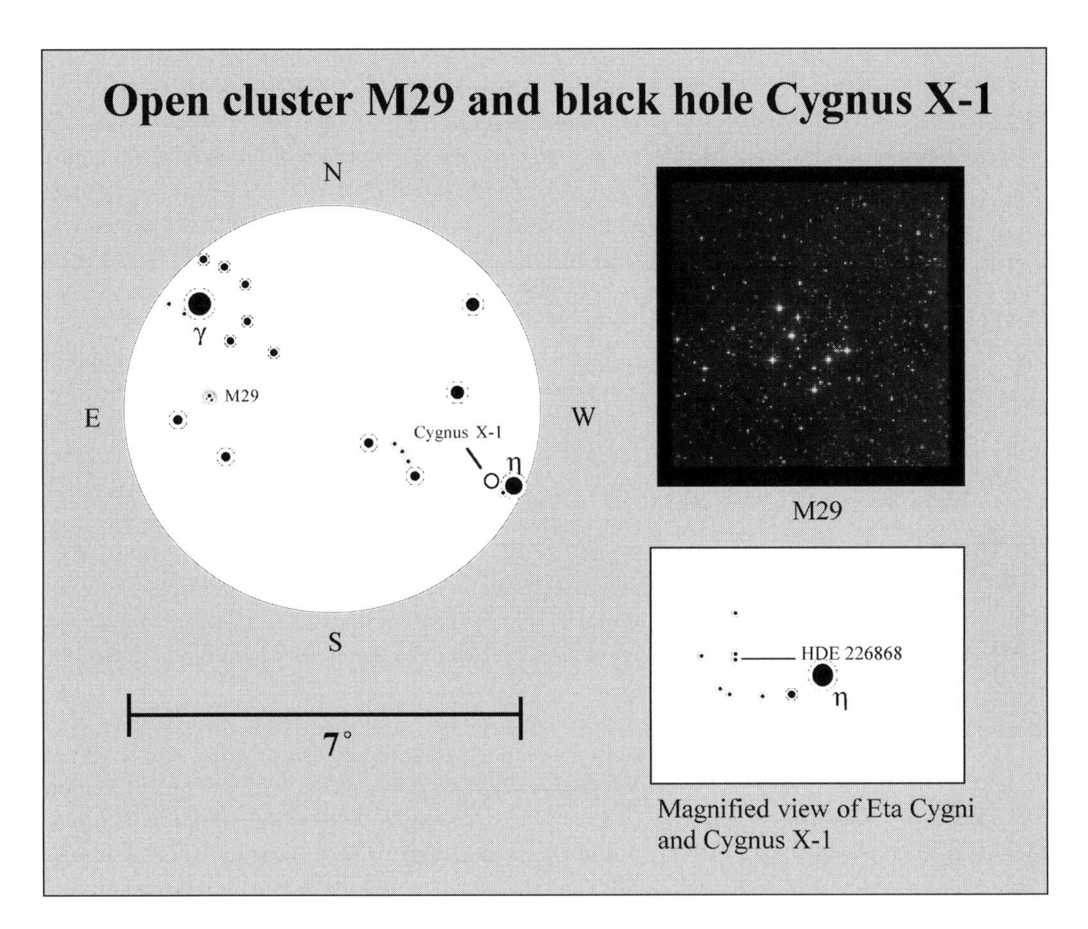

Magnified view of Eta Cygni and Cygnus X-1

stars seen in the chart above. Astronomers believe a black hole is orbiting HDE 226868 because the star is one of the brightest X-ray sources in the sky. The most likely explanation for that fact is that matter is flowing from the supergiant star into the black hole, emitting X-rays as it does. And since this was the first X-ray source discovered in Cygnus (in 1971), the black hole was designated Cygnus X-1. Black holes form when the cores of supermassive stars collapse . . . forever. The hole's gravitational pull is so powerful that light trying to rush out of the hole cannot escape its grasp and is doomed to orbit it forever; since that light never reaches our eyes, the object appears "invisible."

To see a star that ended its life not as a black hole but as a supernova explosion, you will have to be under a dark sky. You then have to get into a comfortable position and breathe rhythmically as you look for two thin wisps of nebulosity known as the Veil Nebula (see chart on page 73). The two brightest pieces of the nebula are separated by 3.5°; the brightest segment lies near, and all but kisses, 4th-magnitude 52 Cygni, which is about 3° south of 2.5-magnitude Epsilon – the star marking the Swan's eastern wing tip. They are extremely thin and difficult wisps. But several amateur astronomers swear that the Veil Nebula is easy to see under very dark skies in 7 × 50 binoculars. The task is a mighty challenge, especially for a beginner, but the more you observe, the better you will become at seeing fainter and fainter objects. If you do get to glimpse these wispy wonders, you are seeing the remains of a cataclysmic stellar explosion that occurred about 15,000 years ago. It heralds the death of a supergiant star – one that started life with at least eight times the mass of the Sun, a star like Antares in Scorpius! In the final moments of its life, the outward pressure that supported the star ceased. The star succumbed to the force of gravity. Within seconds the supergiant's core collapsed into a sphere merely six miles across. The rest of the star rushed inward, then rebounded off the core. A quarter of a second later, the star ended its life in a cataclysmic explosion that shined with the brilliance of a billion suns and scattered its remains across space at speeds measured in thousands of miles per second. The Veil Nebula we see today is the expanding ghostly shell of these stellar remnants.

Just $2\frac{1}{2}°$ southwest of Eta Cygni is the giant, carbon-type, Mira-type variable Chi (χ) Cygni (see chart on page 73). This beautiful star shines at magnitude 3.5 when at maximum light, making it an obvious part of the Swan's neck, before it plunges into deep invisibility at magnitude 14. The star then returns to maximum brightness 400 days later. When brightest, Chi has a pumpkin orange light. The star does not always achieve 3rd magnitude. During many maxima it gets as bright as only about 5th. The odd variability is most likely due to the amount of carbon dredged up from its interior, which can form an opaque outer surface layer, causing the star to dim.

Our next star hop is straight down the Swan's neck to the tip of its nose (see chart on page 73), where we come to beautiful Beta (β) Cygni (Albireo). Albireo is one of the most popular telescopic double stars in the sky. But if you own a pair of 10 × 50 binoculars and can find a way to hold them steady, the single golden light of Albireo will split into two extremely close and

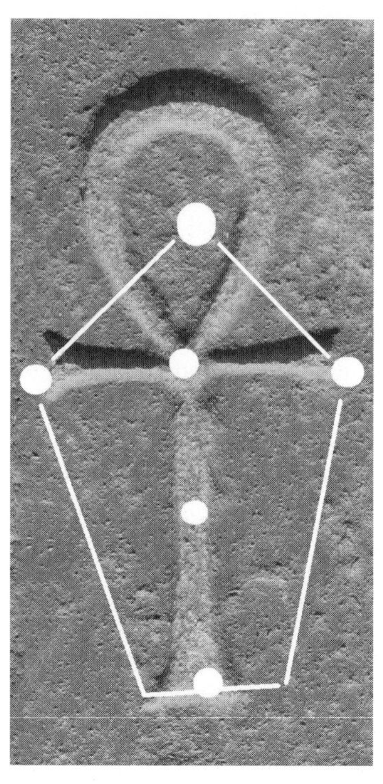

colorful objects – a 3rd-magnitude, golden *K*-type giant and a 5th-magnitude, *B*-type dwarf only 34″ apart! Try looking for the challenging aquamarine companion in late astronomical twilight, about 45 minutes after the Sun has set. With imagination, the two can be visualized as a distant view of our Sun and its orbiting water world . . . Earth! The stars actually lie about 380 light years distant and orbit one another with a period of about 75,000 years.

We finish our tour of Cygnus at the "tail end" – at Alpha (α) Cygni (Deneb), one of the most magnificent *A*-type luminaries in the Milky Way. Take a moment to look at the three stars in the Summer Triangle once again. See how brightly they all shine with a crisp blue light? Indeed, all shine at nearly the same brightness. But while Vega and Altair average only 25 light years distant, Deneb is more than 100 times farther away. It shines so prominently because it is a blue supergiant some 200 times wider than our Sun and 160,000 times more luminous! Deneb was only 7° from the north celestial pole about 18,000 years ago. It will claim that position again in less than 10,000 years.

By the way, take a look at the Northern Cross again, but this time through the eyes of an ancient Egyptian. Here we have a perfect ankh, the symbol of life. Indeed, the symbol for the Christian Cross of Calvary might have been derived from this powerful Egyptian symbol. Another interpretation of the ankh is that it represents the sunrise: The loop represents the Sun (luminous Deneb), which is rising above the horizon (represented by the crossbar [Epsilon, Gamma, and Delta Cygni]); the vertical axis below the crossbar represents the Sun's way. Indeed, the long axis of the Northern Cross almost parallels the ecliptic in this region. Equally interesting is that the ankh, as a symbol of life, refers not only to life on Earth but also to life in the Underworld. This fits nicely with our interpretation of Lyra as Isis (see page 70), for the cross can also represent Osiris in his ankhu (sarcophagus) – a possessor of life.

6 September

Torrent of light and river of the air
Along whose bed the glimmering stars are seen
Like gold and silver sands in some ravine

Henry Wadsworth Longfellow, *The Galaxy* (1875)

The Summer Triangle is still high overhead when September opens. In the south, the Milky Way retains its luster and its near vertical position. As the month progresses, though, the southern heart of the Milky Way will begin to dip toward the southwestern horizon, toward its autumn twilight. So let's return to the glory of the Milky Way, to its high southern skirt, where we'll find a little community of three constellations (Delphinus, Sagitta, and Vulpecula), in a section of sky that I call the Summer Suburbs.

Delphinus, the Dolphin

Face south and find Altair, the bright star marking the Summer Triangle's southern apex. A little more than 10° (a fist) east and slightly north of that blue beacon, you'll see a crisp and conspicuous diamond of four 3rd- and 4th-magnitude stars, packed into an area spanning only 3° of sky. This is part of Delphinus, the Dolphin – the mythical creature that saved the fabled musician Arion from drowning. While sailing to a musical contest in Sicily, a band of pirates robbed Arion of his gold and treasure, then

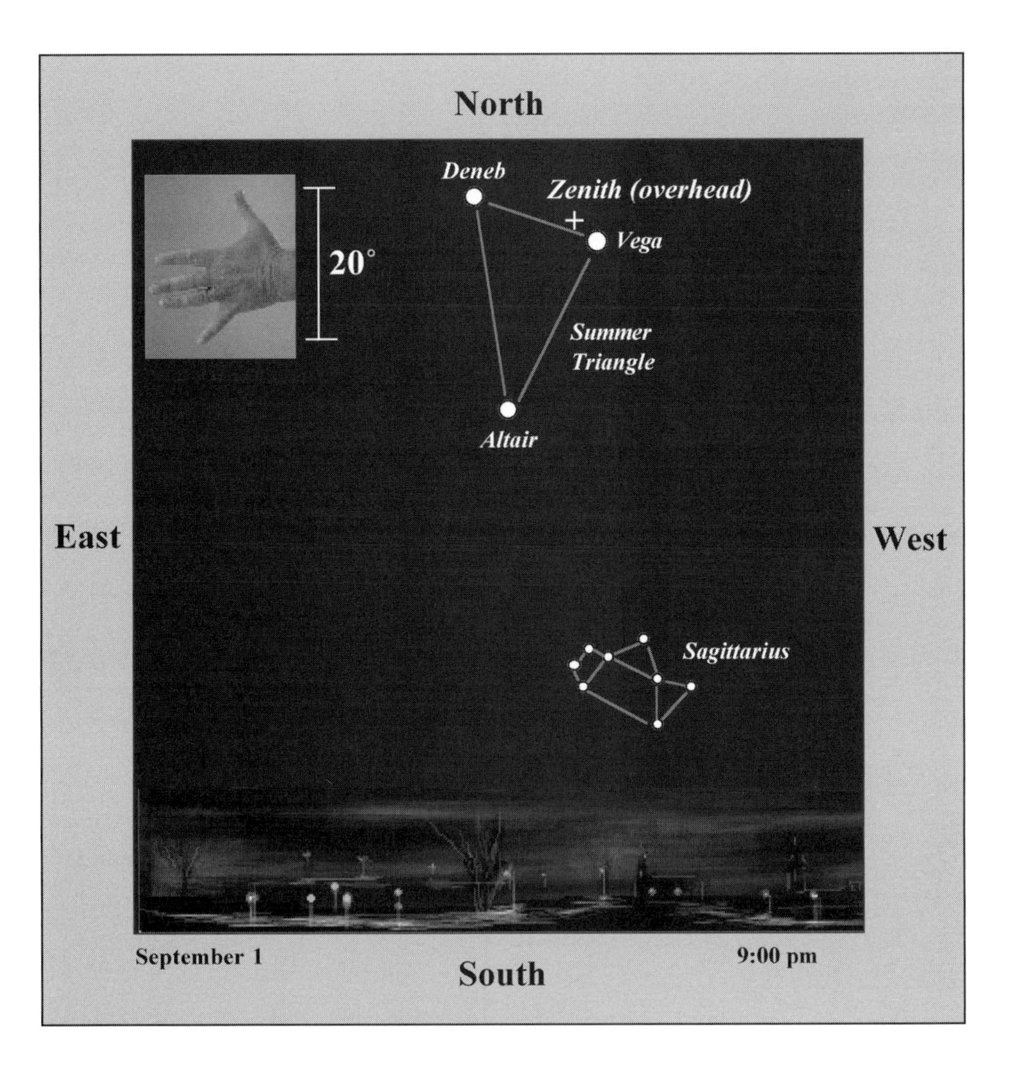

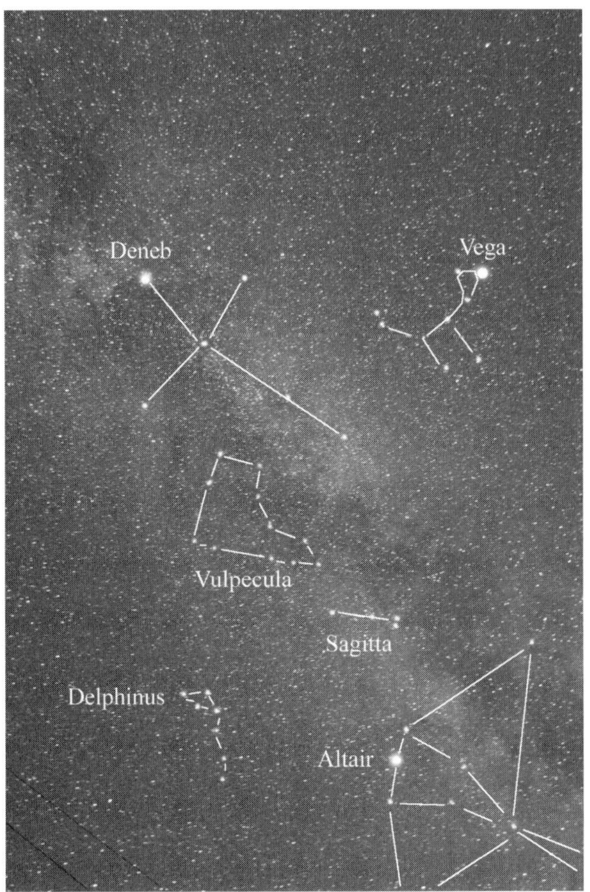

tossed him and his lovely Lyre into the sea. Fortunately, Delphinus had been lured to the boat by the Lyre's lilting music. Seeing Arion in distress, the Dolphin offered the musician a ride on its back (Arion was perhaps the first "poor" musician to hitch a ride) and carried him safely to shore. Arion then helped to capture the pirates, and his wealth and happiness was restored.

The Dolphin's diamond is one of my wife's favorite asterisms of summer (see the chart on the next page). Though small, it's a visual sensation – that pod of stars "magnetically" draws the eye to it, especially if you happen to be casually scanning the region. In your binoculars, all the major stars of the Dolphin fit nicely in the field of view, including the beautiful sapphire gem Alpha (α) Delphini and its distant companion, as well as lemon-yellow Gamma (γ) Delphini (a gorgeous telescopic double).

The diamond is yet another sarcophagus; actually it's a coffin, Job's Coffin. The name is somewhat of a mystery. I suppose there could have been a centuries-old sailor or explorer named Job, whose burial at sea seemed fitting to represent in the sky. Job may also be a misprint of the biblical Jonah who was swallowed by a whale. In ancient times, the shapes of whales and dolphins were not well known. And people fear what they do not know. So if something was big and in the water, it was a beast or leviathan – and leviathans were metaphors for Satan. For instance, in Chapter 41 of The Book of Job, God

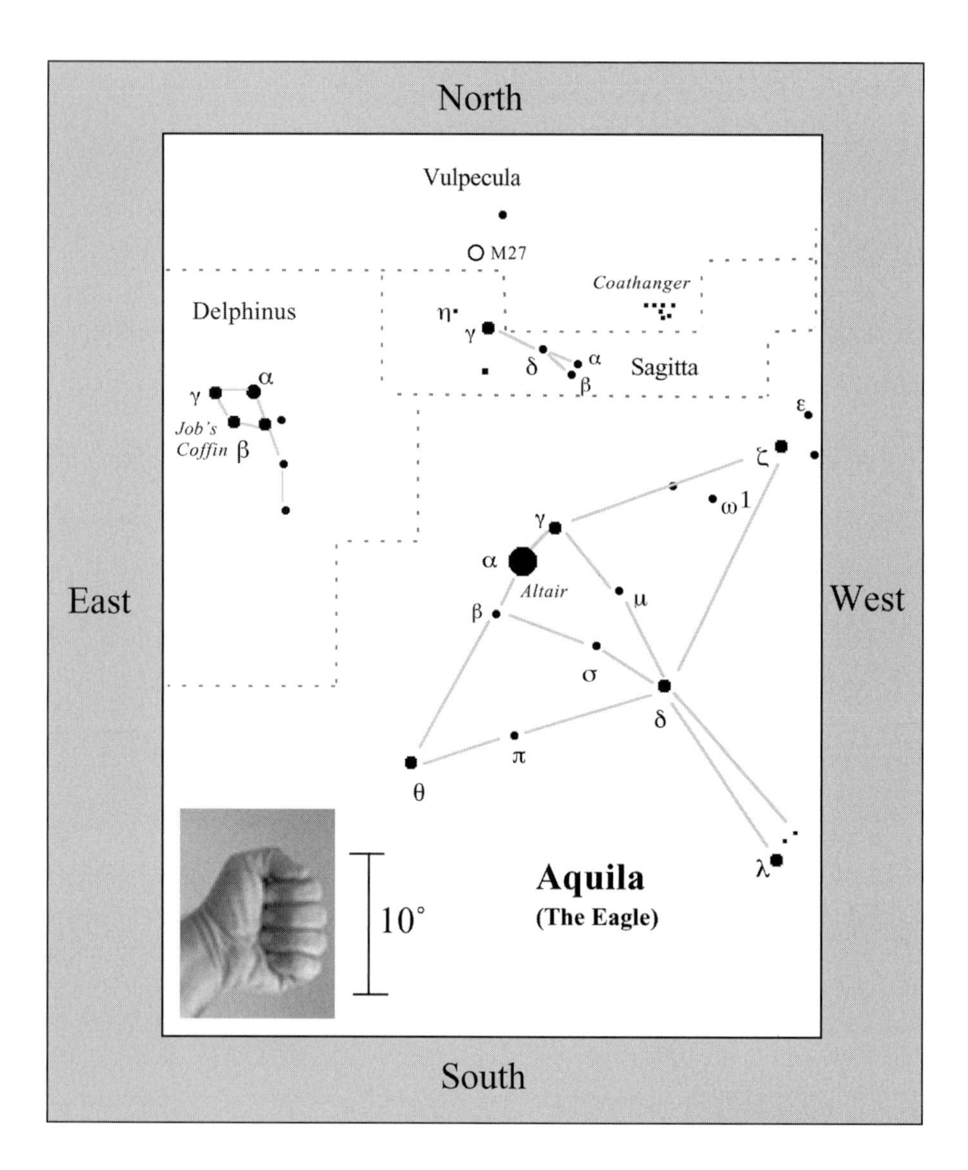

demonstrates to Job that He alone has power over "the leviathan" (Satan) and will slay him on Judgment Day. But the biblical story doesn't make any mention of a "coffin." Most likely the coffin is a metaphor for one of two things: (1) the confining hedge that God had placed around Job and his land to protect him: "Hast not thou made a hedge about him, and about his house, and about all that he hath, on every side? Thou hast blessed the work of his hands, and his substance is increased in the land." Job is rich in livestock, in health, and in his love of the Lord. But God allows Satan to put Job's faith to the test. As with Arion, Job loses all, but keeps his faith. In the end, both Arion and Job are rewarded with the return of their health and riches. Or (2), the coffin could also be a metaphor for the womb: "Naked I came from my mother's womb, And naked shall I return there," Job says.

The Dolphin has one other curiosity. The proper names for the Alpha (α) and Beta (β) stars are Sualocin and Rotanev, respectively. These names are not ancient; they date to 1814, when they appeared in the *Palermo Catalogue*. The names, when reversed, read "Nicolaus Venator," which is the latinized name of Niccolo Cacciatore, the assistant to the great Italian astronomer Guiseppe Piazzi.

Sagitta, the Arrow, and Vulpecula, the Fox

Center Altair in your binoculars. See how wonderfully it is flanked by two reasonably bright and colorful stars? Banana-peel Beta (β) shines at 4th magnitude and is a G-type subgiant 45 light years distant with characteristics like our Sun; 3rd-magnitude Gamma (γ) Aquilae is a beautiful golden K-type luminary 460 light years distant. I find Gamma's light most appealing, especially when seen next to the cool icy glow of Altair – like fire and ice. If you place Gamma Aquilae at the bottom of your field, Alpha (α) and Beta (β) Sagittae, the feather stars of Sagitta, the Arrow (the third smallest constellation in the sky), will tickle the top of your field. Move your wrist, and the entire Arrow will shoot into view.

Here is the arrow that Hercules used to kill the eagle that ate the liver of Prometheus – our ancient Dr. Frankenstein.[1] Prometheus was a titan who enjoyed molding men out of clay and bringing them to life with fire he had

[1] Mary Shelley, the author of the 1816 book, *Frankenstein*, entitled her work *Frankenstein, or, The Modern Prometheus*.

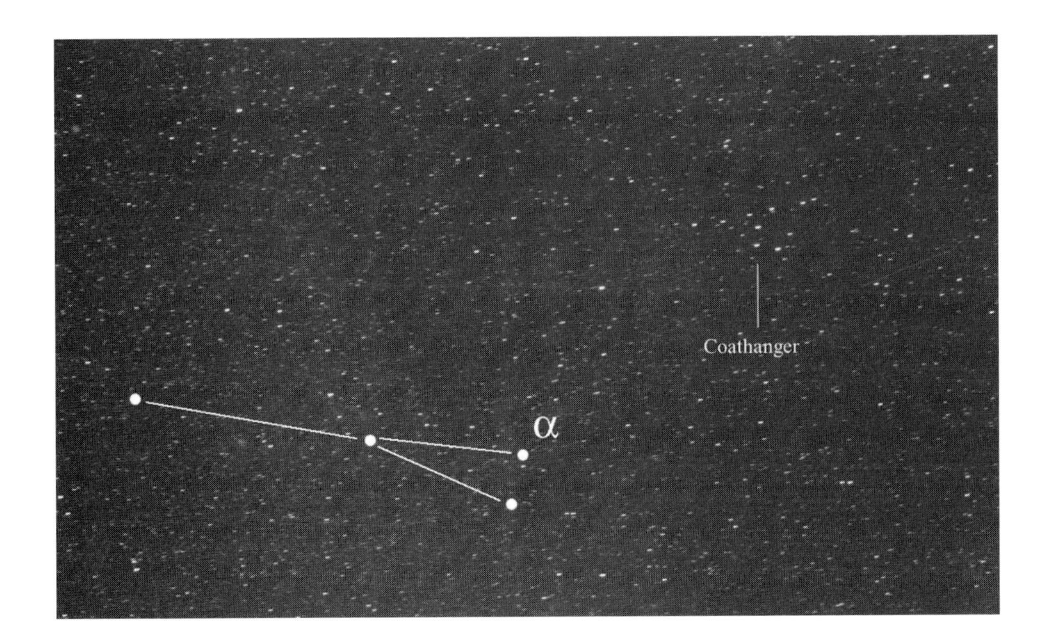

stolen from Jupiter. But all-knowing and all-seeing Jupiter was not amused by Prometheus's sneaking and thieving. The great god put an end to it by chaining the titan to a mountain top and exposing his liver so that the Eagle who lived there could feast on it for eternity; you see, since Prometheus was immortal, every time the Eagle ate his liver, a new organ would grow back, giving the Eagle food for another day. Hercules put an end to the torture by using one of his poison arrows to kill the Eagle and free Prometheus. In return, Jupiter placed both the Eagle and the Arrow in the sky. He also conveniently pointed the arrow away from the Eagle, so that it will forever miss its target, freeing the bird, like Prometheus, from an eternal doom.

Take a moment to study Alpha Sagittae in your binoculars, here is an unusual G-type giant. Although its temperature of 5,400 K is similar to that of our Sun, its diameter is 20 times larger and shines 340 times

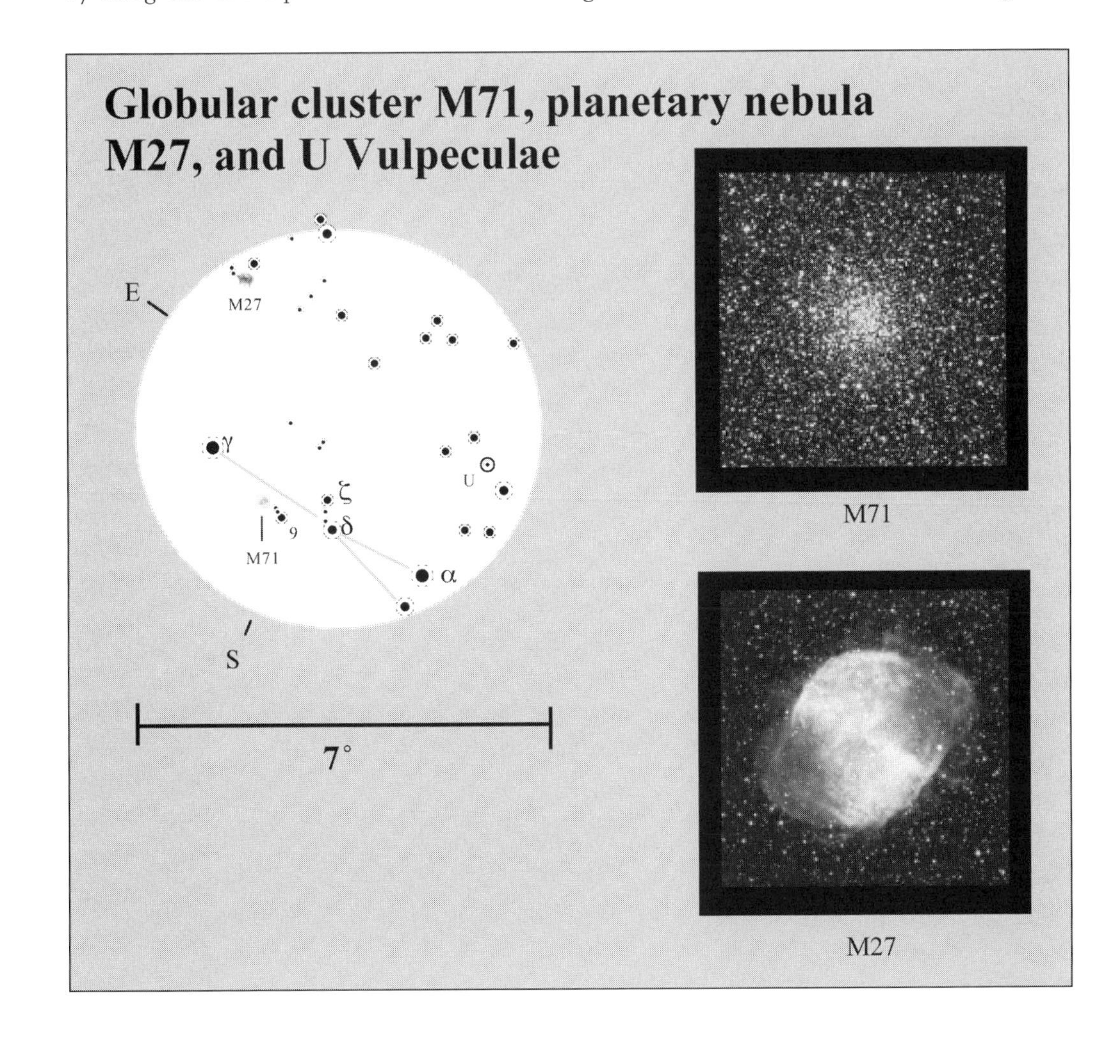

brighter. Its apparent dimness is a reflection only of its distance – 475 light years, which is only slightly more distant than golden Gamma Aquilae.

If you place Alpha and Beta Sagittae in the lower left of the binocular field, you will see a most delightful binocular asterism, Collinder 399, popularly known as the Coathanger, near the middle of the field.

Persian astronomer Al-Sufi (AD 903–86) included this naked-eye grouping in his *The Book of the Fixed Stars* (c. 964), calling it ". . . a little cloud [or cloudy patch] situated to the north of the two stars of the notch of Sagitta." Dalmiro F. Brocchi later independently discovered it during the 1920s or 1930s, so the asterism is sometimes referred to as Brocchi's cluster. But the stars are not physically related, all of them being a chance alignment, which is just as marvelous. The group is quite large, spanning three Moon diameters across. Five of the stars shine between magnitude 5.2 and 6.3, making them visible to the unaided eye under a dark sky, and seven of them are brighter than 7th magnitude – making the group a dazzling sight in binoculars. While the Coathanger moniker has stuck, the asterism has also been seen as a snail "on the move." By the way, although the Coathanger is close to the brightest stars of Sagitta, it actually lies in the western portion of Vulpecula, the Fox.

Hop over to golden Delta (δ) Sagittae at the base of the Arrow's shaft, then look for 9 Sagittae, which forms the eastern apex of a little triangle with Delta and Zeta (ζ) Sagittae. Just 20′ northeast of 9 Sagittae is the pale and tiny glow of the loose globular star cluster M71. This little puff of light is an enormous ball of bristling starlight 90 light years in extent but seen 13,000 light years distant.

Continue your gaze along the Arrow's shaft to 3.5-magnitude Gamma (γ) Sagittae. Here is one of the few naked-eye M-type giants. It lies at 275 light years and burns with a radiance 640 times that of our Sun. We can now use Gamma to lead us to M27, the famous Dumbbell Nebula in Vulpecula – a planetary nebula that gives us a glimpse of the ultimate fate of Gamma. Here is an expanding shell of gas 815 light years distant, which was blown from a Sun-like star during one of its death throes some 4,000 years ago. The gas glows from excitation by ultraviolet radiation emitted by the hot central star – now a tiny white dwarf (actually, in this case, a blue dwarf). That little diffuse blob you see in your binoculars is a swelling sphere measuring 1 light year in diameter. The shell is expanding at velocities ranging from 9 to 20 miles per second, or about 7″ per century. A telescopic favorite among skywatchers of all ages, M27 is only a small but condensed (8′ × 6′) 7th-magnitude glow in 10 × 50 binoculars.

If you return to Alpha Sagittae, then look about 2° due north, you'll see the pulsating yellow supergiant star U Vulpeculae, which varies from magnitude 6.7 to 7.5 (almost a full magnitude) every eight days. U Vulpeculae belongs to a class of pulsing supergiants called Cepheid variables, named after the class prototype Delta (δ) Cephei (see page 88).

CHAPTER 4

The fall stars

7 October

When, in the gold October dusk, I saw you near to setting,
Arcturus, bringer of spring,
Lord of the summer nights, leaving us now in autumn,
Having no pity on our withering;
Sara Teasdale, *Arcturus in Autumn* (1926)

Summer has come to an end. The autumn air is turning cool. As the north winds begin to blow, leaves fall from the trees. By month's end, that wonderful odor of crisp leaf decay will fill the air. Our friend the Great Bear is no longer high overhead in the north, as it was when we started our springtime journey; it is now plodding feet first across the northern horizon, where it will "hang low" for the next few months. In the southwest, the Teapot of Sagittarius is tilting down its spout, as if to empty its contents before being "shelved" for winter. A new zodiacal constellation now rules the low southern skies. It's time to meet Capricornus, the Sea Goat.

Capricornus, the Sea Goat

What is a Sea Goat? Good question! A Sea Goat is a purely imaginative form of a purely imaginative creature. It represents Pan (a fertility deity) in disguise. Usually seen as a creature that's half man and half goat, Pan, as Capricornus, has the upper body of a goat and the lower body of a fish – actually it's the coiled lower body of a serpentine fish. According to legend, Pan turned himself into this curious creature to hide from the giant Typhon – the largest and ugliest creature ever to be seen. Taller than the mountains, his fuzzy head was seen to touch the stars. He had the upper torso of a man but his shoulders sprouted the heads of a hundred hissing dragons. A trunk of

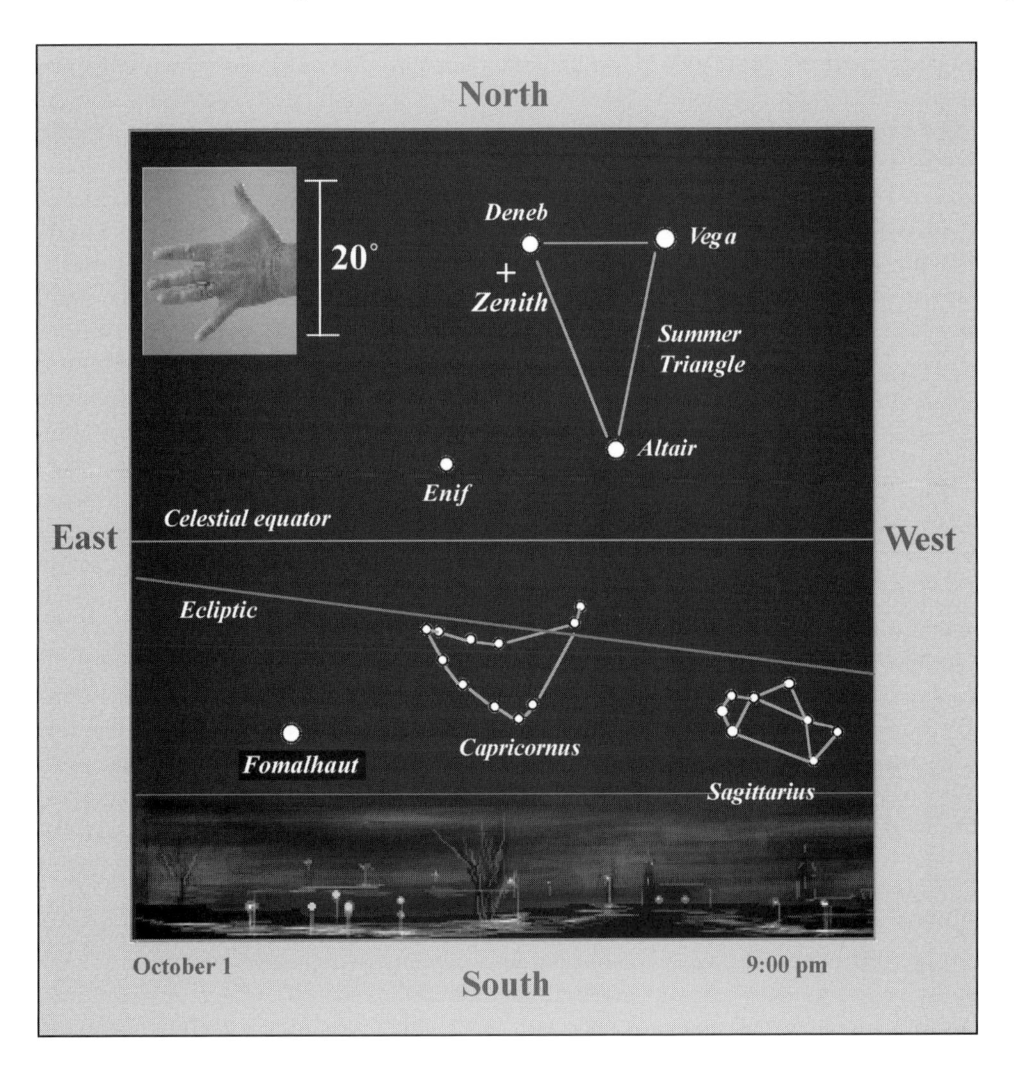

coiling snakes formed his legs. And his mouth dripped lava and molten spittle. When enraged, he would rip mountains apart and throw them at the gods in the sky. (You shouldn't need any convincing that we're actually dealing with a metaphor for a volcano here.)

Anyway, one day, Typhon was getting ready to hurl Sicily's Mount Etna (a volcano) into the sky, when Jupiter, tired of Typhon's tantrums, zapped the ugly beast with a hundred thunderbolts. *Zap! Zap! Zap!* . . . Stunned, Typhon lost his grasp on mighty Etna, and the mountain fell back, pinning him forever to the Earth. Today, we can still see Typhon's fiery rage as he vents his volcanic fury on the summit and slopes of Etna.

It's interesting that Pan would run from the land to the sea, and then transform himself into another beast to hide from the volcanic fury of Typhon. These actions parallel one of the many Hawaiian myths dealing with Pele, the Hawaiian goddess of volcanoes (see photo above). During one of Pele's furies, she directed her lava flows toward her lover Kama-puaa (the Pig God), who saved himself by fleeing to the shore, jumping into the ocean, and transforming himself into his fish form – the *humuhumunukunukuâpua'a* (Hawaii's state fish); that word was made famous by the familiar 1933 song "My Little Grass Shack in Kealakekua Hawai'i" by Bill Cogswell, Tommy Harrison, and Johnny Noble. On another interesting note, the Greeks later linked Typhon with the Egyptian deity Seth – the brother who killed Osiris (see page 70), because both were evil forces and sons of the Earth that attacked the main gods.

Capricornus looks like none of the creatures mentioned above when we see its somewhat inconspicuous form looming over the southern horizon. If anything, it could be likened to a dented sarcophagus, one damaged after Seth had tossed it into the Nile. So the casket now rests in its watery Underworld, dark and deep beneath the celestial equator – the domain now presided over by Osiris. The outline of the constellation also looks like a foreshortened version of an Egyptian solar boat – the kind that carried Re, the Sun god, through the Underworld on his way from setting to rising. With a modern eye, several amateurs have likened its form to that of the B2 Stealth Bomber.

Alpha (α) Capricorni (Algedi) is one of the major showpieces in the Sea Goat. Although it's not the bright-

est star in the constellation, it is a beautiful double comprised of two nearly equal 4th-magnitude stars separated by about 5′. Most casual stargazers immediately notice the star's duplicity with just a glance . . . and a little averted vision. The chance pairing of this optical double is remarkable in itself. $Alpha^1$ Capricorni (the more westerly of the two) is a type G3 supergiant 690 light years distant, while slightly brighter $Alpha^2$ Capricorni is a regular G8 giant 109 light years distant. While $Alpha^2$ Capricorni is more than six times closer, it shines with a luminosity nearly 900 times less than that of $Alpha^1$ Capricorni. So not everything is what it seems in our two-dimensional sky.

Through 10 × 50 binoculars I find both stars pouring forth warm rays of yellow light, though $Alpha^1$ Capricorni is slightly more orange. The binocular view is enhanced by the appearance of a roughly 6th-magnitude star a little more than 20′ west–northwest of $Alpha^1$ Capricorni, and a 5th-magnitude star about 40′ southeast of $Alpha^2$ Capricorni. Seen together, they make a lovely bracelet of starlight. $Alpha^1$ and $Alpha^2$ Capricorni were not always so easy to see with the unaided eye. In fact early skywatchers didn't take notice of the duplicity until the seventeenth century, when the stars' independent proper motions carried them far enough apart.

Now lower your binoculars a bit to view 3rd-magnitude Beta (β) Capricorni (Dabih), which is another beautiful binocular double and a great naked-eye challenge. Beta's 6th-magnitude companion is nearly 3.5′ away to the west, which is a snap to resolve in binoculars but it's not so easy to see with the unaided eye; the dimness of the companion relative to the brightness of the primary makes it difficult to discern. Through binoculars, I see a yellowish primary and an opalescent secondary. As Garret P. Serviss tells us in his 1923 book *Astronomy with an Opera-Glass*, when he saw the secondary, he exclaimed to himself, "Why, the little one is as blue as a bluebell." Others have likened the pair's colors to yellow–orange and sky blue. These stars form a true binary pair 330 light years distant and take about a million years to complete a single orbit around their common center of gravity.

If you place Beta Capricorni in the upper right of your binocular field, a pretty triangle of three 5th-magnitude stars will appear near the center. The brightest of the three is Rho (ρ) Capricorni and is itself a pretty double star. And if you place Rho Capricorni at the right edge of your

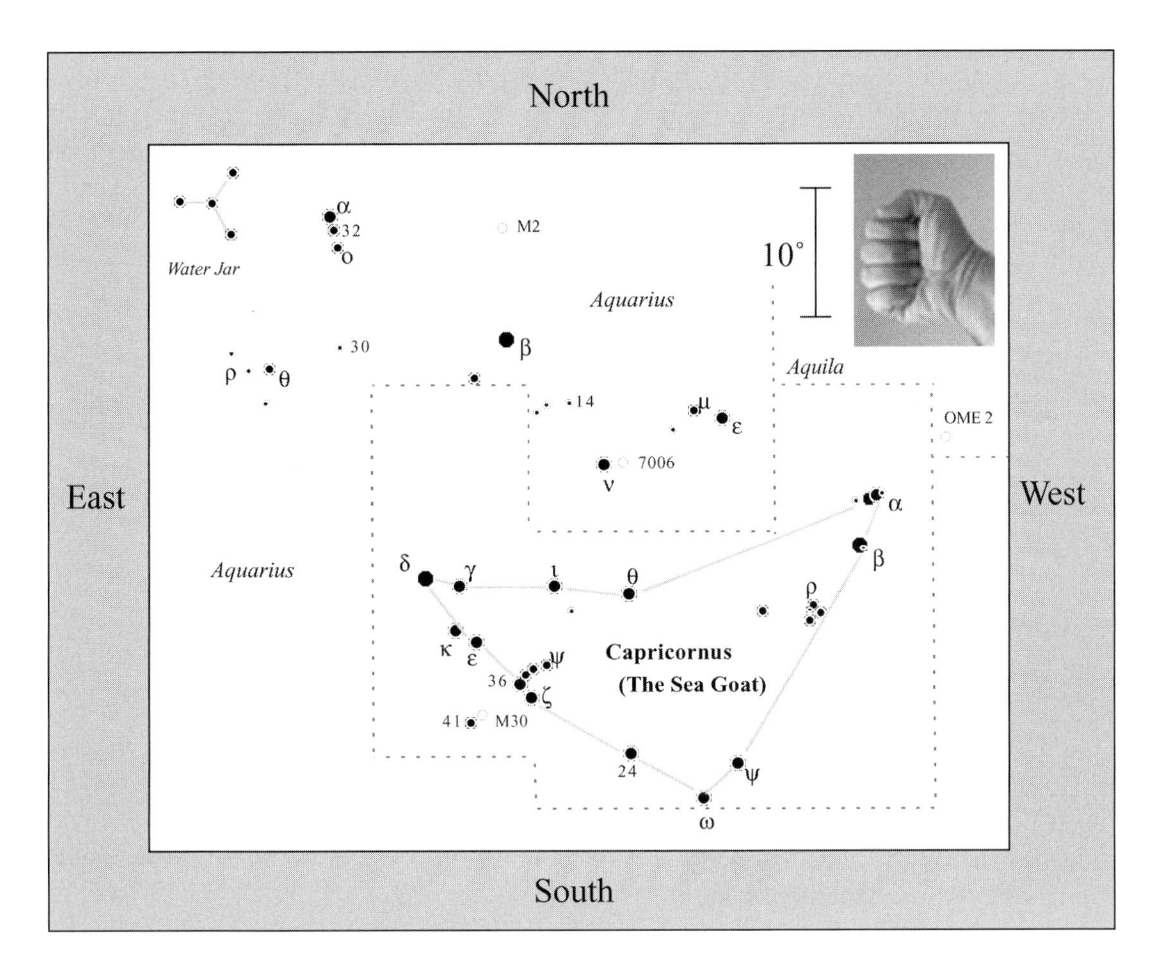

binocular field, you can use the chart below to hop over to 6th-magnitude 19 Capricorni, where I found a neat little cascade of dim stars immediately to its south.

Be sure to sweep your binoculars all along the perimeter of the Sea Goat, where you will find a multitude of quaint stellar arrangements – all pleasing to the eye, all worthy of your attention. For instance, Zeta (ζ) Capricorni is part of a 3°-long fishhook of stars including 36 and Phi (ϕ) Capricorni. Zeta itself is a remarkable star, being a type G4 supergiant 122 light years distant. It's nearly 500 times more luminous than our Sun and is nearly 60 times wider. Now place Zeta and 36 Capricorni in the right side of your field of view and use the chart on the next page to find the little observed globular star cluster M30; it will be near the center of the field, just west of 5th-magnitude 41 Capricorni. Shining at 7th magnitude, the cluster is surprisingly obvious to see under a dark sky. Look for a round glow about 12′ in extent that's moderately condensed. Messier 30 is nearly 27,000 light years distant and its true physical extent spans more than 90 light years of space.

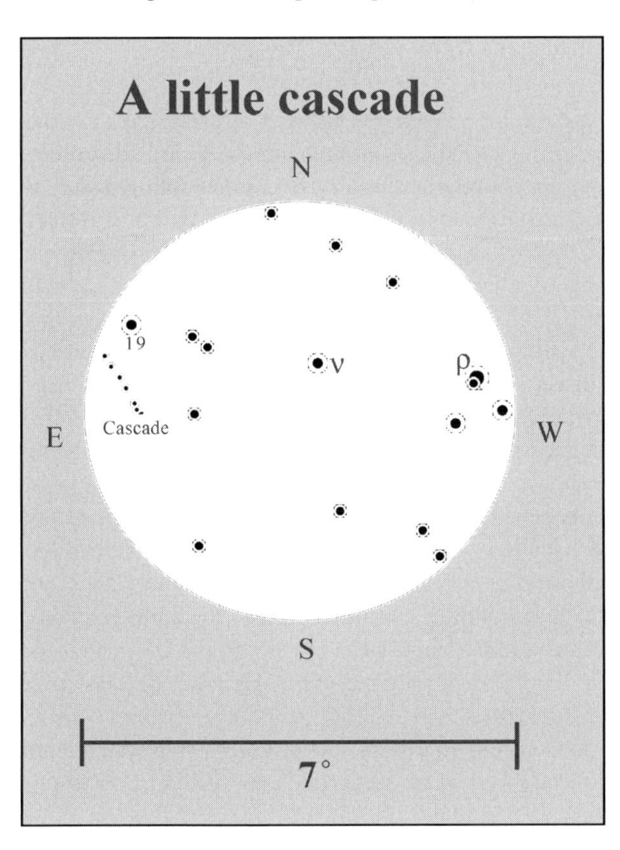

For a challenge and a most interesting sight, return to Alpha and Beta Capricorni and place the stars in the lower left side of the field of view. Now use the chart at the bottom of page 84 to find O'Meara (OME) 2, a curious asterism or open star cluster I independently discovered on October 16, 2006, while making observations for this book. I say that I *independently* discovered it, because the object is not plotted on any star atlases. I recently learned, however, that an advanced amateur astronomer in Sao Paulo, Brazil, had discovered it before me sometime between 1997 and 1998. Its official designation is Alessi J20046–1030, and it is now cataloged as such in the 2003 publication *Star Clusters* (Virginia, VA: Willman-Bell, Inc.), by Brent A. Archinal and Steven J. Hynes. The cluster is not in Capricornus, but lies just over the border in Aquila the Eagle. But what a mysterious object. It is a challenge

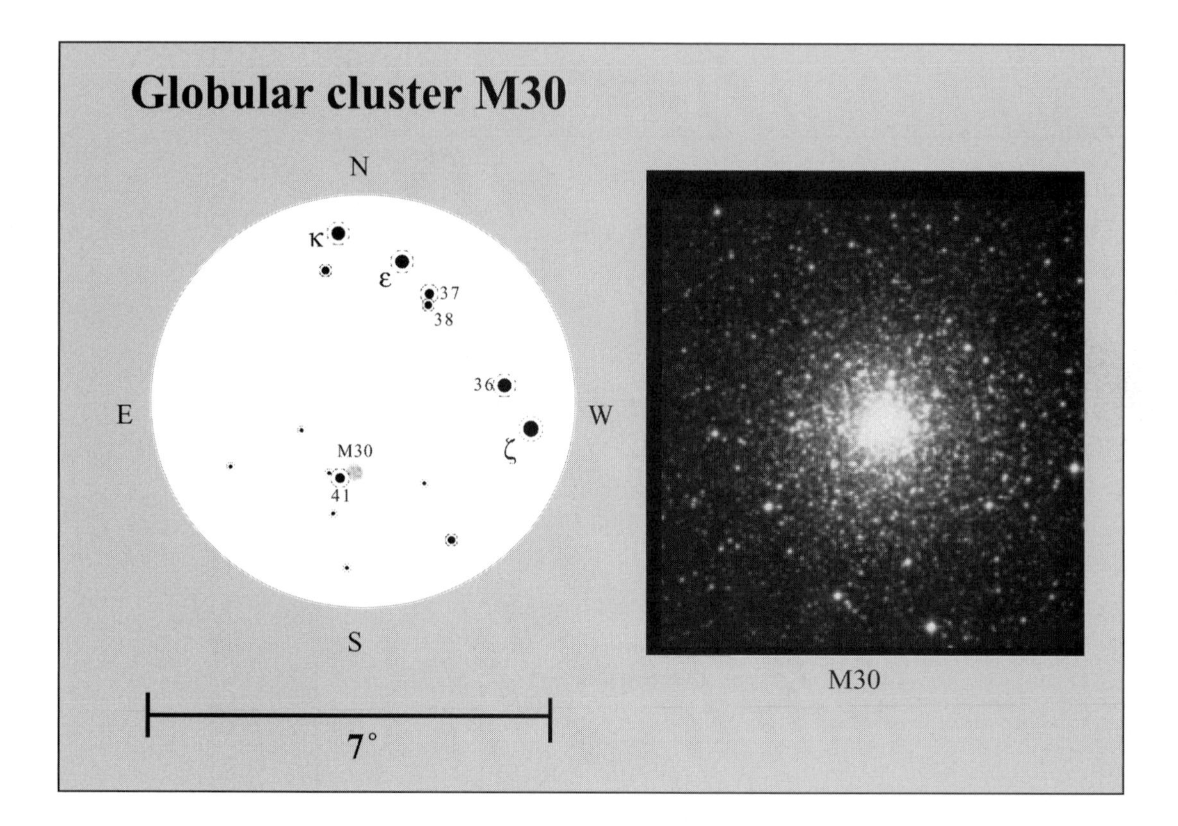

M30

to see because it is large (20′) and somewhat dim (7.5 magnitude). So you need to be under a clear dark sky to see it. The grouping contains about 20 stars, the brightest of which shines at magnitude 8.8. Indeed, as seen with averted vision through 10 × 50 binoculars, the diffuse glow breaks apart into little sparks of light. Research by Bruno Alessi has shown that, based on the proper motions of these stars, the members may be physically related.

Before we leave Capricornus, consider how appropriate it is for Pan, a fertility god, to patrol this region of the zodiac. When the Sun sails through Capricornus, it's on its way out of the celestial Underworld (see the chart on page 81). It is as if Pan impregnates the Sun with life, helping it swim toward a breach. The Sea Goat – a creature that's part upper-air animal and half fish – then could symbolize that the Sun is still "half in the water."

OME 2 = Alessi J20046-1030

N

OME 2

E

W

α

β

S

7°

Aquarius, the Water Bearer

Aquarius, the Water Bearer, has a fuzzy past. It may be Ganymede, the cup-bearer of the gods, who serves them their nectar and ambrosia. Jupiter was so taken by Ganymede's beauty that he disguised himself as Aquila, the Eagle (as you can see the same constellation figures were used repeatedly in different myths) to . . . well, swoop him off his feet. Ancient Egyptians believed that the setting of the Water Bearer's urn stars caused the Nile to rise.

Aquarius is a large and relatively dim constellation that fills a void north and east of Capricornus. The figure shown on the next page is one adapted from a version created by H. A. Rey and depicted in his wonderful 1952 book, *The Stars*. It shows the Water Bearer lunging toward the east and pouring water from a jar, whose contents flow south into the mouth of Pisces Austrinus (the southern fish); the fish's mouth is marked by the bright 1st-magnitude star Fomalhaut, which is the Alpha star of that latter constellation.

Take a moment to look at that bright *A*-type star, which is only 25 light years distant and 200 million years young.

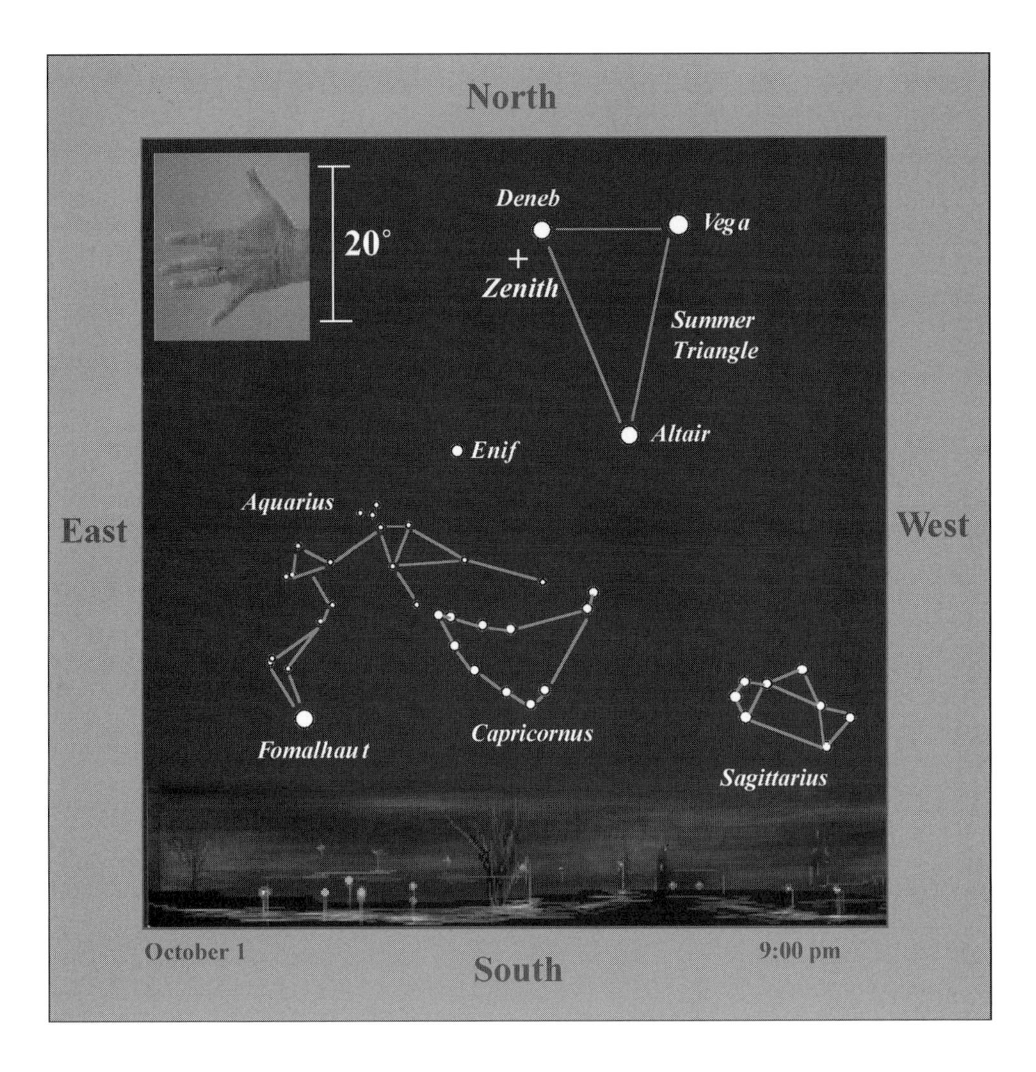

Fomalhaut is surrounded by an enormous dust disk some four times the extent of our Solar System. Recent observations have also revealed a central gap, or clearing, in the disk, which may be caused by a Saturn-sized planet orbiting the star. As you look at Fomalhaut, imagine that you are seeing a phase in the star's life that was much like our Sun's 4.3 billion years ago.

The most interesting binocular objects belonging to Aquarius lie in its more northerly realms – around the Water Bearer's head and torso. Let's start with his urn – the urn as depicted in old star charts, which is the tight, Y-shaped asterism of 4th-magnitude stars that I show as the Water Bearer's head (see chart on page 83). Commonly known as the Water Jar of Aquarius, this keen little grouping lies almost two fists northeast of Delta (δ) Capricorni. It is a fine sight in binoculars, appearing as a sprinkling of four diamond chips spread across 4° of sky. Each star has a sharp blue sheen. Move your binoculars slightly to the right and 2.5-magnitude Alpha (α) Aquarii (Sadalmelik) will come into view. Sadalmelik marks the northern end of a little line of three stars, the other two being 32 and Omicron (o) Aquarii. Alpha is a G-type yellow supergiant, 750 light years distant. It is 3,000 times more luminous than our Sun and 60 times wider.

Third-magnitude Beta (β) Aquarii (Sadalsuud), which lies about a fist to its lower right, is a near twin of Sadalmelik. Here is a G-type yellow supergiant 600 light years distant, with a luminosity 2,200 times solar and a width 50 times that of the Sun. Indeed, it's believed that these suns formed together in a loose aggregation, but, because they were not tightly bound by gravity, moved apart over the course of tens of millions of years.

If you place Beta in the upper left of your field and look toward the center, you'll see 7th-magnitude 14 Aquarii, which forms the heart of a tiny Y-shaped asterism (not shown on the star chart) only about $\frac{1}{4}°$ in length. I see 14 Aquarii, an M-type star, shining with a squash-colored light.

Now lower your sights to 4th-magnitude Theta (θ) Capricorni and place it in the field as shown on page 86. You want to look north to 4.5-magnitude Nu (ν) Aquarii, then $1\frac{1}{4}°$ due west to what appears as an 8th-magnitude star. It is a star, but one about twice the mass of our Sun that has cast off luminous shells of gas, which, as the Hubble Space Telescope image accompanying the chart shows, is an organism of visual complexity. Cataloged as NGC 7009, this planetary nebula, which has two linear lobes that extend from the main ellipse like Saturn's rings seen edge on, is commonly called the Saturn Nebula. That tiny pinprick of light you see in your handheld binoculars is the dying gasps of a star 1,400 light years distant. University of Washington astronomer Bruce Balick and his colleagues believe that the lobes are gas jets fleeing the nebula's dark central cavity, like streams of water

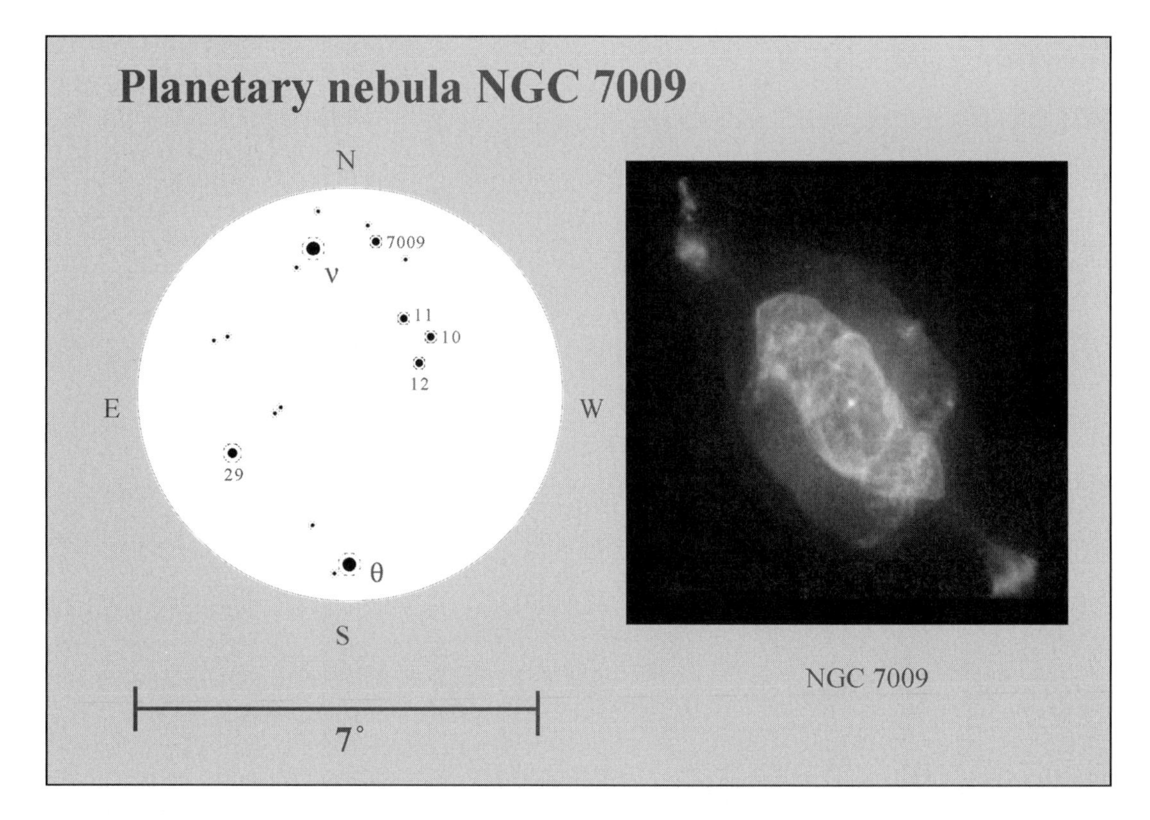

NGC 7009

from the nozzle of a garden hose. "The dense tips of the jets might be material that the jets have plowed ahead of them as they push forward," Balick says, though more observations are needed to confirm this theory.

If you return to Beta Aquarii and place the star at the bottom of your field, you will see a very noticeable, condensed glow near the top of the field. You have just found the magnificent 6th-magnitude globular cluster M2 – a 170-light-year-wide swarm of 100,000 suns, some 13 billion years old and 37,000 light years distant.

Center 4th-magnitude Theta (θ) Aquarii in your binoculars. Theta marks the southeast corner of a near equilateral triangle with Alpha and Beta Aquarii; the star should be easy to identify because the 5th-magnitude star Rho (ρ) Aquarii is immediately to its east. Now use the charts on pages 83 and 87 to identify the 6.5-magnitude star HD 210277 Aquarii. A Jupiter-sized object circles that G-type dwarf roughly every 1.2 years in an orbit that's just 10 percent larger than Earth's orbit around the Sun.

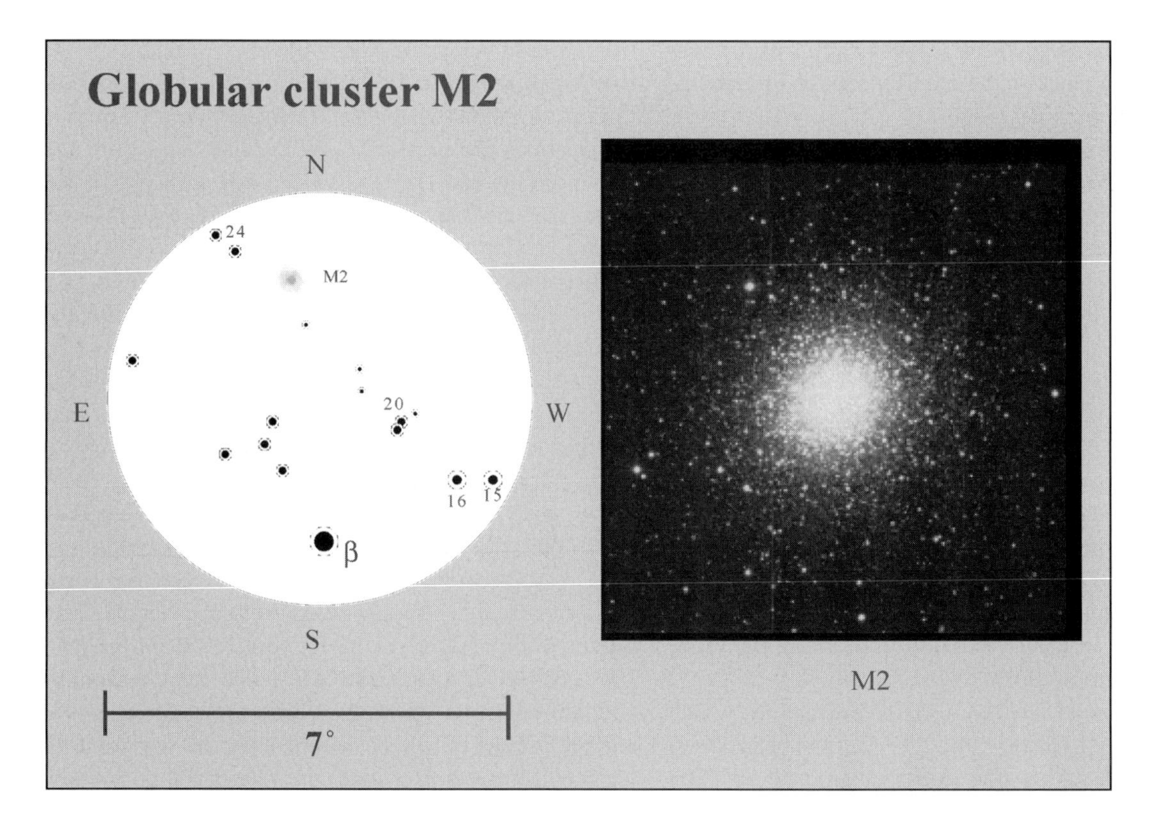

M2

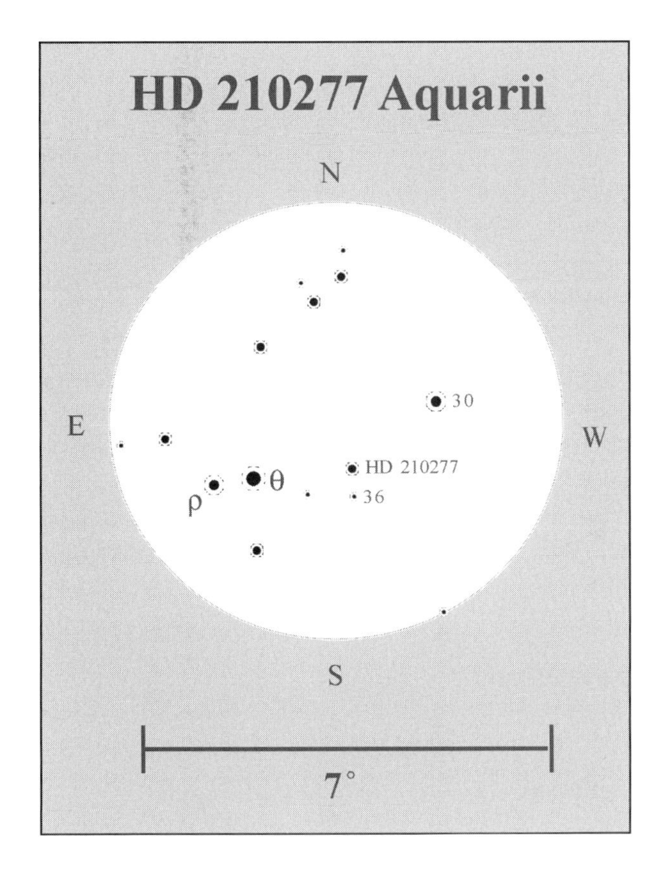

Cepheus, the King

Turn to face north, then look directly over the north celestial pole for a pattern of moderately bright stars that form the stick figure of a house. This is the main body of Cepheus, the King of Ethiopia and husband to Queen Cassiopeia. Cepheus plays a bit role in one of the most famous myths of the heavens surrounding the sacrifice of his daughter, Andromeda, to the sea monster, Cetus, the Whale. But we will encounter that legend in more detail next month with the introduction of Cassiopeia.

For now, you want to locate Gamma (γ) Cephei (Errai), the 3rd-magnitude star at the tip of the House asterism, which is about 10° (a fist) directly above Polaris. Confirm Errai by first identifying the rest of the House asterism, whose major axis (from the tip of the house to its base) measures about 20° or two fists. Gamma Cephei shines with a warm yellow light. Indeed, it is a K-type subgiant that is on its way to becoming a red giant. But Errai is a wonder for another reason: it is one of the brightest stars in the night sky with a suspected planet. The planet is at least 1.6 times more massive than Jupiter. It orbits Errai with a period of $2\frac{1}{2}$ years at an average distance of 189 million miles – twice that between the Earth and the Sun. Interestingly, Errai is a known double star, making

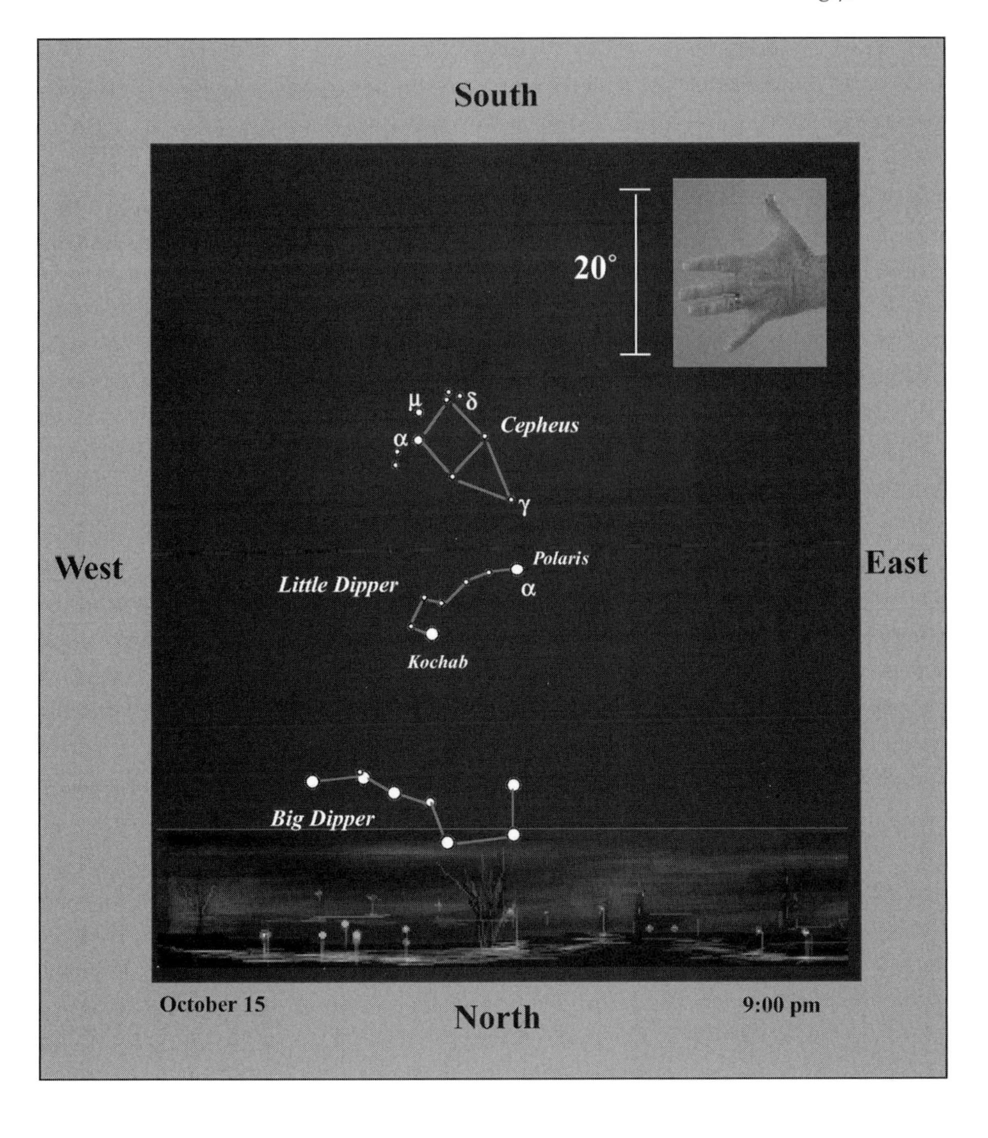

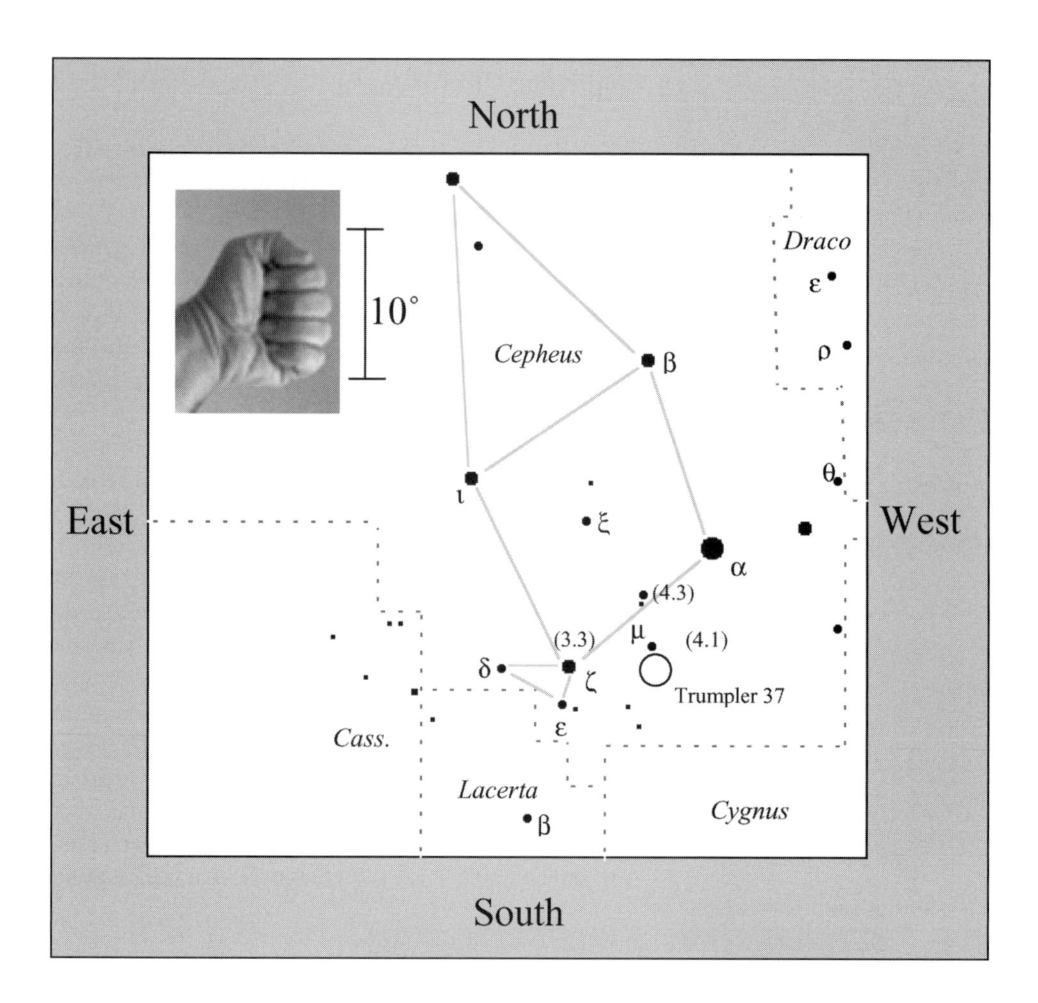

its planet one of the few known to exist in a double star system. Errai's stellar companion is a red dwarf that lies about as distant as Uranus is from our Sun.

Follow the stars in the House until you identify the two stars marking its base, 2.5-magnitude Alpha (α) and 3rd-magnitude Zeta (ζ) Cephei: Alpha is bright-white *A*-type star 49 light years distant, while Zeta is a fantastic *K*-type supergiant 725 light years away. Now look between these two stars and a bit southwest for 4th-magnitude Mu (μ) Cephei. Dipped in the shallows of the autumn Milky Way, you cannot mistake Mu Cephei because it is one of the reddest binocular stars in the night sky – so rich it looks like a drop of spilled blood on black velvet.

Also known as Herschel's Garnet star (after the eighteenth-century astronomer Sir William Herschel who likened the star's color to that gem), Mu Cephei is an *M*-type red supergiant – one of the most brilliant of all stars of its class in the entire galaxy. Lying some 2,400 light years distant, the star measures some 2.4 billion miles across. If it replaced our Sun, its outer shell would extend beyond the orbit of Saturn. Mu Cephei is also a slow, semiregular variable whose fluctuations are easy to detect with the unaided eye or binoculars. When at maximum brightness, the star shines at magnitude 3.4, then dips to magnitude 5.1 irregularly over the course of years. Mu is nearing the end of its life, and may hang around for another few million years – not long in astronomical terms. It will eventually explode as a supernova, leaving a remnant with a neutron star or a black hole.

Do not rush away from this star or this region. Mu Cephei is also swimming on the fringes of a maelstrom of starlight – the enormous open star cluster Trumpler 37, which spans three full-Moon diameters in apparent size. The cluster is situated near the galactic plane at the core of what's known as the Cepheus *OB*2 Association – an extensive region rich in bright-rimmed star clouds distributed over an area of 3°. The region that contains Trumpler 37 is three million years young! The cluster contains 480 members that span 46 light years of space. Trumpler 37 then is one of the youngest star clusters known in our galaxy. The brightest member of the cluster is a hot, *O*-type star that shines at magnitude 5.5. This star is the main source of excitation of a nebula that surrounds the entire region. Under the darkest skies, it may be possible to see this nebula in your binoculars as a faint frost covering the cluster. See what you think.

Return now to Zeta Cephei and look 2° due east for 4th-magnitude Delta (δ) Cephei, one of the most famous (and important) variable stars in the night sky. Note that Delta Cephei also marks the eastern apex of a little triangle with Zeta and Epsilon (ε) Cephei. Delta Cephei is the prototype Cepheid variable – a class of yellow giant or supergiant stars whose clocklike period of variability is directly related to its visual magnitude. The longer a Cepheid's pulsation period, the more luminous the star by a constant factor. Because of this precise relationship, astronomers can use Cepheid variable stars as distance indicators. The relationship was first discovered in 1912 by Henrietta

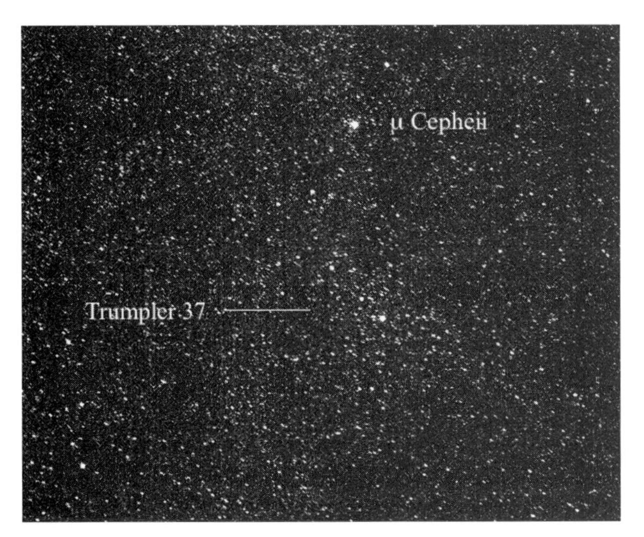

Leavitt at Harvard: "A straight line," she said, "can be readily drawn among each of the two series of points corresponding to maxima and minima, thus showing that there is a simple relation between the brightness of the variable and their periods." The relationship is especially important since the discovery of a Cepheid variable in another galaxy tells astronomers the distance to the galaxy. Delta Cephei, which lies at a distance of 950 light years, varies between magnitude 3.5 and 4.4 with a period of 5.4 days. It is a marvel to watch in binoculars, especially since the star has a slight color shift, changing from an F-type star to a cooler G-type star. Use the chart on page 88 to estimate the star's nightly change in brightness; the stellar magnitudes on this chart are given in parentheses.

Equally fascinating is that Delta is a much under-observed binocular double star. The yellow variable has a pretty blue, 7th-magnitude companion 41″ away. The pair, which looks like a distant view of the Earth and Sun, may be physically related. If so, they will take at least 500,000 years to complete an orbit. As seen from a hypothetical planet orbiting Delta, the companion would burn as a rich-blue beacon – one as bright as Venus in our nighttime sky.

8 November

There is no light in earth or heaven
But the cold light of stars;

Henry Wadsworth Longfellow,
The Light of Stars (1839)

Winter is closing in. The Summer Triangle is slipping ever westward, and the fallow starfields of autumn shine forth in peaceful austerity. Gone is the milky blaze of glory

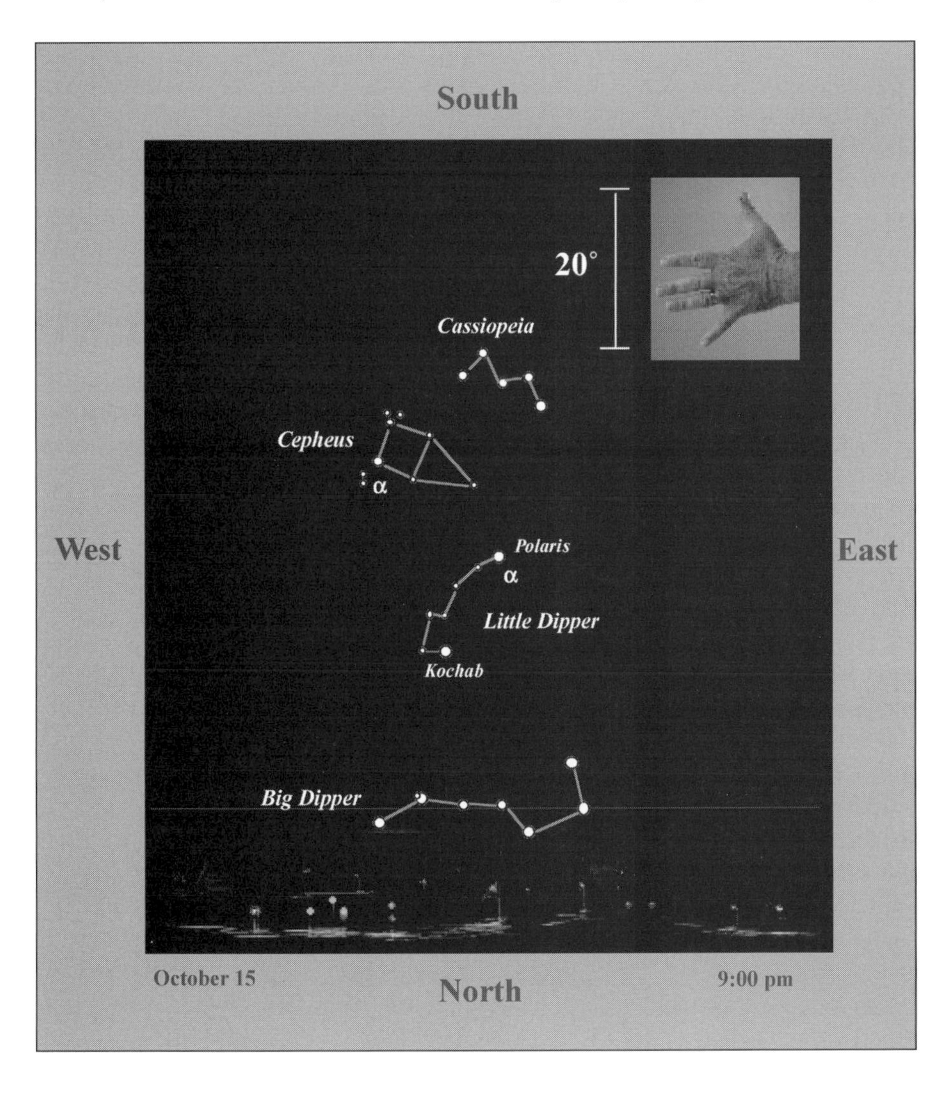

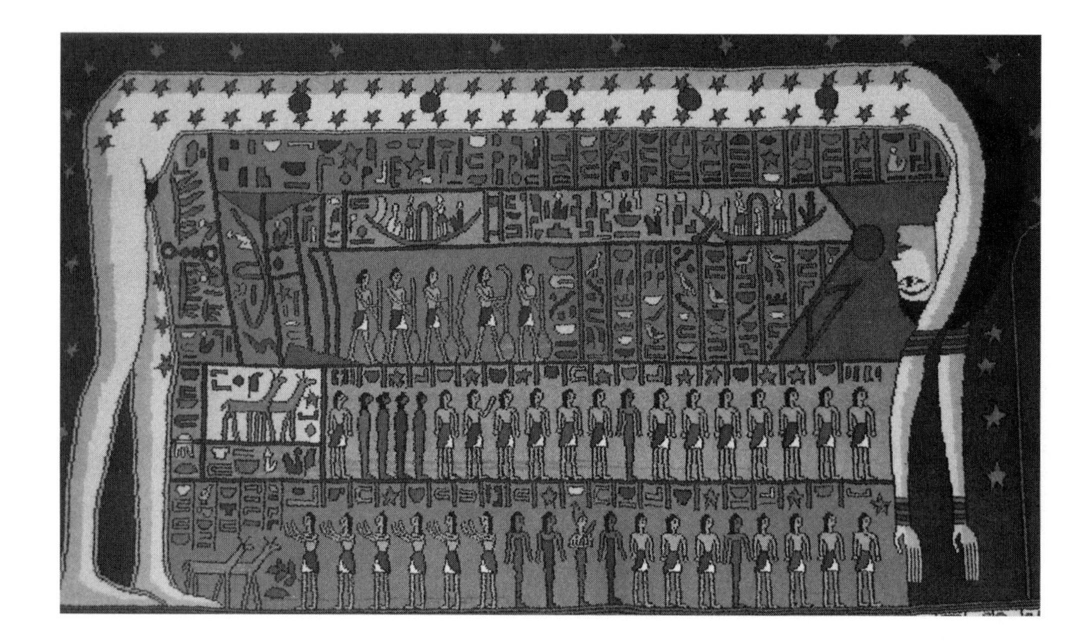

sweeping through Sagittarius. The western horizon now hides it from view. But if you face north and look high overhead, you will see the Milky Way arching over the north celestial pole from west to east like a snow-covered trestle, or the Egyptian goddess, Nut, the very "firmament strewn with the stars" who arches over the sacred pole.

The Milky Way band's intensity appears greatly diminished, as if that river of stars has begun to frost over. The disparity in optical density is real. We are no longer looking toward the bright and dense center of the galaxy but in the opposite direction – where the plane of the Milky Way is less congested and visually subdued. The most prominent star pattern now ruling the high north is the stunning M-shaped asterism of five 2nd- and 3rd-magnitude stars in Cassiopeia, the Queen.

Cassiopeia, the Queen

Arguably the most famous of all the classical sky legends is the stirring tale of Andromeda, the Woman Chained, and the vain antics of her Ethiopian mother, Queen Cassiopeia. Cassiopeia loved the way she looked in a mirror and made no attempt to hide her vanity. But one day she went too far and dared to boast that she was fairer than the Nerieds, who were the sea's most exquisitely beautiful nymphs. When the Nerieds heard the flaunts, they sent a sea monster (Cetus) to dine on the inhabitants of Ethiopia. As the monster terrorized these people, the Queen looked to her husband, King Cepheus, for help. Powerless to fight the beast, Cepheus decided to consult the Oracle of Ammon, which told him that the only way out of the situation was to sacrifice his beautiful daughter to the hideous leviathan. With great reluctance the King gave in to the prophesy. He stripped the princess and bound her by the arms and waist to coastal stone. And there under the hot Sun poor Andromeda awaited her fate.

Fortunately for Andromeda, Perseus – the son of the comely mortal Danae and great Zeus (*Per Zeus*, "fathered" by Zeus) – saw the chained maiden from his aerial perch. At first, he nearly mistook her pure white form for marble, except that he noticed her flaxen hair blowing in the gentle breeze, and her brilliant eyes welling with tears. Flying to her on winged sandals Perseus pleaded for the virgin, to explain her dire situation. At first, the maiden, in her naked form, could only blush in shame. But fearing that the powerful youth might wrongfully consider her a sinner, she revealed the truth.

As Perseus listened to Andromeda's woeful tale, his muscles tensed. The youth was already feeling quite the champion, because he had just slain the Gorgon Medusa, whose gruesome head (with those I'll-turn-you-to-stone eyes) he had in a leather sack. Perseus whispered some words of hope to Andromeda. The enterprising young man also made her a bargain: if he slew this beast, she would be his bride. Andromeda agreed and sealed the promise with a kiss.

In his book, *The Heroes* (Cambridge: Macmillan, 1855; Boston, MA: Warner, 1855), Charles Kingsley continues the story in a most illustrious fashion:

> Perseus laughed for joy, and flew upward, while Andromeda crouched trembling on the rock, waiting for what might befall.
>
> On came the great sea-monster, coasting along like a huge black galley, lazily breasting the ripple . . . . His great sides were fringed with clustering shells and sea-weeds, and the water gurgled in and out of his wide jaws, as he rolled along, dripping and glistening in the beams of the morning sun.
>
> At last he saw Andromeda, and shot forward to take his prey, while the waves foamed white behind him, and before him the fish fled leaping.
>
> Then down from the height of the air fell Perseus like a shooting star; down to the crests of the waves, while Andromeda hid her face as he shouted; and then there was silence for a while.
>
> At last she looked up trembling, and saw Perseus springing toward her; and, instead of the monster a long, black rock, with the sea rippling quietly about it.

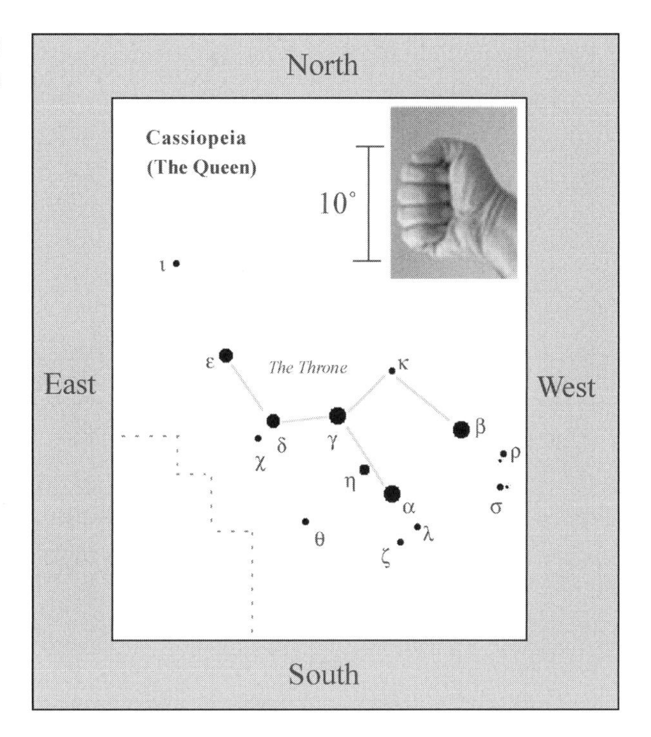

Perseus had held the blood-freezing head of Medusa before the monster and turned it to stone. The story has a rare happy ending: the Gorgon and the Sea Monster are slain, the beautiful maiden is saved by the handsome hero (in some versions of the story, the pair ride off on the winged horse Pegasus, who was born from a vile mixture of sea foam and the blood that dripped from Medusa's freshly severed head), and the lovers marry. They become the great grandparents of yet another Hero – the mighty Hercules (see page 48).

And what of Cassiopeia? She was ultimately seated in her throne and placed in the high northern heavens, where, for half the night, she must hold her head downward, to give her a lesson in humility – or, as Jack Horkheimer of the Miami Space Transit Planetarium says in his book *Stargazing with Jack Horkheimer* (Peterborough, NH: Cricket Books), "Cassiopeia must ride endlessly around the North Star, holding on to her throne for dear life to keep from falling off."

The main stars of the celestial M (or W, depending on the asterism's position above or below the pole), form the framework of the Queen's famous throne. The Ethiopian Queen's famous throne may also be a derivative of the Throne of Egyptian goddess Isis. In hieroglyphs, the throne is either the headdress of Isis or a symbol for Isis. Then again, it might also represent the throne of any ruler.

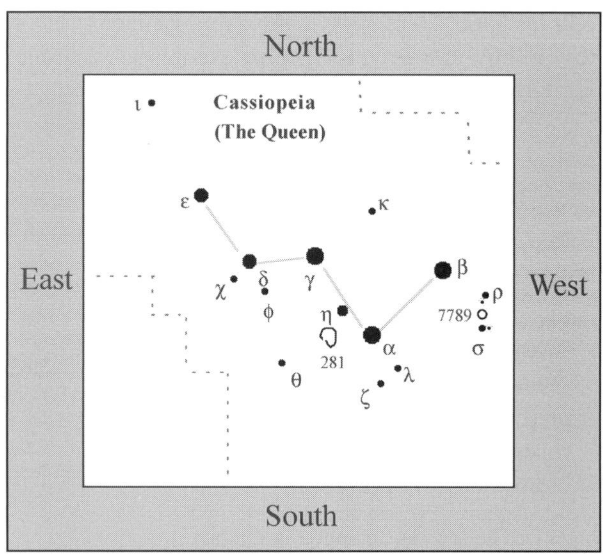

But there may be evidence that the throne does indeed belong to Isis. In ancient Egypt, when Cassiopeia rose high over the mystical north, the waters of the Nile flooded, bringing life-sustaining waters to the parched desert lands. The hieroglyph for water is a series of connected letter M's – like the M-shaped asterism of Cassiopeia when it is directly above the north celestial pole.

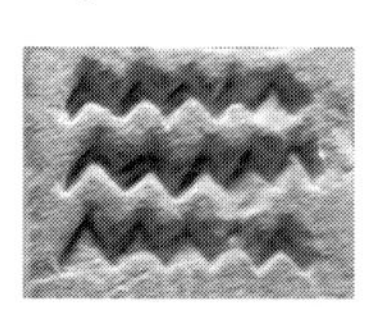

As R. R. Sellman writes in his 1960 book *Ancient Egypt* (New York: Roy Publishers), "Since life centered on the annual flood, the calendar had the practical purpose of predicting its arrival and arranging the farming year. To the peasant the ability of the priests and officials to tell when the Inundation was due probably gave the impression that they actually controlled it, though the official explanation was that it came from the weeping of Isis over the body of Osiris." At this crucial time of the year, Isis is seated in her throne, facing down, her teardrops falling into the Underworld. Certainly Cassiopeia's prominent position in the river Milky Way gives it watery rights.

If you sweep your binoculars across the five major stars in the celestial M or W, you'll find most gleaming with blue light, except for Alpha (α) Cassiopeiae (Shedir), which is a bright K-type orange giant 230 light years distant. Shedir is 90 light years farther away than similarly bright and colorful Alpha Ursae Majoris (Dubhe). When I compare the light of these two stars, I find Shedir more on the yellow side of orange, glowing like a tulip under the golden light of the setting Sun.

Under a dark sky, try to see the relatively bright emission nebula NGC 281, also known as the Pacman Nebula, because of its distinctive form in photographs. This irregular patch of glowing gas lies only $1\frac{3}{4}°$ east of Alpha Cassiopeiae; its Moon-size glow forms the southeastern apex of a triangle with Alpha and yellowish Eta (η) Cas. The nebula NGC 281 is a vast cloud of ionized hydrogen gas excited by a compact cluster of *OB* stars (Index Catalogue 1590), which some recent studies say formed only about 3.5 million years ago. The glowing gas and dark obscuring clouds are actually part of a larger complex of atomic and molecular clouds forming an 880-light-year-wide ring around NGC 281, which is expanding away from the nebula at 14 miles per second. The entire complex was probably first formed by supernovae explosions, which triggered subsequent episodes of star formation, including those continuing today.

Now slide over to 2nd-magnitude Delta (δ) Cassiopeiae, and scan the area to the east–northeast in your binoculars. If you sweep carefully and spend the time to relax your gaze, you might spy several open star clusters, three of which are of special note. First, there's 7th-magnitude M103, which lies just 1° northeast of Delta. Note that it is very small and compact (6′) with three dim stars knitted across its face. Just $1\frac{1}{2}°$ to the east–northeast is the equally bright but larger (15′) NGC 663. This cluster is very easy to see, appearing as a bright and irregularly round, mottled patch; with careful scrutiny it looks like a double "nebula." Lying at a distance of 7,200 light years NGC 663 is about a thousand light years closer than M103. Finally, look about 2° southwest of Delta Cassiopeiae for NGC 457, which hides in the glare of 5th-magnitude Phi (φ) Cassiopeiae. This cluster (of an indeterminate distance) is commonly called the ET Cluster because when seen through a telescope its form looks like that big-screen "Extra terrestrial," with Phi and its colorful companion representing the creature's eyes. In binoculars, though, the cluster is a phantom glow, like the smoke from a burning candle (Phi).

Return to Beta (β) Cassiopeiae, the westernmost star in the celestial W. If you place the star in the field of view so

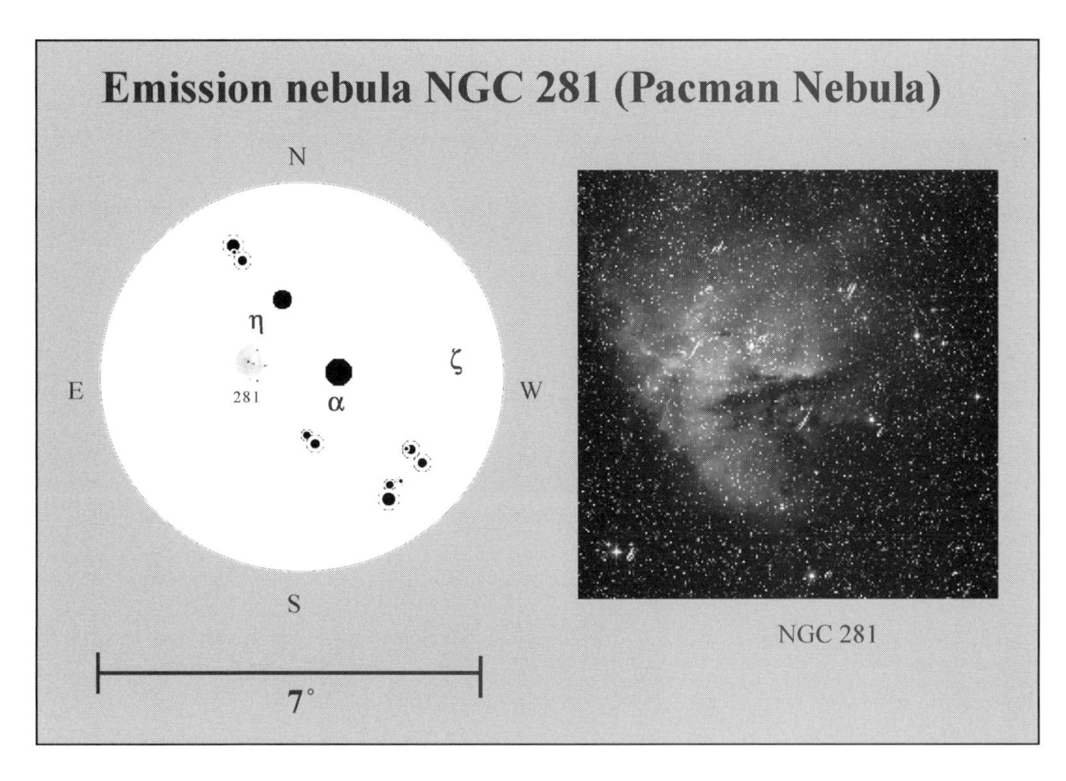

NGC 281

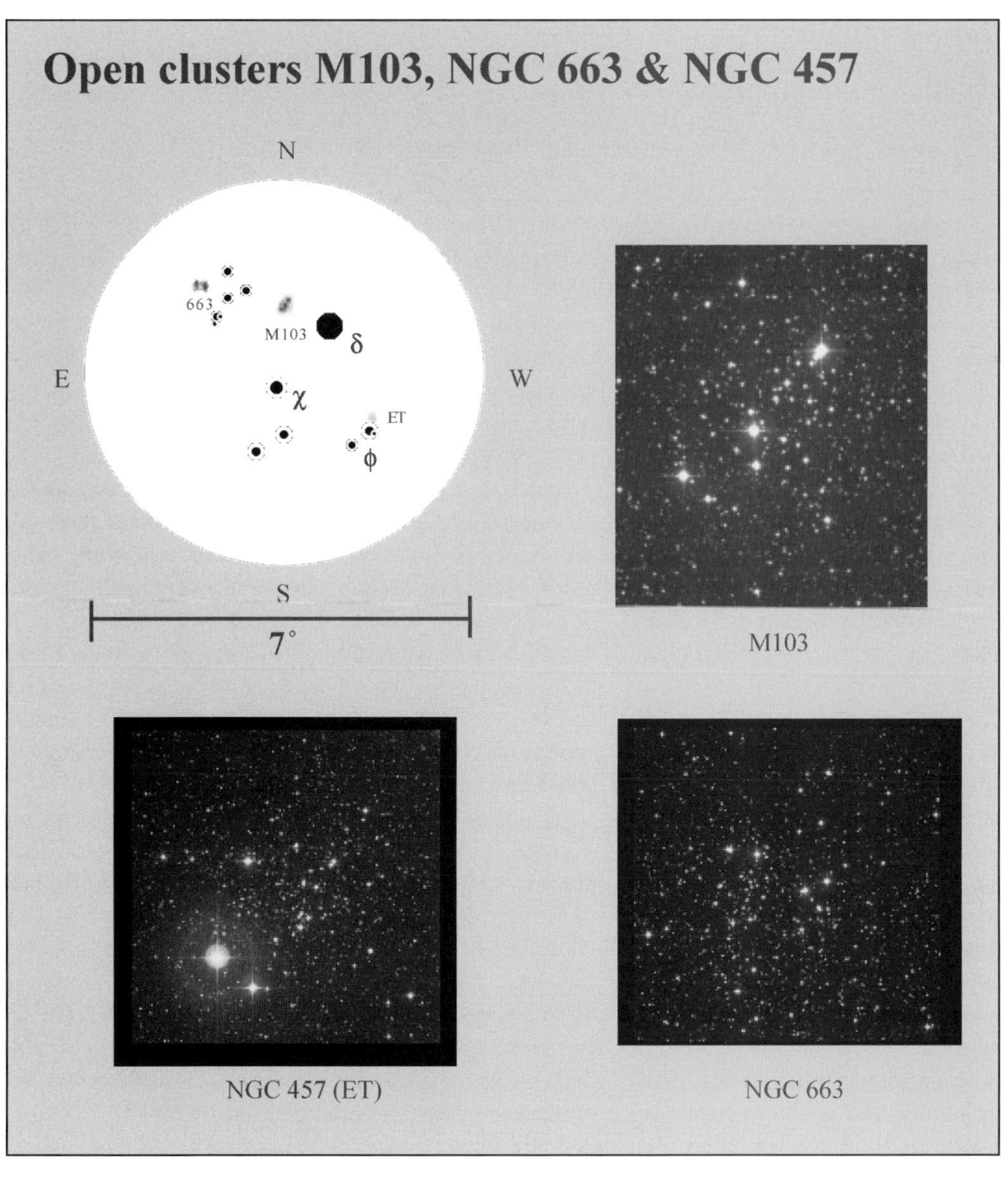

M103

NGC 457 (ET)

NGC 663

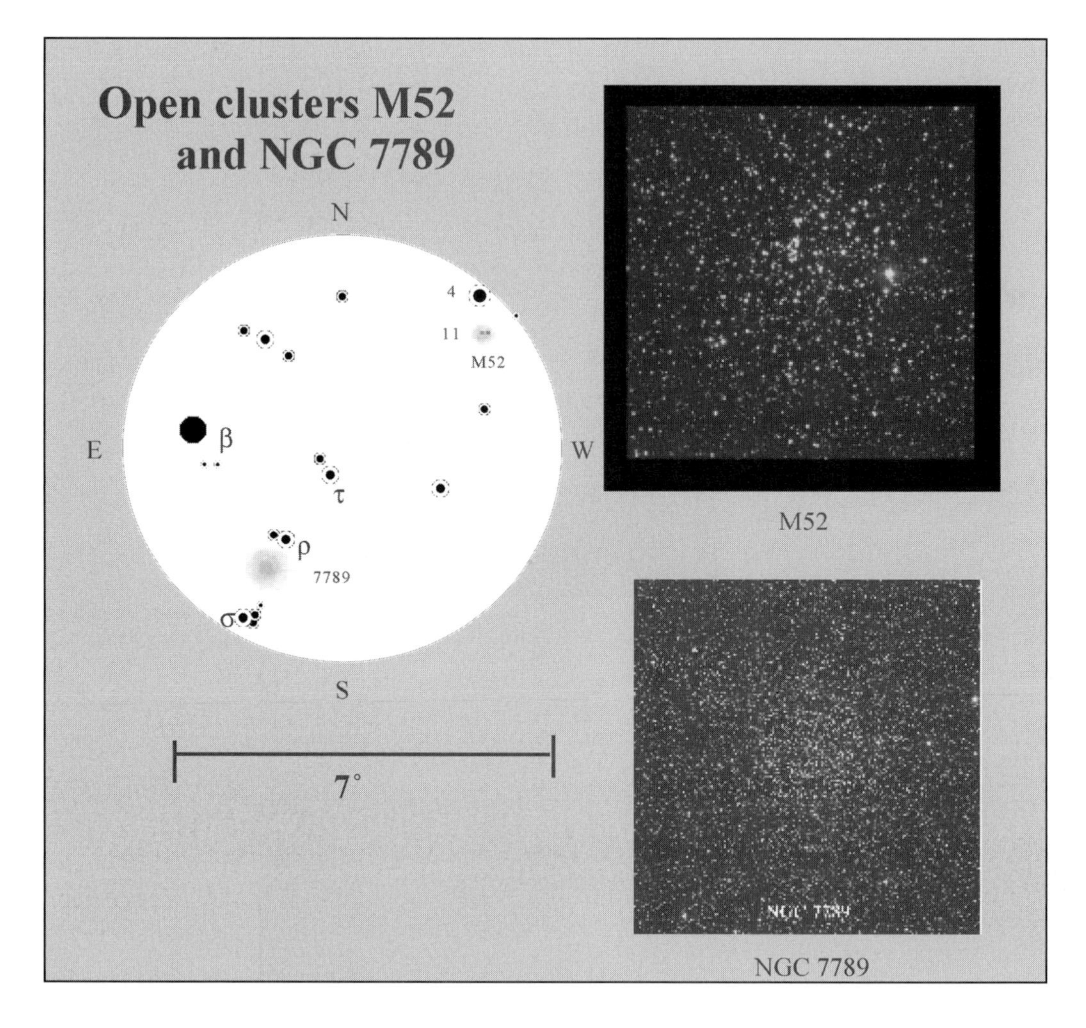

that it matches that in the chart above, you will see two large and dynamic open star clusters at the extreme and opposite edges of the field. Try first for 7th-magnitude M52, which lies about 6° northwest of Beta and just $\frac{1}{2}$° south of 4 Cassiopeiae, a reddish 5th-magnitude star. In binoculars, the cluster looks like a double star in a haze. This is a rich cluster of stars, 20 million years young, with 200 members spread across 24 light years of space. Just 3° south of Beta Cassiopeiae trapped between 4.5-magnitude Rho (ρ) and Sigma (σ) Cassiopeiae, is NGC 7789, a dynamic 6.5-magnitude open star cluster, that requires more than binoculars to appreciate fully. That's because, through a telescope, it is one of the finest and richest open star clusters in the night sky, a superb collection of nearly 600 little specks of starlight 6,000 light years distant. Still, the binocular view under a dark sky is magnificent, the cluster being nearly as large as the full Moon and just as round. And though no stars pop clearly into view at a glance, the object is mottled with some flecks of dim starlight eking into view with averted vision. But overall, it looks like the head of a tailless comet drifting through the Milky Way.

Of special interest is NGC 7789's 4.5-magnitude attendant, Rho Cassiopeiae – a hypergiant star some 8,000 light years distant. This G-type star, while as hot as our Sun, is about a *million* times more luminous! Rho is, in fact, in the very last stages of its evolution and could go supernova in as little as 50,000 years. In 2000, the star underwent the largest stellar mass ejection ever recorded, ejecting the equivalent of 50 Earths per day for 200 days. About five percent of a solar mass was expended during that event, which is roughly a thousandth of Rho's mass. Since the star cannot withstand many more such mass ejections and remain stable, it's believed the end is near for Rho, which in 2007 I found had a delicious pumpkin color.

To see another doomed star, turn your binoculars to 4th-magnitude Kappa (κ) Cassiopeiae, in the Queen's throne. Kappa is a cleverly disguised B-type supergiant, 3,500 light years distant. I say cleverly disguised because, despite its great distance, the star appears fainter than it should because a dense cloud of dust lies between it and us, reddening the star's light. Born with a mass of around 40 Suns, Kappa is priming itself to explode as a grand supernova whose central core, when it finally collapses, could turn into a black hole.

If you look about $1\frac{1}{2}$° northwest of Kappa (κ) Cassiopeiae, you will see the *location* of a supernova that erupted in 1572. (You'll have to use your mind's eye to see this one through binoculars.) The event was first witnessed by W. Schuler on November 6th, 1572, and possibly others in the following days, but the object is often referred to as Tycho's Star – in honor of the Danish astronomer Tycho Brahe (1546–1601) who not only independently discovered the surprise star, but provided modern astronomers with a detailed account of its brightness and longevity.

In his 1978 *Celestial Handbook*, Robert Burnham, Jr., provides us with a loose translation of Tycho's discovery observation:

> "On the eleventh day of November in the evening after sunset . . . I was contemplating the stars in a clear sky . . . I noticed that a new and unusual star, surpassing the other stars in brilliancy, was shining almost directly above my head; and since I had, from boyhood, known all the stars of the heavens perfectly, it was quite evident to me that there had never been any star in that place in the sky, even the smallest, to say nothing of a star so conspicuous and bright as this. I was so astonished at this sight that I was not ashamed to doubt the trustworthiness of my own eyes. But when I observed that others, on having the place pointed out to them, could see that there was really a star there, I had no further doubts. A miracle

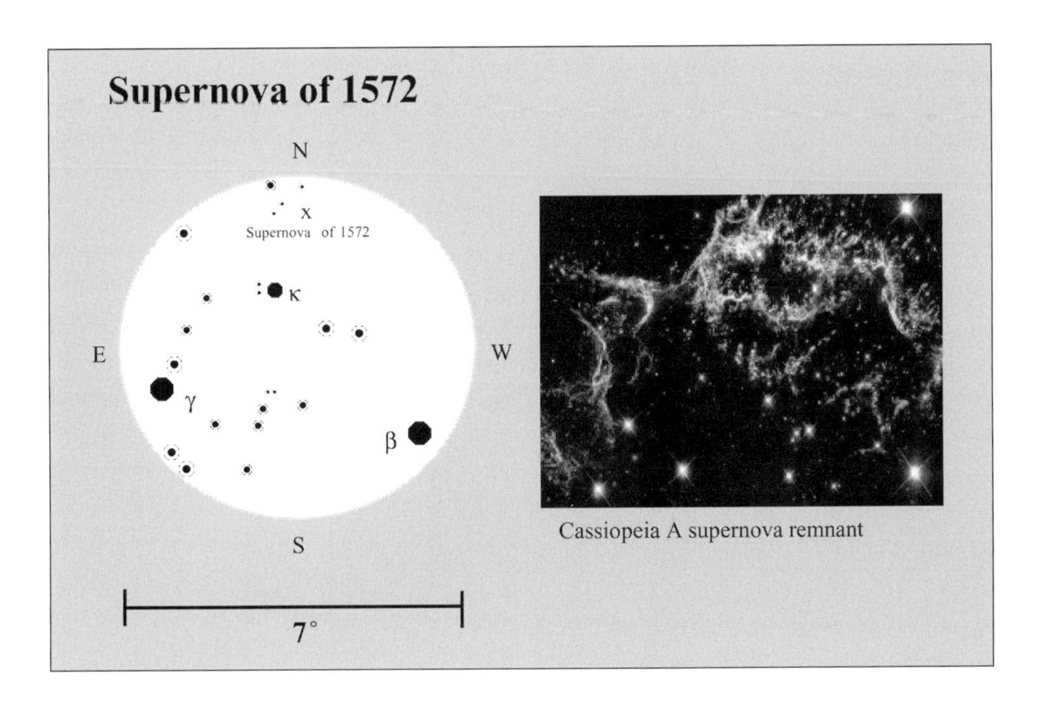

indeed, one that has never been previously seen before our time, in any age since the beginning of the world."

The new star equaled Venus in brightness and remained brighter than all the stars in the nighttime sky for two weeks and did not disappear from view until 1574; when brightest, the star was also seen in broad daylight. The photo illustrations on page 95 show how that new star would have appeared to a naked-eye observer in the sixteenth century.

Today astronomers are still observing the shredded remains of that cataclysmic stellar explosion, called Cassiopeia A. The Hubble Space Telescope image below, for instance, shows only a portion of the expanding shell of debris, which is arranged into thousands of small, cooling, knots of gas. Each small clump, originally just a small fragment of the star, is tens of times larger than the diameter of our Solar System!

The exploded star must have been on the order of 15 to 25 times more massive than our Sun. Such massive stars live fast and die young; Cassiopeia A's progenitor probably burned through its fuel supply in just a few tens of millions of years – that rate of consumption is about 1,000 times faster than our Sun's. In the end, these heavy stars begin a complex chain of events that lead to its explosive finale.

When I first learned of Tycho's Star in the mid 1960s, I mounted a campaign to discover a nova using binoculars. I encourage you to try doing the same. You can read about this exciting adventure in Appendix B.

Pegasus, the Winged Horse

Turn to face south, then look about halfway up the sky for Pegasus, the Winged Horse. Actually, you can start by identifying autumn's hallmark asterism: the Great Square of Pegasus. The Square's four corner stars, which measure about 15° apart, all shine brightly at 2nd and 3rd magnitude. The stars stand out because the Square's interior is bleak. So you are looking for a near-perfect square of four stars that will frame your fist held at arm's length with room to spare. Some skywatchers judge the clarity of autumn nights by counting naked-eye stars within the Great Square. If they see a dozen or so stars, for instance, the naked-eye limit is about magnitude 6. The reason this area of sky looks so bleak is because we're looking high above our galaxy's star-studded plane, where a thin veil of starlight hardly obstructs the heavens beyond.

You must do a little mental gymnastics to see the mythical Horse because it's upside down. Also, only the Horse's upper body is visible. In the mind's eye of our ancient ancestors, Pegasus was seen rising out of a mixture of sea foam and blood, so the Horse's hind quarters have not yet risen. It's like seeing a snapshot of the creature halfway through its birth. Note that the Square also has two Alpha (α) stars. That's because it's an asterism – like the Summer Triangle, which is comprised of three Alpha stars. To complete the Square, we must draw upon Alpha Andromedae to mark its northeast corner. The Horse's faint legs stretch to the northwest from Beta (β) Pegasi,

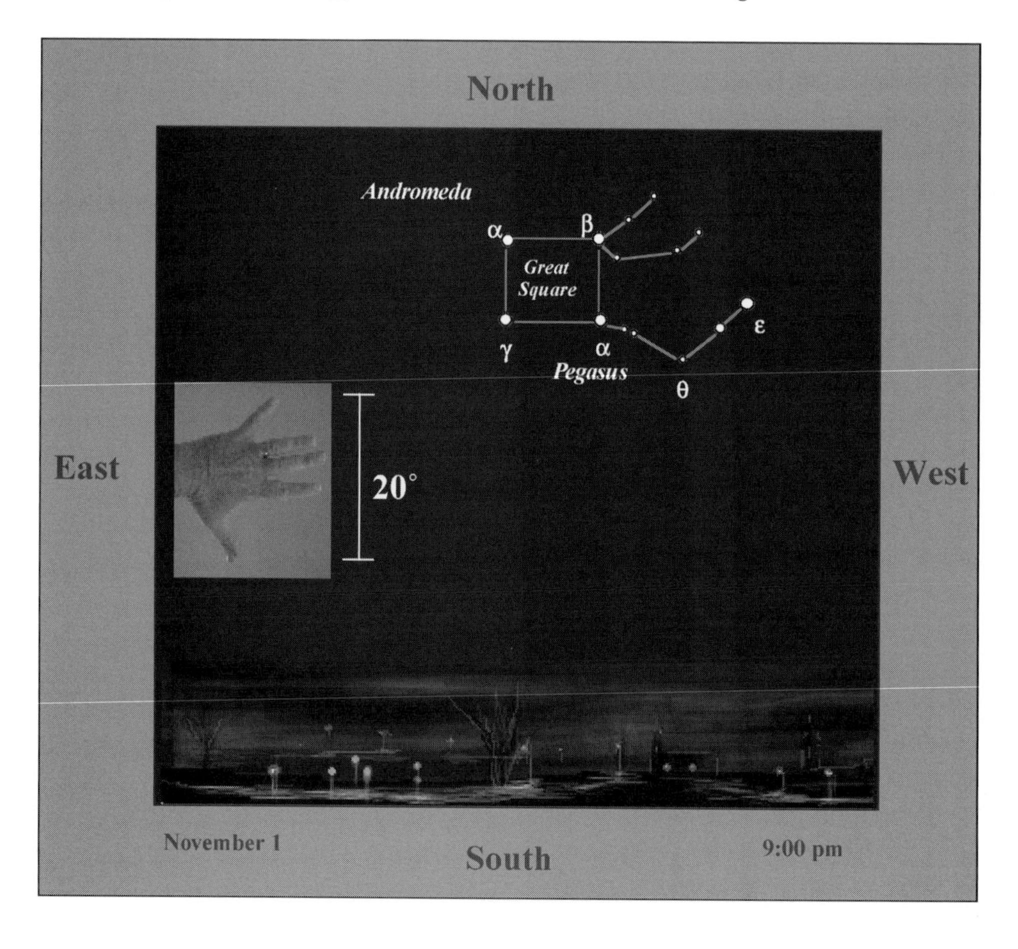

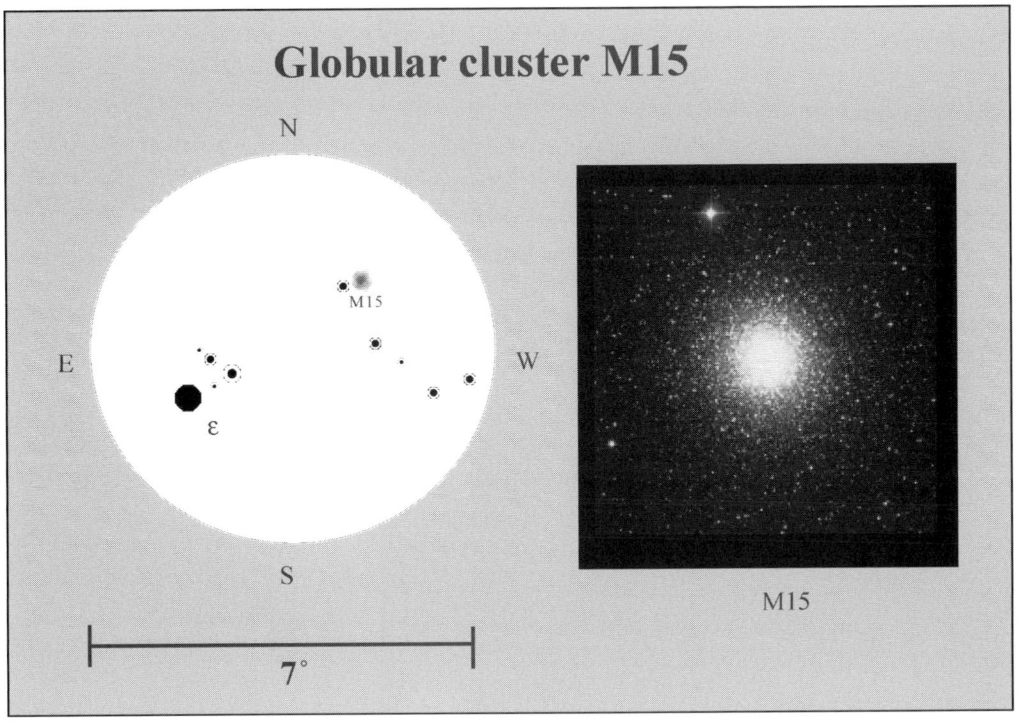

M15

while the Horse's neck and head are formed by Alpha, Theta (θ), and Epsilon (ε) Pegasi.

The Great Square is autumn's Grand Celestial Station. You can depart from any of the Square's corners for different celestial destinations. Let's start at Alpha (α) Pegasi (Markab), a white-hot *B*-type 140 light years distant; it's similar to Alpha Leonis (Regulus) but nearly twice as distant. Now use your eyes or binoculars to "ride the train" down the Horse's neck, making the dogleg turn at Theta (θ), before ending your ride at 2nd-magnitude Epsilon (ε) Pegasi (Enif).

Compared to the crystal clearness of Markab, Enif shines with a titillating marigold light. While both stars shine at 2nd magnitude, Enif is a *K*-class supergiant 10 times more massive than our Sun and nearly five times more distant than Markab. The star might have just enough mass one day to explode as a supernova. Now compare the golden hue of Enif to the warmer orange glow of 2nd-magnitude Beta (β) Pegasi (Scheat). Scheat is an *M*-type red giant 200 light years distant. Typical of other *M*-type stars, it is also a variable star, whose light fluctuates by half a magnitude (from 2.3 to 2.7) with an unpredictable period.

Return to Enif and center it in your binoculars. Just 4° northwest should be a small but brilliant, 6th-magnitude orb of diffuse light, with a highly condensed center. This is the famous globular star cluster M15, a magical blizzard of hundreds of thousands of stars seen some 30,000 light years distant. This little sphere of light is actually a monstrous 160 light years in true physical extent. Be sure to use averted vision, which will cause the orb to swell to twice its visual diameter as seen with direct vision.

From M15, take the return "train" to Enif, then ride the rail back to Alpha Pegasi. Now use the chart above to find 5th-magnitude 51 Pegasi, which lies midway between Alpha and Beta Pegasi and a tad west. Visible to the unaided eye under a dark sky, 51 Pegasi is a *G*-type star like our Sun 50 light years away, making it another great birthday star. More importantly, 51 Pegasi has a planet with a mass about half that of Jupiter orbiting it only five million miles away, which is some 19 times closer to its sun than Earth is to ours. It takes the planet only four days to complete an orbit. That's got to be one wild ride! It also means that if you're 50 years old on Earth, you'd be 4,562 years old on this planet!

Pisces, the Fish

Lower your glass and use your unaided eyes to once again look at the Square of Pegasus. Just a fist southeast of Alpha (α) Pegasi is a dim but distinct circlet of 4th-magnitude stars called the Circlet of Pisces – the western fish in the zodiacal constellation Pisces, the Fish. The rest of the constellation is faint because the stars are dim and spread out. On the northeastern end, the dim stars reach toward the constellation Andromeda, which we will discuss next month. The ancients saw Pisces as two fish tied together by a long cord. Its mythology is related to that of Capricornus (see page 81). The two fish represent Venus, the goddess of love and beauty, and her son, Cupid. To escape the wrath of Typhon, Venus grabbed Cupid and fled to the banks of the Euphrates River. There they hid among the reeds and awaited the storm to pass. When the winds began to blow, the surrounding reeds rattled, causing Venus to panic. Afraid, she called to the sea nymphs for help. In one version of the story, two nymphs came to the rescue and carried Venus and Cupid off to safety on their backs;

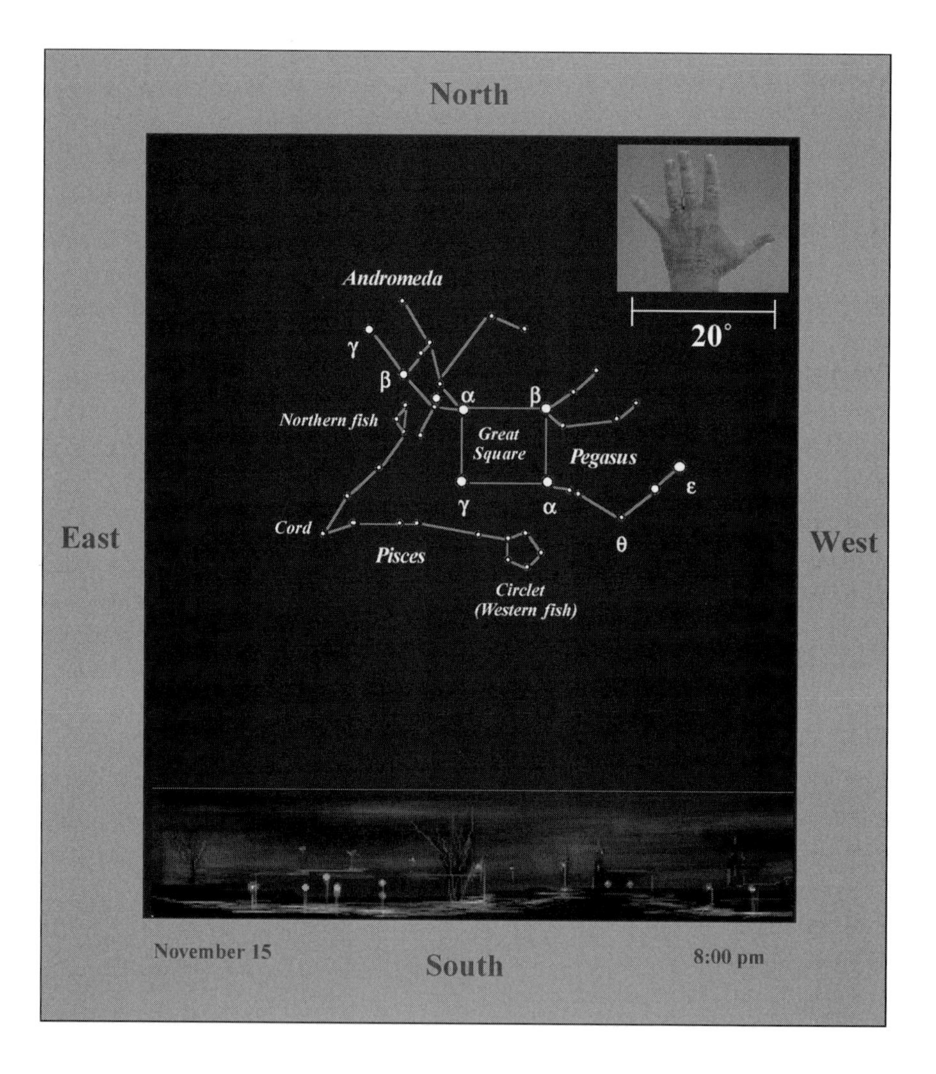
North
20°
Andromeda
γ
β
α
β
Northern fish
Great Square
Pegasus
ε
γ
α
East
West
Cord
Pisces
θ
Circlet (Western fish)
November 15
South
8:00 pm

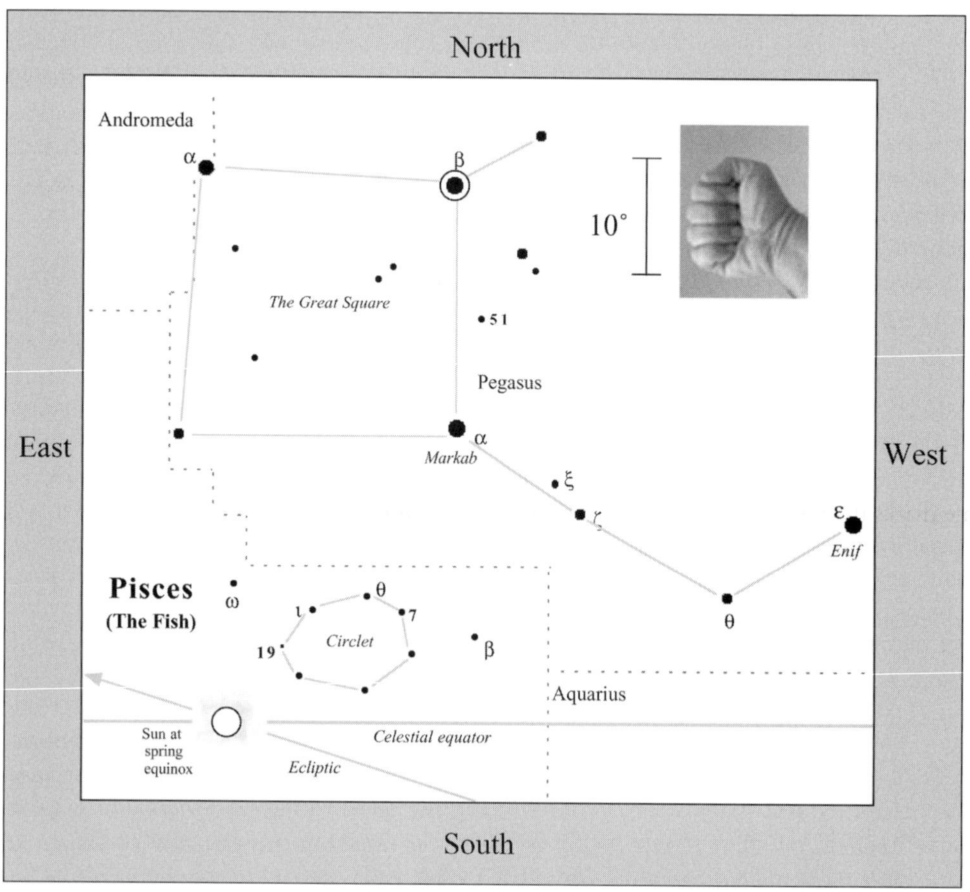
North
Andromeda
α
β
10°
The Great Square
51
Pegasus
East
α
Markab
West
ξ
ζ
ε
Enif
Pisces (The Fish)
ω
θ
ι
7
Circlet
19
β
θ
Aquarius
Sun at spring equinox
Celestial equator
Ecliptic
South

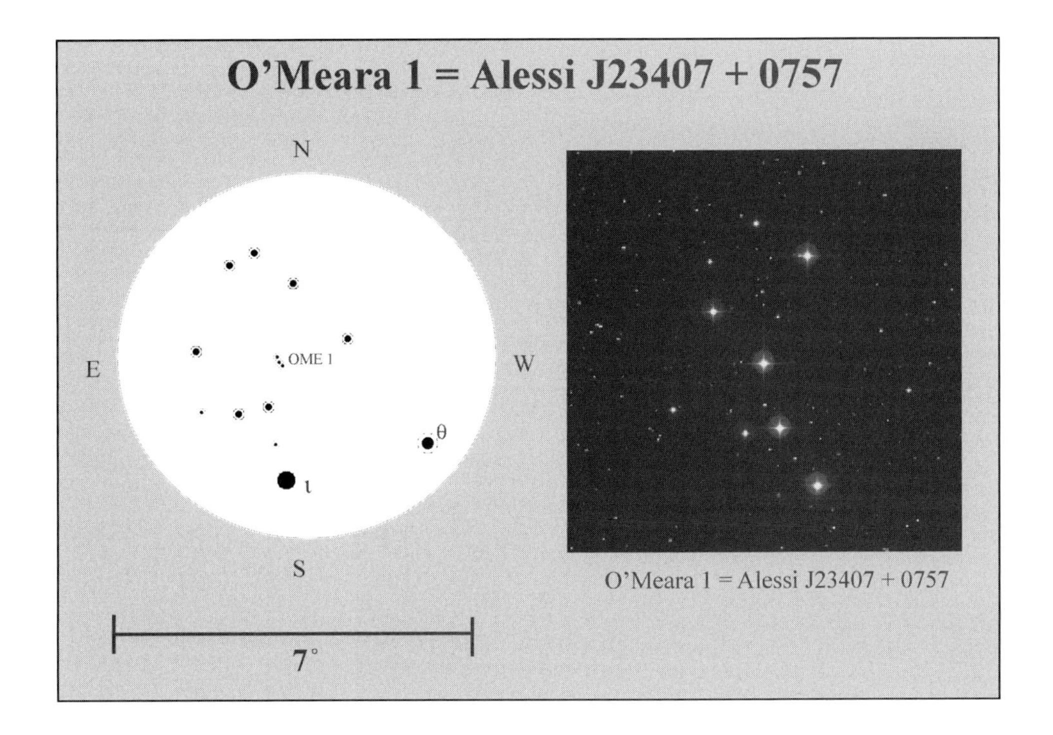

other versions have the deities changing into fish to flee Typhon's rage. As for the cord, no one knows why the fishes are tied together in such a fashion; the symbolism has been lost over the ages. But it has been suggested that Venus and Cupid tied themselves together so as to not lose each other in the dark waters of the Euphrates.

No matter, the most attractive part of the constellation is its Circlet, which fills the binocular field of view. A bit south of Lambda (λ) and Kappa (κ) Piscium is the point where the Sun finally crosses the celestial equator on its way out of the Underworld – the point of the spring equinox, which we celebrate in the third week of March.

Since both fish are leaping out of the dark waters of the Underworld, I have to wonder if the mysterious cord is not symbolic of the bond that unifies the deities (who in their human form obviously live in the open air) with their saviors from the sea. It may also symbolize the tight leash that the Sun is on; although the Sun leaves the Underworld in Pisces, it is destined to return. Interestingly, in Catholic religion, the fish is a symbol of Christ who rose from the dead. Like the Sun, Christ ascends from darkness into the light, just as springtime marks the renewal of life and daylight's triumph over night.

Finally, if you'd like a challenge, turn your binoculars to Iota (ι) Piscium in the Circlet and look $2\frac{1}{4}°$ north for the asterism I call the Little Ladle (O'Meara 1). I independently discovered it on the evening of January 20, 2003, though it was first discovered by the advanced Brazilian amateur astronomer Bruno Alessi years earlier; the asterism is now formally known as Alessi J23407+0757. Although the Ladle is a telescopic object, its brightest stars form a curious fuzzy string of dim stars visible in binoculars. See what you think.

9 December

O night divine! the stars are brightly shining,
It is the night of the dear Saviour's birth:
Placide Cappeau; *O Night Divine* (1847)

Andromeda, the Chained Maiden

If you go outside on December 1 around 7:00 pm and look south, the Great Square of Pegasus will be in the same place it was on November 1 at 9:00 pm. But now we can look for regal Andromeda, the Chained Maiden. Like Pegasus, we see this member of the Royal Family upside down. Alpha (α) Andromedae (Alpheratz) at the northeast corner of the Great Square marks the woman's head. Gamma (γ) and 51 Andromedae are her feet, while Beta (β) Andromedae (Mirach) and two attendants to the northwest (Mu and Nu) form the chain that girdles the Maiden's waist. There she stands, arms outstretched, pinned to invisible rocks, waiting for Cetus to devour her; we will look at Cetus shortly. The constellation of Andromeda, from Alpha to Gamma should span the width of your fully splayed fingers held at arm's length.

Of all the objects in the Chained Maiden, none is fairer than M31, the Great Spiral Galaxy – a naked-eye and binocular wonder and one of the grandest spectacles in the heavens. To find this, the nearest large galaxy to our own, use Autumn's Grand Celestial Station. You depart from the northeastern corner of the Great Square (Alpheratz) and travel two stops eastbound to snapdragon-yellow Delta (δ), then to golden Mirach. Stop here a moment and take the time to study Alpheratz and Mirach in your binoculars. While both stars shine at 2.1 magnitude, they

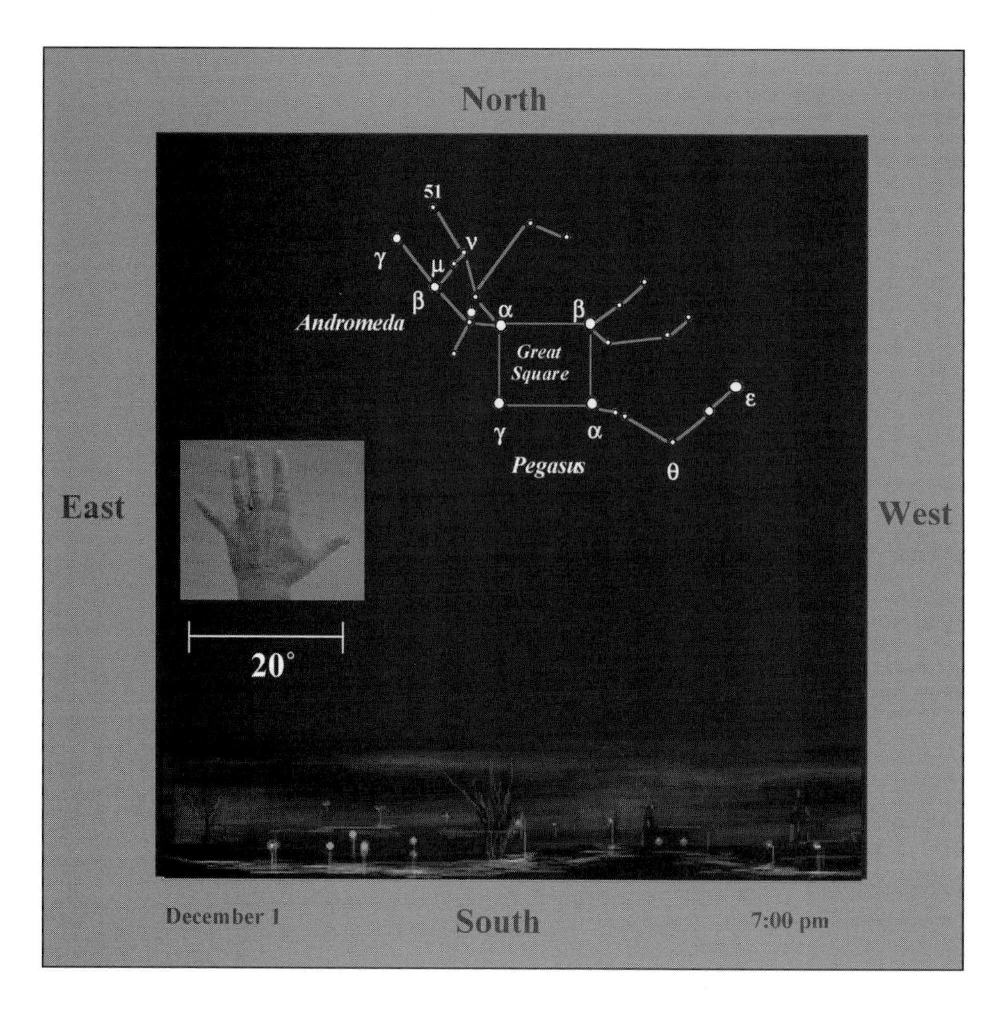
North
51
ν
γ
μ
β
Andromeda
α
β
Great
Square
ε
γ
α
Pegasus
θ
East
West
20°
December 1
South
7:00 pm

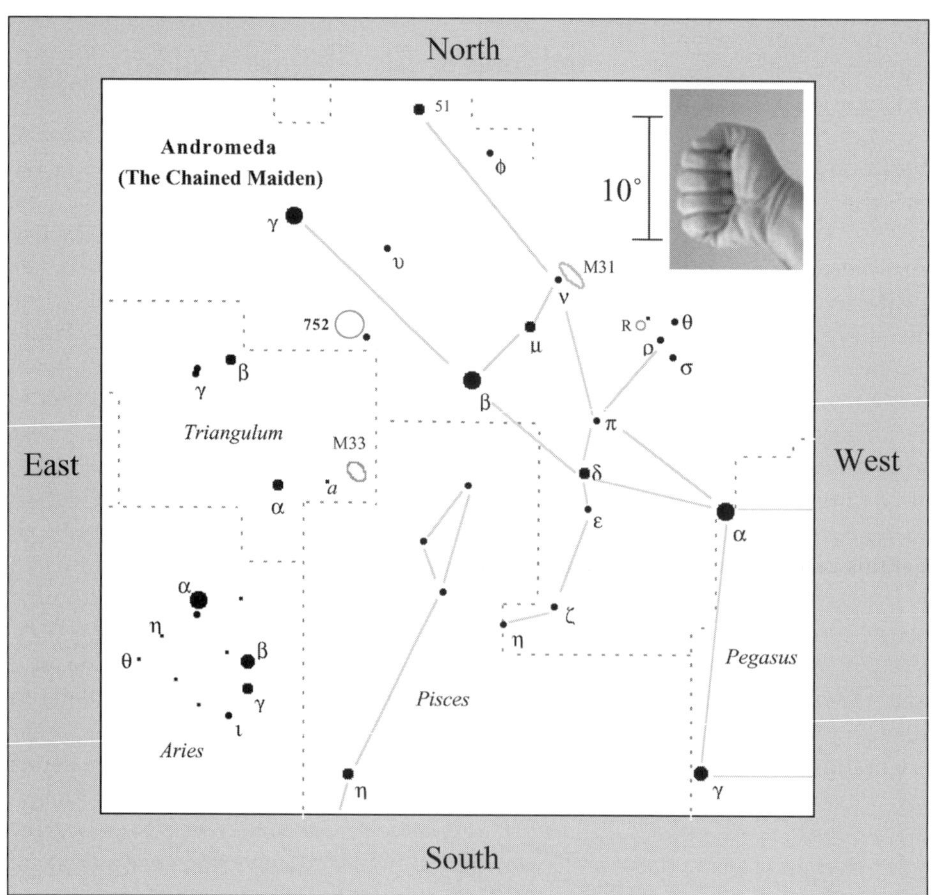
North
51
φ
Andromeda
(The Chained Maiden)
10°
γ
υ
M31
ν
752
μ
R
θ
ρ
σ
β
γ
β
Triangulum
π
M33
δ
α
a
ε
α
East
West
α
η
ζ
θ
β
η
Pegasus
γ
Pisces
ι
Aries
η
γ
South

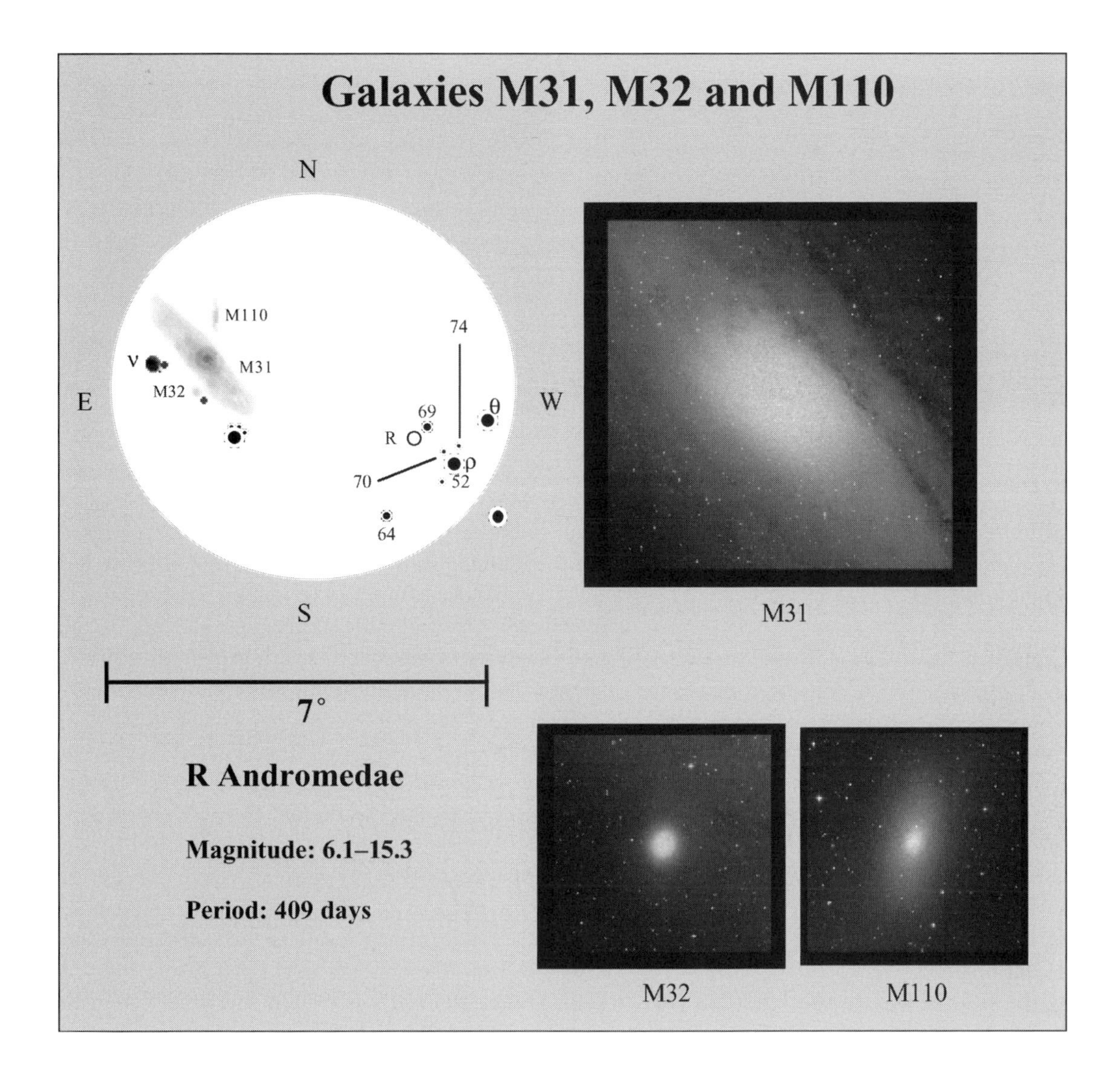

are vastly different. Alpha is a hot B-type star 97 light years distant and about 200 times more luminous than the Sun. Mirach, on the other hand, is a cool, M-type giant twice as distant as Alpheratz and nearly 2,000 times more luminous than the Sun. Now return to Mirach and head northbound, past 4th-magnitude Mu (μ) and Nu (ν) Andromedae, until you reach your destination 1° west of Nu.

In binoculars M31 stretches more than 3° in extent and is quite apparent as a diffuse spindle of light that gradually, then suddenly, brightens to a sharp nucleus. The galaxy is an enormous pinwheel of dust and gas mingling with some 300 billion stars all splashed across 130,000 light years of space. When we look at this island universe, we are peering beyond the forest of stars in our own galaxy, across a lonely 2-million-light-year void of space, at a galaxy that might just mirror our own. So if you were an alien stargazer living on a planet in the Andromeda Galaxy, our Milky Way would look very much like M31 in their night sky. Indeed, M31 is one of the most distant objects visible to the unaided eye.

If you look carefully at the galaxy's borders, you may be able to detect that its northwestern rim has a sharp edge to it. This edge marks the location of a prominent dust lane slicing through that part of the galaxy. By contrast, the galaxy's southeastern rim diffuses gradually into the sky background.

The first recorded sighting of the galaxy dates to the tenth century when Persian astronomer Al-Sufi called it "the little cloud" in his *Book of Fixed Stars* (AD 964). The light reaching our eyes from M31 left that galaxy two million years ago, when cosmic rays and ultraviolet radiation streaming from a supernova explosion 130 light years distant in the Scorpius arm of the Milky Way may have caused a wave of extinctions here on Earth. Interestingly, in August 1885, a new star erupted near the nucleus of M31. The new star achieved a maximum brightness of magnitude 6, making it just visible to the unaided eye. Designated S Andromedae, the *nova stella* was the first supernova known to erupt in a galaxy outside of the Milky Way. The drawings on page 102 from the 1898 *Atlas der Himmelskunde* (Vienna: A Hartleben's Verlag) show how the star altered the galaxy's appearance.

Binoculars will also reveal two of M31's companion galaxies: M32 and M110. Messier 32 is a small (8′) and highly condensed dwarf elliptical that looks like a slightly swollen 8th-magnitude star; it all but kisses the galaxy's southern rim. Messier 110 is a similarly bright, though

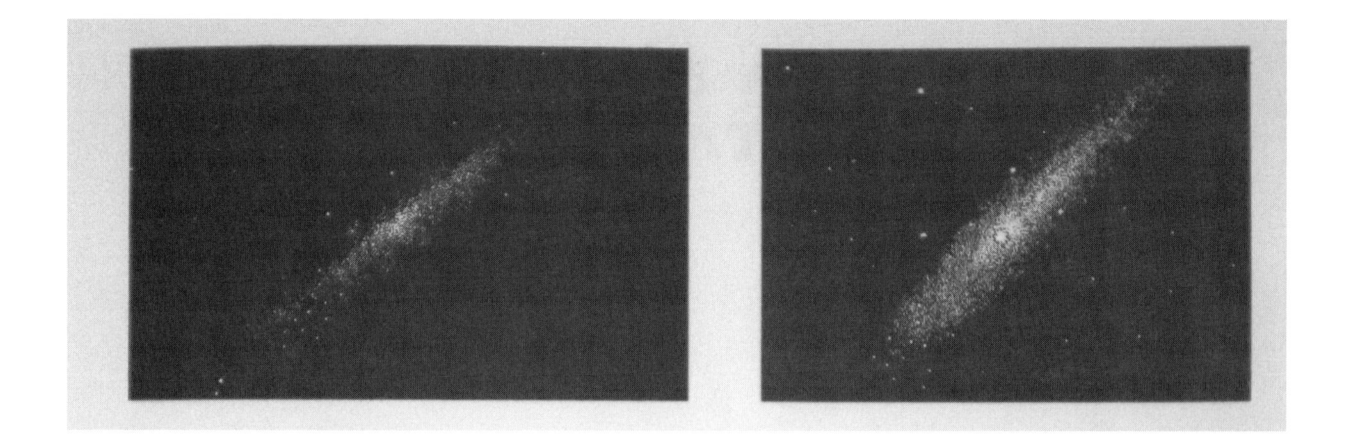

much larger and more diffuse, spindle (22′ × 11′) 37′ northwest of M31's nucleus. Messier 32 is much easier to see under city lights, while M110 is easily washed out by them.

In the same binocular field of M31 is the Mira-type variable star R Andromedae, whose light varies between a bright magnitude 6.1 and a dim magnitude 15.3 every 409 days. When at maximum light, R Andromedae shines as brightly as S Andromedae did when it was at maximum light, so you can imagine how it would have been to see that rare event, especially through binoculars. You should keep an eye on M31 with your binoculars for future events, because, when it comes to supernovae in external galaxies, lightning can strike twice . . . thrice . . . or more times in our lifetimes.

Before leaving the area, be sure to check out Nu (ν) Andromeda, the 4.5-magnitude star east of M31; it is a beautiful binocular triple. Nu is a luminous B-type dwarf that appears as a white spark with two tight, lemon-tinted companions to the west and northwest. The brightest shines at magnitude 7 and lies 6′ west of Nu. The other is 8th magnitude and 4′ northwest of its neighbor. Seen together, the three stars look like a bent matchstick starting to ignite.

It's quite a challenge, though not impossible, to see Nu and its attendants with the unaided eye. I call it my Invisible Man Challenge, because the dimmer stars will weave in and out of view at the limit of perception. Trying it myself one night, I thought of film historian David J. Skal's comments on the 1933 film adaptation of H. G. Wells's 1897 novel, *The Invisible Man*. "The idea of an invisible man is a wonderful metaphor for the outsider," Skal says, "a person who exists on the fringes of society." To the naked eye, Nu's attendants are the perfect metaphorical outsiders.

Take the "train" back to Mirach, then head about a little more than 10° (a fist) northeast to 2nd-magnitude Gamma (γ) Andromedae (Almach). A beautiful telescopic double, the star is nonetheless a gorgeous imperial yellow K-type giant now in the act of dying. Almach lies at a distance of 355 light years and shines about 2,000 times more brightly than the Sun. If Almach were our Sun, its disk would be large enough to swallow the planets out to the orbit of Venus. Look for a pretty trapezoid of 7th-magnitude suns 1° east–southeast of Gamma and a tight trio of stars comprised of a single 7th-magnitude sun and two 9th-magnitude companions about 1° to the south–southeast; neither of these star groups are charted.

If you place Gamma in the upper left of your binocular field a big and sparse open star cluster, NGC 752, will come into view. Although the cluster shines at 6th magnitude, it is also very large (75′) making it appear more than twice as large as the full Moon. Under a dark sky, it is a beautiful splash of 8th-magnitude and fainter suns. Unlike most open star clusters, which are tens or hundreds of millions of years old, NGC 752 is about 2 *billion* years old, making it one of the oldest open star clusters known. The cluster's very open nature hints at its age. Unlike globular star clusters that tend to remain tightly bound together for billions of years, open star clusters are more loosely bound and tend to disperse much more quickly as they age; in the end, an open star cluster's stars – especially those lining the cluster's edge – become lost to the gravitational pulls of passing stars, molecular clouds, and other clusters. These vagabonds then blend

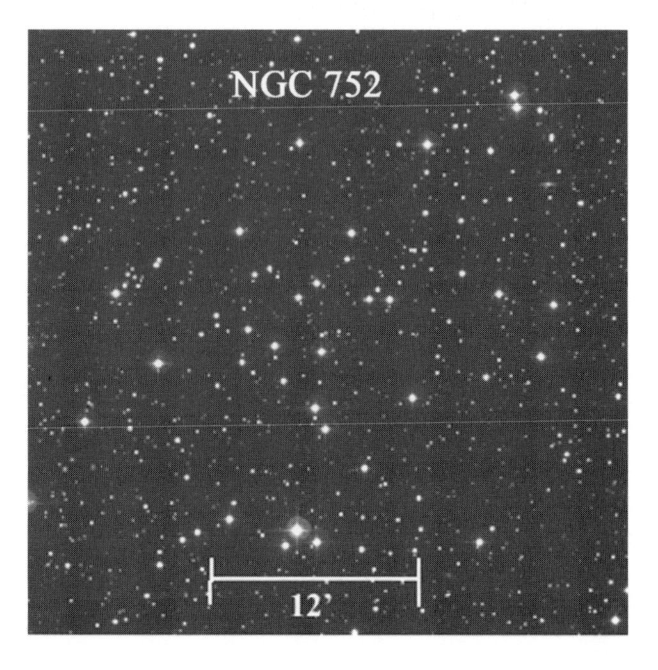

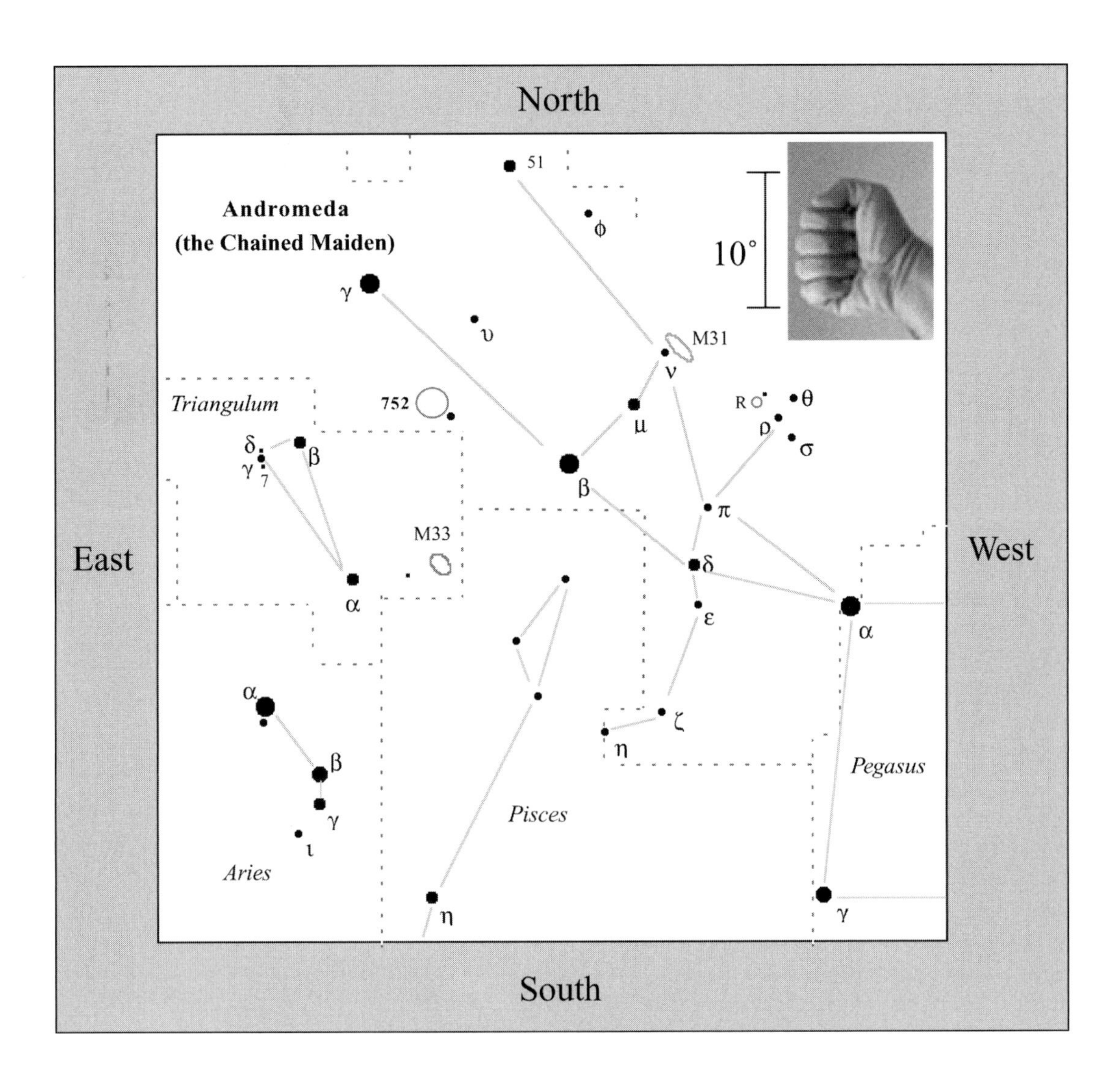

into the background Milky Way like drifters on a busy New York City street, where the records of their origins are essentially erased.

Aries, the Ram, and Triangulum, the Triangle

Use the chart above to find the small and somewhat minor constellations, Aries, the Ram, and Triangulum, the Triangle. You can start by once again returning your gaze to Gamma (γ) Andromedae, then looking about 10° (a fist) due south for the Triangle, which is comprised of two 3rd-magnitude stars – Alpha (α) and Beta (β) Trianguli – and 4th-magnitude Gamma (γ). Despite its dimness, the Triangle is a very ancient constellation. Although it appears acute in the sky, its name was originally derived from Deltoton, from the capital Greek letter Delta (Δ). Other myths have it representing the Nile Delta and the triangular-shaped island of Sicily. Apparently Ceres, the daughter of Saturn, who presided over agriculture, pleaded to Jupiter that he honor Sicily – an important agricultural land in early classical times – with a place in the heavens.

Since the volcano Etna is on Sicily, it is possible to see the mountain's smoke drifting away from the island in the form of M33, the Great Triangulum Galaxy. To find it, refer to the chart on page 104 and simply place Alpha (α) Trianguli (Mothallah) in the lower left of your binocular field; 6th-magnitude M33 will be near the center. Look for a large ghostly glow more than twice the apparent diameter of the full Moon. It should appear slightly oval with a strange "silver" quality to it – a light that makes the galaxy somehow appear three dimensional. Like the Andromeda Galaxy, M33 is 2 million light years distant and may be a satellite galaxy to that larger wonder. Popularly known as the Pinwheel Galaxy, M33 spans 50,000 light years and is 15 times less massive than the Andromeda Galaxy. But it is still a binocular wonder and a great naked-eye challenge. The ease of its visibility fast made it a barometer for the clarity of one's observing site.

Slip over to 4th-magnitude Gamma (γ) Trianguli, which is part of a very pretty coincidental triple star; the other members being 5th-magnitude Delta (δ) Trianguli to the north and similarly bright 7 Trianguli to the southwest. (A dimmer companion lies to the south–southwest.) If you want to impress your friends with the power of the three-dimensional sky, show them this triple through your binoculars. Delta is 35 light years distant,

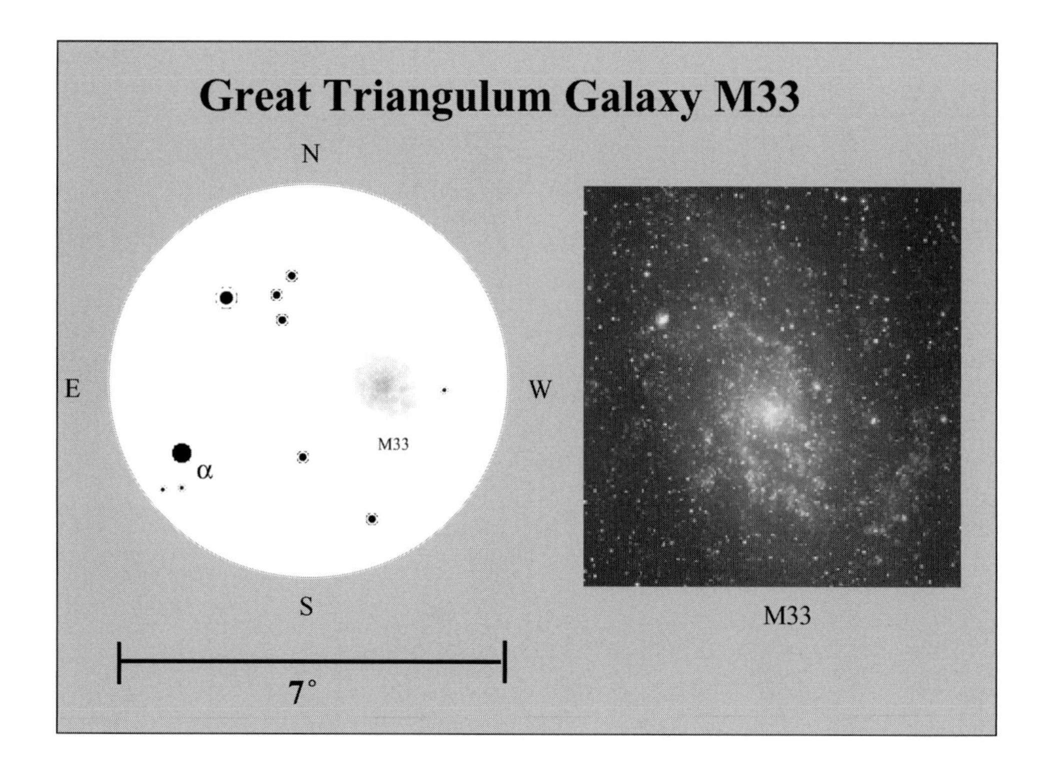

Gamma is 118 light years distant, and 7 Trianguli is $2\frac{1}{2}$ times farther away from us than Gamma.

From Gamma Trianguli, drop your sights another 10° to 2nd-magnitude Alpha (α) Arietis (Hamal). Hamal, together with 3rd-magnitude Beta (β) Arietis (Sheraton) and 4th-magnitude Gamma (γ) Arietis (Mesartim), form the constellation that represents the ram that wore the golden fleece – the object of Jason and the Argonauts' desire. Hamal is, in fact, a golden K-type giant 66 light years distant. But I see it as looking canary yellow. When Homer wrote about Jason's epic adventure in the seventh century BC, the spring (or Vernal) equinox – the point in the sky occupied by the Sun as it crosses the celestial equator on its journey northward – occurred in Aries, at a point nearly 10° south of Hamal. Known as the First Point of Aries, it became a powerful astrological symbol with strong religious ties.

Although we commonly refer to Aries as being a Ram, it was also at one time depicted as a lamb. Hamal is, in fact, Arabic for "lamb." Around the time of Christ, Aries was the leader of the heavens and Christ was the sacrificial Lamb of God. The Christian celebration of Christ's rising from the grave (Easter) occurs on the first Sunday after the first full Moon following the spring equinox – when the Sun, baptized in water, rises from darkness into light.

Cetus, the Whale

We now plunge back into the Underworld to meet the beast who was to devour Andromeda. To the ancients, Cetus did not look like a whale – like a humpback or sperm whale – but a Sea Monster. As the depiction below shows, it was a hideous amalgam of taxonomical horror. Here is what I call a *Drahorspenteal*: an animal with a dragon's head and neck, a horse's torso, serpent's tail, and leopard-seal like fins; in other words, a Sea Monster by committee. It exemplifies the magnitude of our ancestors' ignorance of the undersea world. It's as if a heap of bones exploded from a museum, washed up on shore, and was pieced together by a child with a vivid imagination.

Actually, that scenario is not far from the mark. The legend of Cetus was born when Pliny the Elder (AD 23–79) recorded the discovery in 58 BC of the "spine of the sea serpent killed by Perseus at Joppa." Seeing the bones displayed in Rome, Pliny must have gasped at the great skeleton's 40-foot length, with each vertebra measuring six feet in circumference. While the bones most

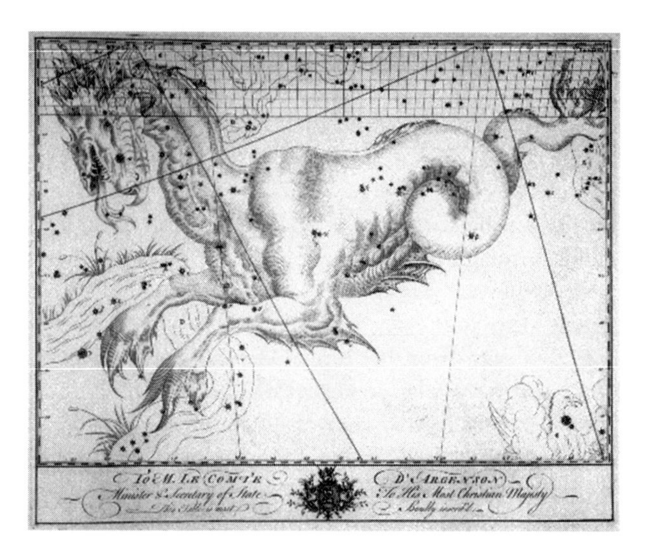

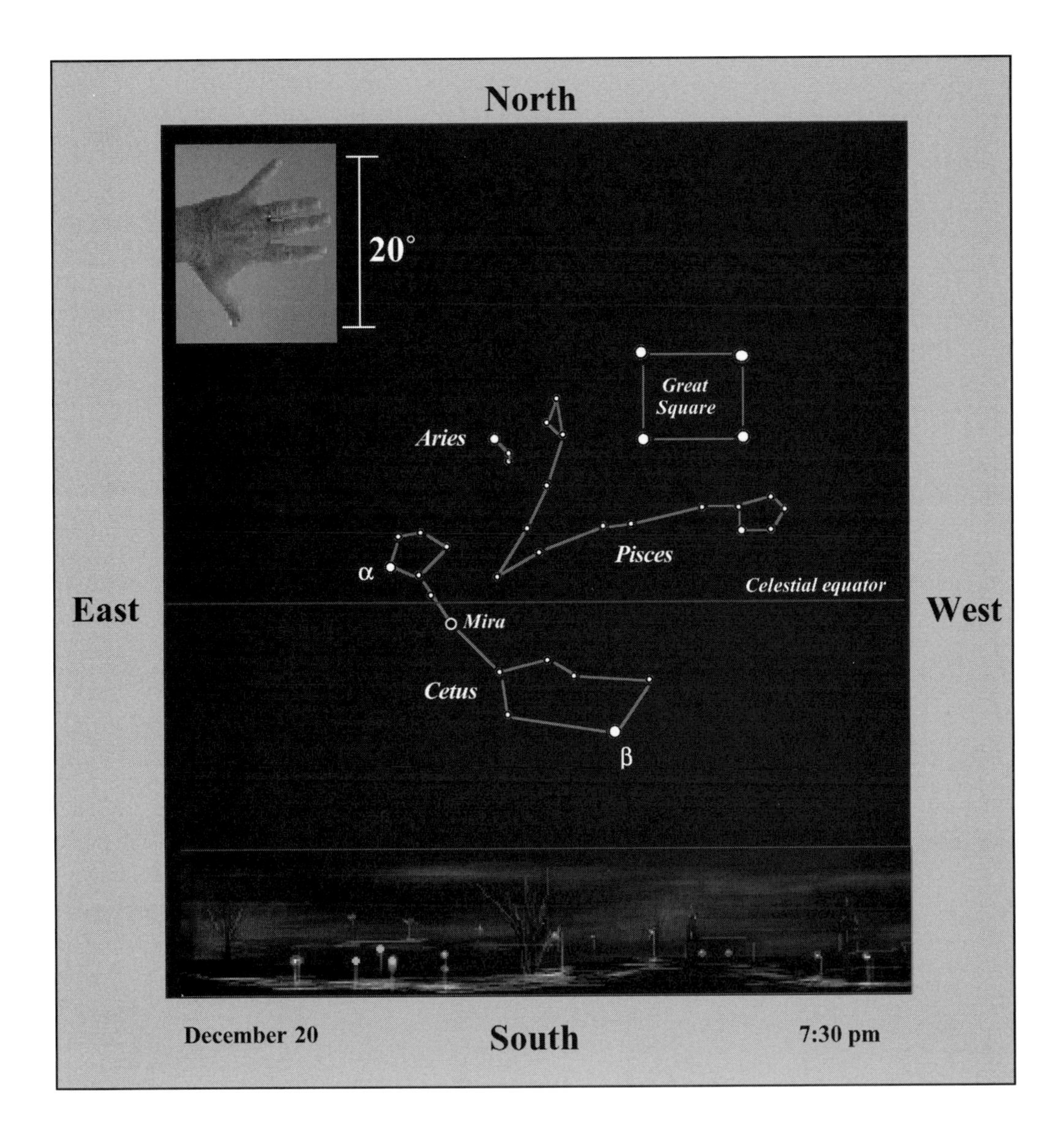

likely did indeed belong to something like a deceased and decayed sperm whale, Pliny truly believed in dragons. In faraway India, he said, these creatures were large enough to prey on elephants by dropping out of trees and strangling them; the crests on their heads were so large that they could sail to Arabia to hunt. (These stories, no doubt, described exaggerated python behavior.) It's also possible that Pliny had heard about, or seen, the rare oarfish, whose snakelike body can measure 30 feet in length and have a red head crest.

Mythology aside, if you scan the south meridian in late December between 7:00 and 8:00 pm, you can see the Sea Monster poking its head above the waters of the celestial equator. The Sea Monster's head (which your fist should comfortably cover) is one of the few easily identifiable parts of this large and rather dim constellation. Like the western fish of Pisces, the head of Cetus is a circlet of stars (more of a squashed circlet) that deserves a scan with your binoculars because there are some interesting star couplings.

The head's Alpha star, Menkar, is the best. This M-type giant lies 220 light years distant and shines like emperors' gold. Seen through binoculars, Menkar has a colorful 6th-magnitude companion, 93 Ceti, to the north, which has a beautiful apple-green sheen. The color contrast is most striking.

While Menkar is the constellation's Alpha star, it is not the brightest. Shining at magnitude 2.5, Menkar is actually half a magnitude fainter than Beta (β) Ceti (Deneb Kaitos), which marks the end of the Monster's coiled tail. In true physical dimensions, Menkar is the Alpha star, being 84 times larger than our Sun, while 96-light-year-distant Beta is a K-type giant with a diameter of only 34 times our Sun. This ranking scheme excludes the Whale's great prototype variable Mira, which on rare occasion can exceed the brightness of either Alpha or Beta Ceti; but we'll come back to Mira.

If you have a clear and low horizon, and are under a dark sky, lower your binoculars a little more than one field to the south. There you will encounter one of the sky's dynamic duos: the spiral galaxy NGC 253 and globular star cluster NGC 288 – both of which reside in the southern constellation Sculptor. The pairing is a chance alignment, with NGC 288 being 27,500 light years distant

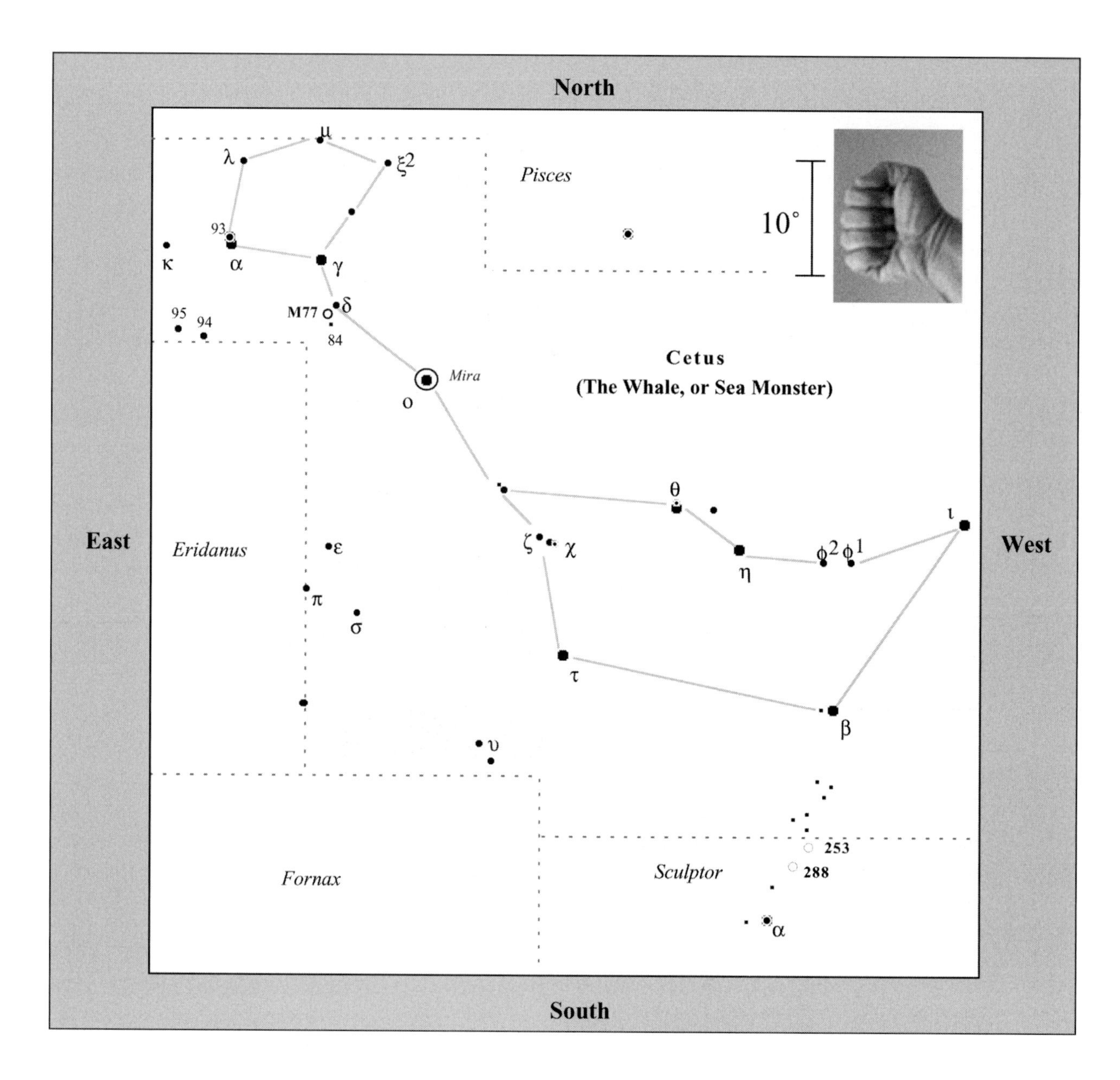

in the halo of our galaxy and NGC 253 being an extragalactic system 9.8 million light years away; but we see them on the two-dimensional sphere separated by only $1\frac{3}{4}$°.

The galaxy NGC 253, the brightest member of the Sculptor Group of galaxies, measures 54,000 light years in diameter and has a total mass 75 billion times our Sun (making it about $2\frac{1}{2}$ times smaller and 4 times less massive than M31 in Andromeda). The Hubble Space Telescope image of it on the next page is a close-up of the galaxy's core; streaming bands of dark clouds nearly obliterate the view of a multitude of stellar furnaces below. Through 10 × 50 binoculars, it looks like a cigar-shaped glow. By comparison, NGC 288 is a round orb of celestial cotton. In true physical extent, it measures 104 light years across and is some 13 to 14 billion years old. Indeed, NGC 288 is among the very oldest globular star clusters in the halo of our galaxy. Note too how NGC 288 lies only 40′ north and slightly east of the *south galactic pole* – so this globular star cluster is the South "Star" of the Milky Way Galaxy.

Now look about 15° east and slightly north of Beta Ceti for 3.5-magnitude Tau (τ) Ceti – a G-type sun only 12 light years away. In 1960 astronomer Frank Drake attempted to detect intelligent radio signals coming from Tau Ceti and Epsilon (ε) Eridani (see page 120), which is just over the eastern border of Cetus in Eridanus, The River. Drake dubbed the search "Project Ozma," after the princess that appears in every Frank L. Baum book except the *Wizard of Oz*. In that bold attempt, Drake pointed the 85-foot radio telescope at Greenbank, West Virginia, toward these stars for a total of about 200 hours over a period of two months to see if ET was going to phone home. Ironically, on the day the project commenced (April 8), Drake picked up an alien signal; long weeks passed before he confirmed that the signal originated from a secret military operation. It was a false alarm. Although no further "alien" signals were detected, the project stirred up so much interest that the National Academy of Sciences organized the first major meeting about the search for extraterrestrial intelligence (SETI), which ultimately gave birth to the modern era of SETI research.

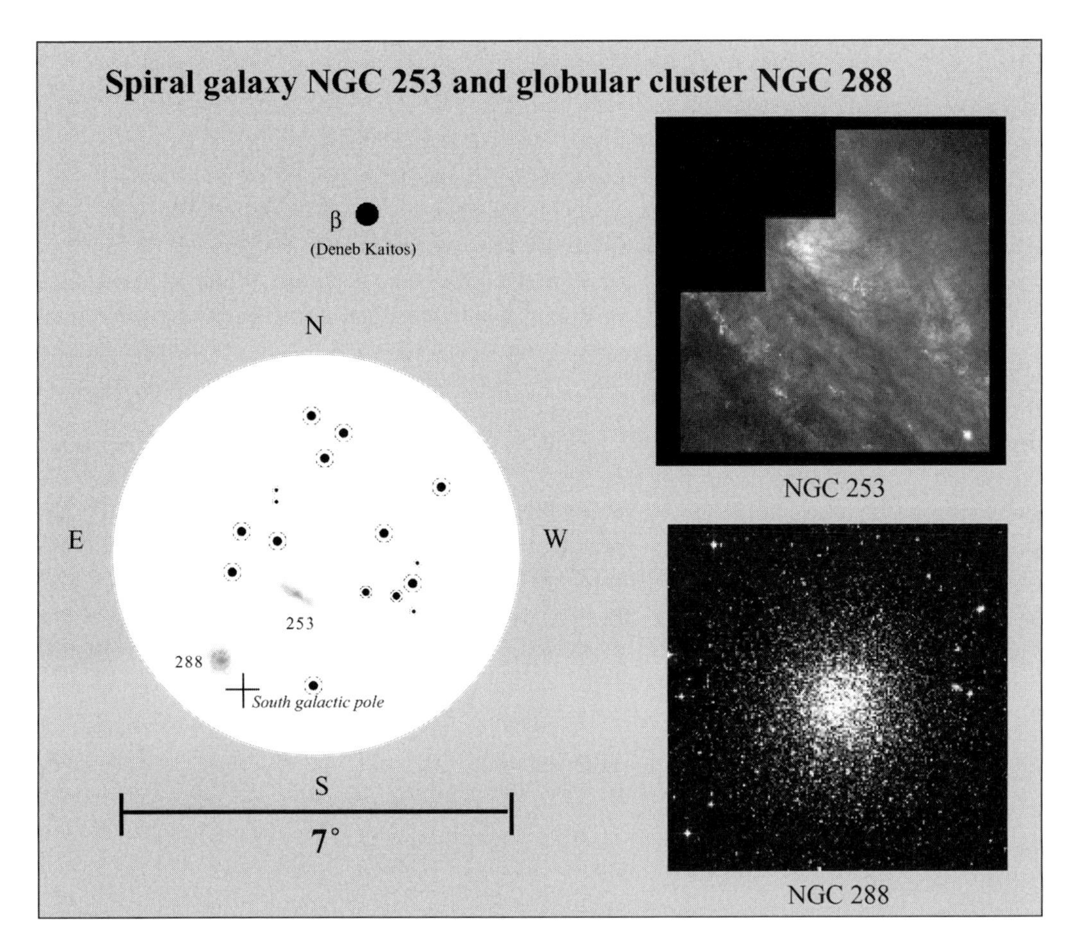

Perhaps Drake was listening to the wrong stars. You see, Cetus has two other stars that we now know have planet-like bodies orbiting them. One of them, 5th-magnitude 94 Ceti (see chart above), is visible to the unaided eye. An F8-type sun, 94 Ceti lies 73 light years away and has a planet with twice the mass of Jupiter in an orbit 120 million miles away – just 27 million miles farther than Earth from the Sun.

From 94 Ceti, swing a little more than one binocular field to the west–northwest, to 4th-magnitude Delta (δ) Ceti. The Seyfert galaxy M77 lies only 1° southeast of that star. Look for a tiny fuzzy pip of light in a row of three stars. This system, 47 million light years distant, looks stellar because the galaxy has an active (Seyfert) nucleus that behaves like a low-energy quasar, meaning it periodically ejects matter at incredible velocities (6 percent

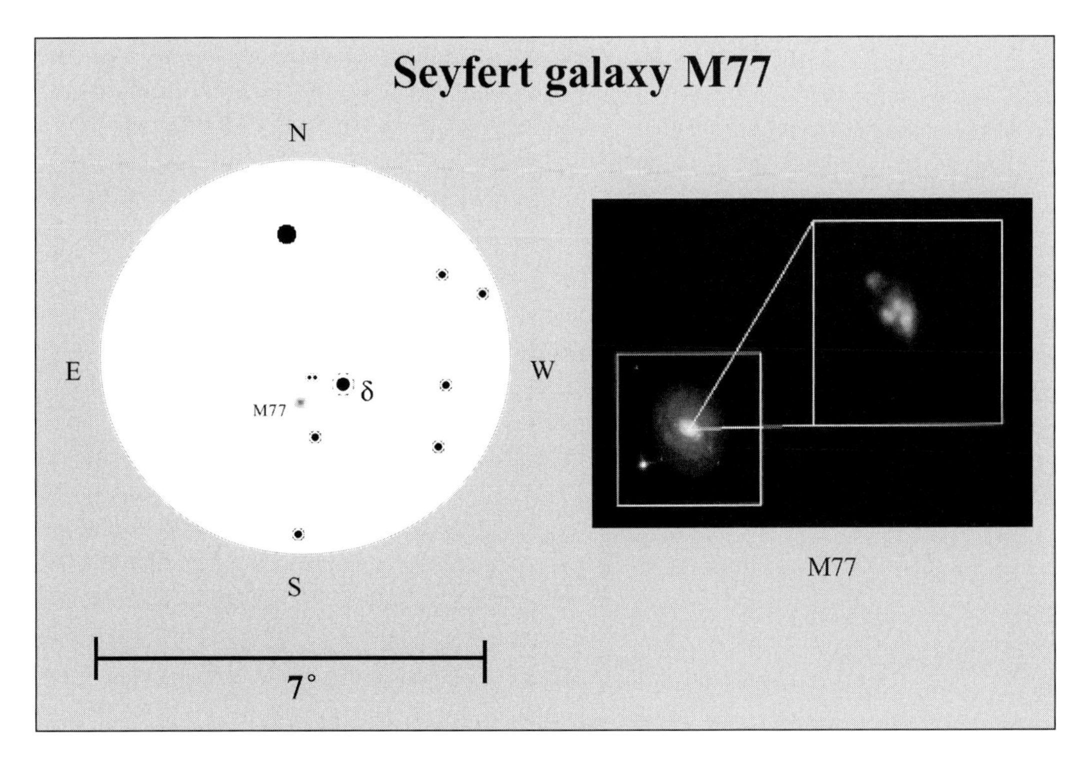

of light speed and up). All galaxies, including our own, might spend about 10 percent of their lifetimes in the Seyfert phase. Timothy Ferris, in his popular book *Galaxies* (New York: Stewart, Tabori and Chang, 1990), draws a fine analogy: Galaxies, he says, are like fireflies that light up for only a fraction of the time that they spend in flight; the rest of the time the fireflies are dark, storing up energy for another flash. In the case of M77, gas clouds, some with the mass 10 million times our Sun, are being blasted away from the galaxy's nucleus with velocities up to 360 miles per second. That's enough energy to power several million supernovae explosions!

Finally, we come to the Grand Poobah of variable stars: Mira, the wonderful. Otherwise known as Omicron (o) Ceti, Mira is the brightest and most famous M-type variable of its kind. All long-period variables of its type are known as Mira-type stars. When at maximum light, the star can be a glorious 2nd-magnitude beacon, though in 1779 it achieved 1st magnitude. Most of the time it attains only 3rd. At minimum light, it dips to 9th magnitude and sometimes 10th – well below naked-eye visibility. The star winks in and out of view with a period of 330 days – almost a full year, making it a great star to follow every season; sometimes it will be a part of Cetus, sometimes it won't. The star's vanishing act also has the effect of decapitating the Sea Monster, leaving its head to float ghostlike well above its body. Mira lies 420 light years distant. The star is nearing the end of its life. When its outer shell expands at maximum light, it pushes out to a point that would equal that distance of our Asteroid Belt from the Sun, a position that would place it between the orbits of Mars and Jupiter.

By the way, many skywatchers who grew up reading H. A. Rey's *The Stars: A New Way to See Them*, have adopted Rey's interpretation of the constellation's star pattern. In his scheme, Beta Ceti becomes the tip of the Whale's snout, and the famous circlet is the Whale's tail. Fourth-magnitude Upsilon (υ) Ceti, then, becomes the bottom of the Whale's belly. See what you think.

CHAPTER 5

The winter stars

10 January

Thou art the star for which all evening waits –
O star of peace, come tenderly and soon,
George Sterling, *Aldebaran at Dusk* (1911)

Winter has arrived. Nights loom long and cold. The trees are bare like bones. And the stars shine like white stones locked in black ice. Pegasus, the Winged Horse, has drifted away from the meridian, carrying Andromeda toward the western getaway. In the north, the Great Bear has left its lowly domain under the north celestial pole and is now standing to the east on its tail. As Cetus backs away from the south meridian, a new constellation has risen to prominence, Taurus, the Bull.

Taurus, the Bull

When with his golden horns in full career
The Bull beats down the barriers of the year.
Virgil, *The Georgics*, as translated by John Dryden

If you go outside at 9:00 pm on New Year's day and look high in the south, you will see two distinct star patterns: a wide V-shaped gathering of suns with a bright orange member, and, higher up in the sky, a tight congregation of glittering jewels in the shape of a little dipper. The V-shaped gathering is the Hyades open star cluster, while the "little dipper" is the Pleiades open star cluster. Both are part of the zodiacal constellation, Taurus, the Bull, and both have mystified skygazers since the dawn of history.

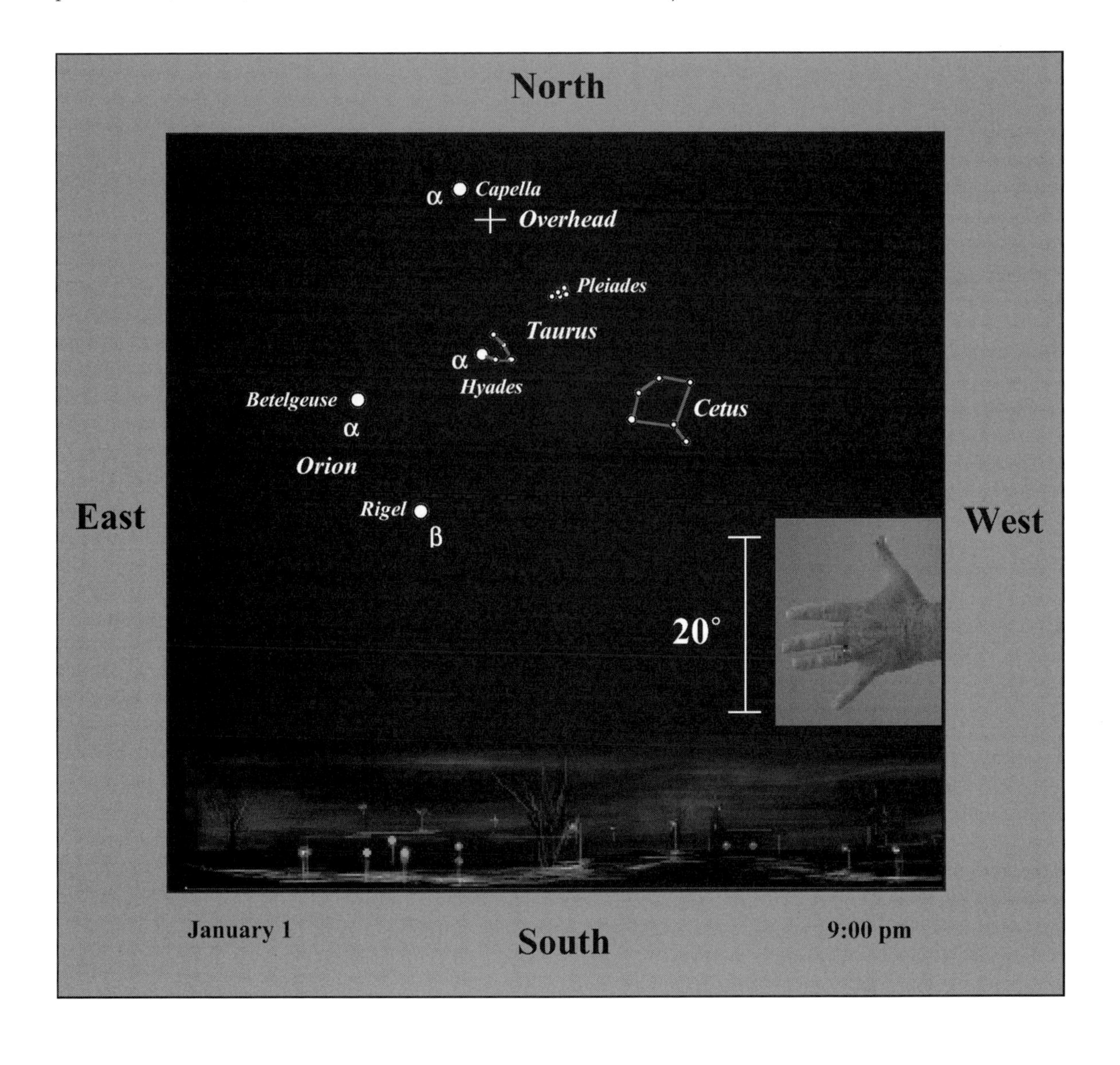

The Hyades represents the Bull's face, with Alpha (α) Tauri (Aldebaran) marking one blazing eye. The Pleiades, depending on the myth, is either a sword wound in the Bull's back or seven doves cowering on the Bull's back. In Greek mythology, seven bright stars in the Hyades (though no one knows which seven) were Hyads – the daughters of Atlas (the Titan condemned to support the Earth and sky on his shoulders for eternity) and Aethra. The Hyads' paternity makes them half sisters of the Pleiades, who were born to Atlas and Pleione; together the Hyads and Pleiades constitute the 14 daughters of Atlantis, or the Atlantides. Jupiter transformed the Pleiades into doves and placed them on the Bull (the protector) to help them escape the lustful advances of Orion, the Hunter; we'll meet Orion next month.

Actually, Jupiter himself was a bull, a white bull, a disguise he concocted to win the affection of Europa, daughter of Phoenician king Agenor. When Europa saw the beautiful bull looking so peaceful by the sea, she went to it, patted its head affectionately, then ventured to mount its back. Suddenly the bull leaped into the sea. Fearing for her life, Europa clung to the Bull as it pushed with powerful strokes toward Crete. Once safely ashore, Jupiter revealed his true form and mated with the princess. (Power corrupts!)

The stars of Taurus have been seen as a Bull since at least 4000 BC, when the Sun resided among them during the spring equinox. The union was a symbol of virility, the birth of spring and agricultural activity, which the earliest peoples depended so much on for survival.

The Bull also figures prominently in Egyptian hieroglyphs. Take, for instance, these words from the *Pyramid Texts* of ancient Egypt, first inscribed by the last ruler of the Fifth Dynasty on the internal walls of his pyramid at Saqqara:

> Your fields are in fright, o jAd-star, before the pillar of stars, when they have seen . . . the Bull of Heaven. How the herdsman of the bulls is overwhelmed with awe before him! . . . Unas is the Unique One, the Bull of Heaven. . . . Ignore not Unas, o Bull of Heaven, you know him because he knows you. Ignore not Unas, o Bull of Heaven.

In ancient Roman times, the Hyades were a bad omen for sailors, because they rose shortly after sunset during the Mediterranean's rainy season. The stars became known as Sidus Hyantis, the Rain-Bringers. The phrase refers to the legendary grief the Hyades felt over the death of their terrestrial half brother, Hyas; the sisters' tears were so profuse that they caused heavy rainfall on Earth. Today, the Hyades rise at the end of the reaping season, and their pumpkin orange star Aldebaran is now an appropriate and vivid reminder of the season.

Aldebaran is a K-type orange giant 60 light years distant. It has a diameter only 40 times that of our Sun. Its brightness varies slightly, between magnitude 0.75 to 0.95. The star might also have a giant planetlike body orbiting it every two years, but the object's true identity remains uncertain; it could be a low-mass brown dwarf star. There's another surprise: despite its key location in the Hyades, Aldebaran is not a member of that cluster, which lies about 150 light years away. What we see is a lovely chance alignment. So the Bull's eye is really popping out of its socket!

Run your binoculars across the face of the Bull, which is chock full of attractive stellar pairs. Most obvious are the Theta (θ) Tauri and Delta (δ) Tauri groupings in the middle of each branch of the V. The Theta pair is surrounded by a circlet of tight little doubles, all pleasing to the eye through binoculars. Actually there are two Theta stars: Theta^1 (θ^1) and Theta^2 (θ^2) Tauri, which shine at 4th and 3rd magnitude, respectively, and are separated by about 6′. Theta^1 is a yellow G-type giant, so be sure to compare its sheen with that of Aldebaran. I find Theta^2 looking more pearly white than orange Aldebaran. The Sigma stars (σ^1 and σ^2) are another pretty pair of chalk-white stars of near equal intensity; they shine at 5th magnitude and are separated by 7′. Now hop over to the western branch of the V, where you'll see Delta^1 (δ^1) and Delta^2 (δ^2) Tauri. Like the Theta stars, the Delta stars are a yellow and white pair but more widely separated (18′); they're also slightly fainter, with Delta^1 shining at 4th magnitude and Delta^2 at 5th. Also of note are the pretty Kappa stars to the north.

If you place Aldebaran in your binoculars as shown in the chart on page 111, you'll find NGC 1647, the Pirate Moon Cluster. Through 10 × 50 binoculars, this open star cluster is a beautiful sprinkling of about a dozen stars seen against a round ghostly glow of unresolved starlight. The cluster is quite large, having an apparent diameter larger than that of the full Moon, which is, in part, why I gave the cluster its moniker; besides, under a dark sky, the binocular view makes the cluster look like a fog-enshrouded moon sailing over the vast ocean of space. Light pollution will drown out the cluster's pale glow, so save it for a clear, dark night. Although NGC 1647 is positioned close to the Hyades cluster, it has no physical relation to it. NGC 1647 lies 1,700 light years distant, covers 20 light years of space, and is 190 million years old – that makes it some

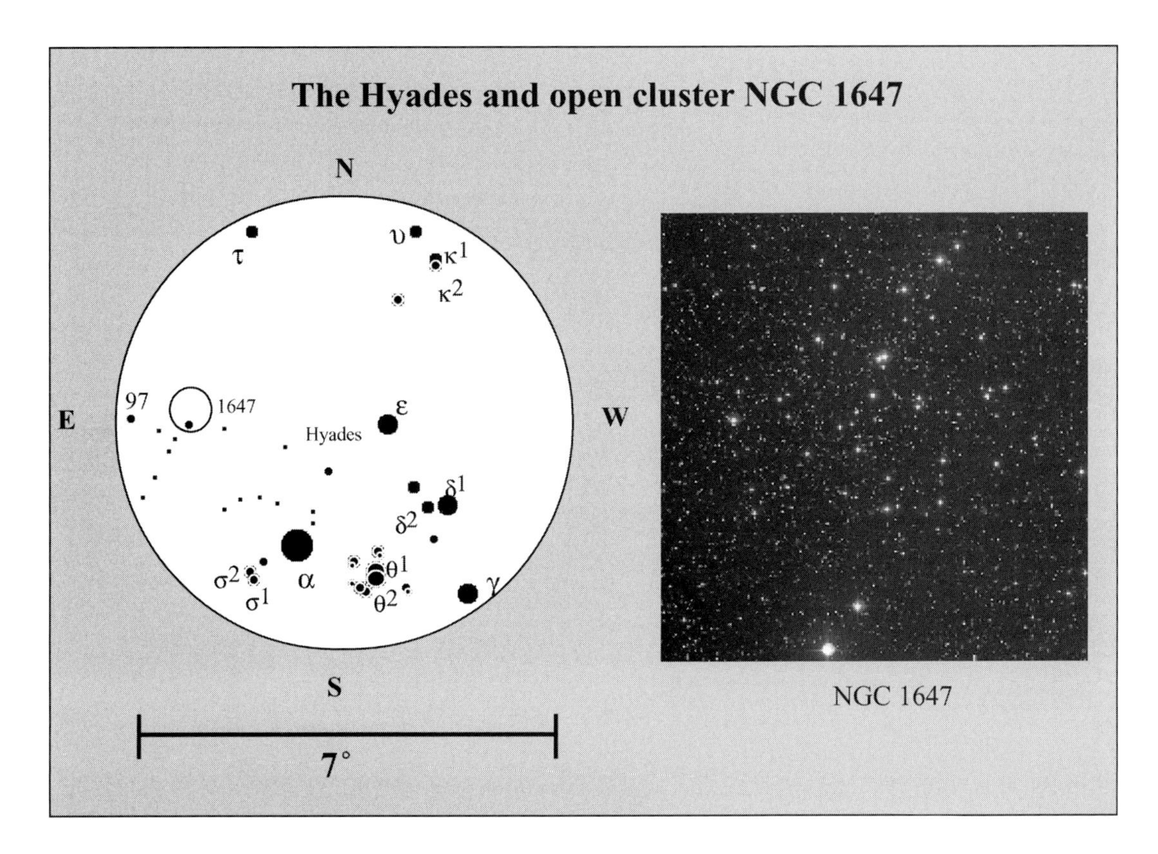

400 million years younger than the Hyades and 11 times more distant.

Now use the chart below to look for 3rd-magnitude Zeta (ζ) Tauri, which will be a little more than a fist to the east–northeast of NGC 1647. If you are under a dark sky, you might be able to see the dim glow of M1, the famous Crab Nebula (a supernova remnant) – the remains of a cataclysmic stellar explosion that occurred in our own Milky Way galaxy. While Zeta Tauri is 420 light years away, the supernova remnant is a whopping 6,300 light years distant. When its light finally reached Earth, Chinese star-watchers in the year AD 1054 recorded its appearance

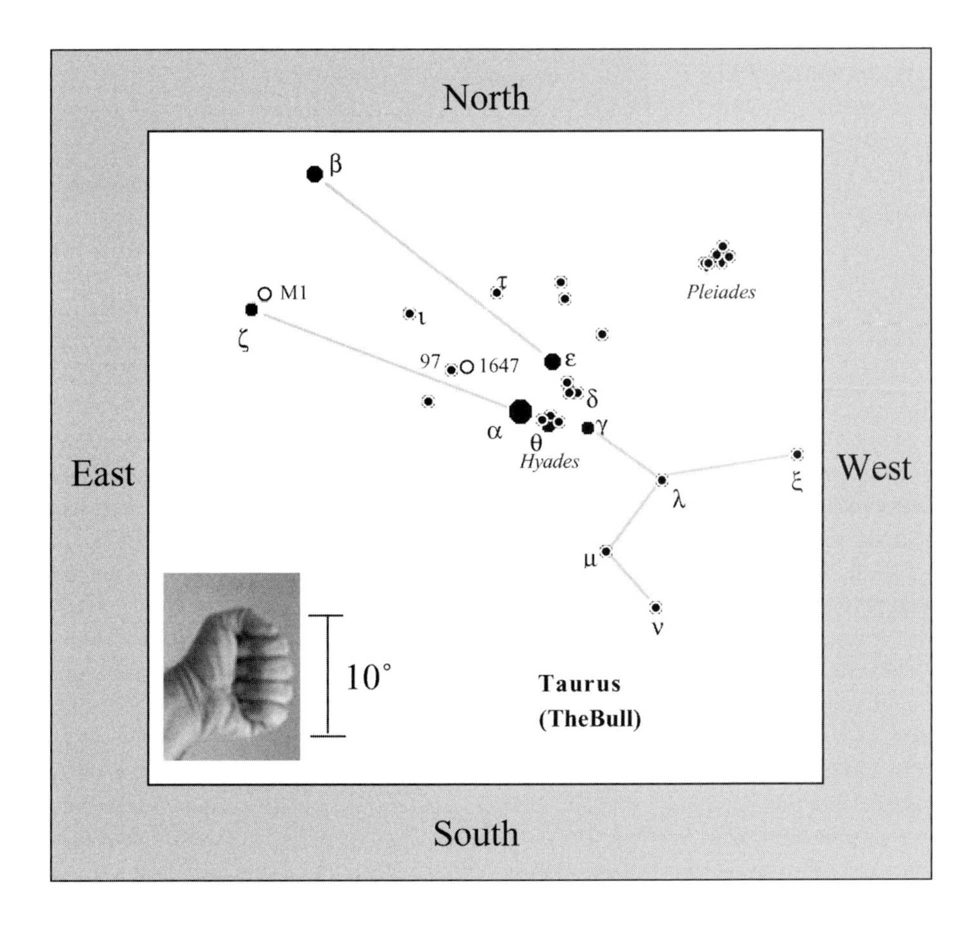

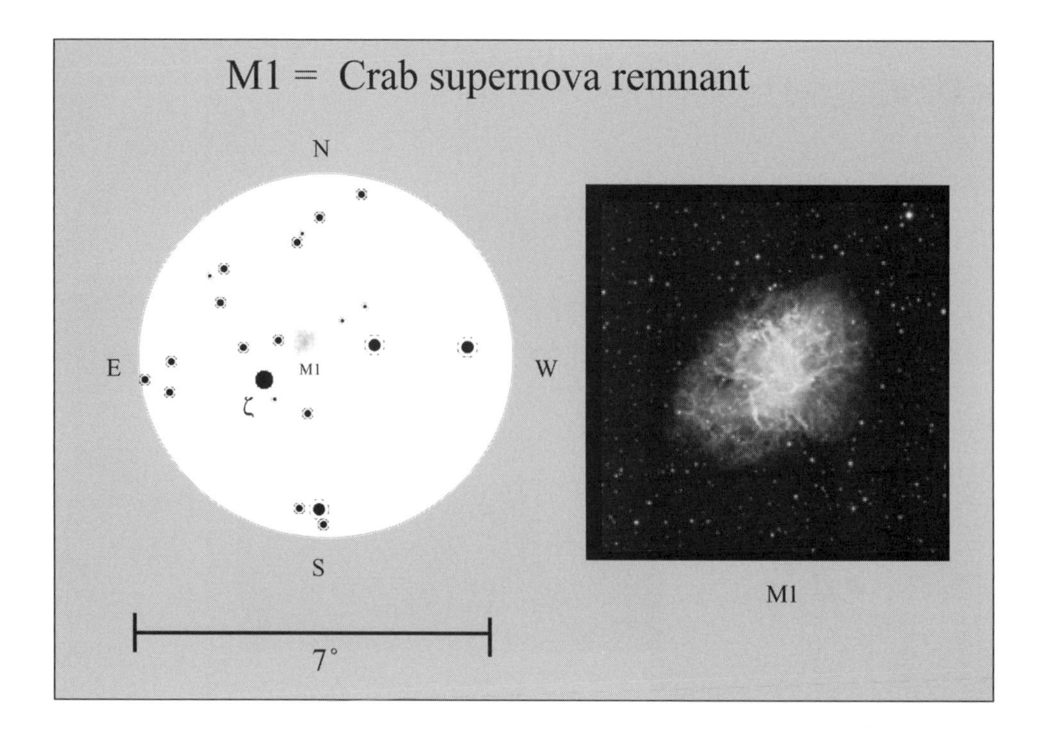

as a "guest star" in the annals of the Sung dynasty. The new star rivaled Venus in brightness and remained visible in the daytime sky for 23 days. The progenitor star of this supernova blast must have had a mass of at least 15 times our Sun and consumed its nuclear fuel in about 10 million years, before imploding, then exploding as a supernova. When Charles Messier encountered this stationary object in 1758, it enticed him to start cataloging other objects like it, "so that astronomers would not confuse these same nebulae with comets just beginning to shine." Thus began the now famous Messier catalog of nebulae and clusters of stars, which first appeared in the French almanac *Connaissance des Temps* for 1801. Messier 1's popular name is the Crab Nebula, a name honoring the long, leg-like filaments that radiate from the main body of the destroyed star; these rays fill a volume of space some 10 light years across. And though the nebula is too distant for you to appreciate, its creepy filaments are expanding some 1,120 miles each second you look.

We now come to perhaps the most beautiful object in the naked-eye and binocular heavens, M45 – the seven misty sisters known as the Pleiades, one of the sky's grandest open star clusters (see photo at right). On cold January evenings, this tangle of starlight stands high overhead in the south, just 10° (a fist) from the Hyades. No one that I know has ever spied the cluster without wondering what it is they're seeing. With any sort of study, a novice might immediately ask whether the group is the Little Dipper, because the stars do appear to form a miniature pot. On the darkest of nights, each member sparkles like a gathering of veiled brides under the moonlight. When seen with west up, they also create the singular form of a woman in white with low arms outstretched in a loving gesture.

The exact number of stars you can see with the unaided eye is variable. Depending on the conditions and clarity of the night, you might see anywhere from 6 to 16 or more. The Pleiades contains 10 stars brighter than 6th magnitude (a general limit for casual stargazing); before the invention of the telescope Kepler's tutor, Maestlin, mapped 11, and keen-eyed stargazers have recorded as many as 18. Why then are the Pleiades referred to as the Seven Sisters? The answer may simply be in the power of "seven" – history's long-enduring mystical number. Traditionally, the number refers to the seven doves that carried ambrosia to the infant Jupiter, or to the seven sisters who were placed in the heavens so that they might forget their grief over the fate of their father, Atlas. Under the darkest night skies, observers can see some of the swaddling nebulosity that still clings to this 20-million-year-young cluster. The nebula shines by reflected starlight radiating from the Pleiades' brightest members.

The classical Seven Sisters are Asterope, Alcyone, Electra, Maia, Merope, Taygeta, and Celaeno; two other stars, Atlas and Pleione, are named for their parents. One enduring (and endearing) myth considers the "missing Pleiad." Of it, Aratus, writes, "Their number seven, though the

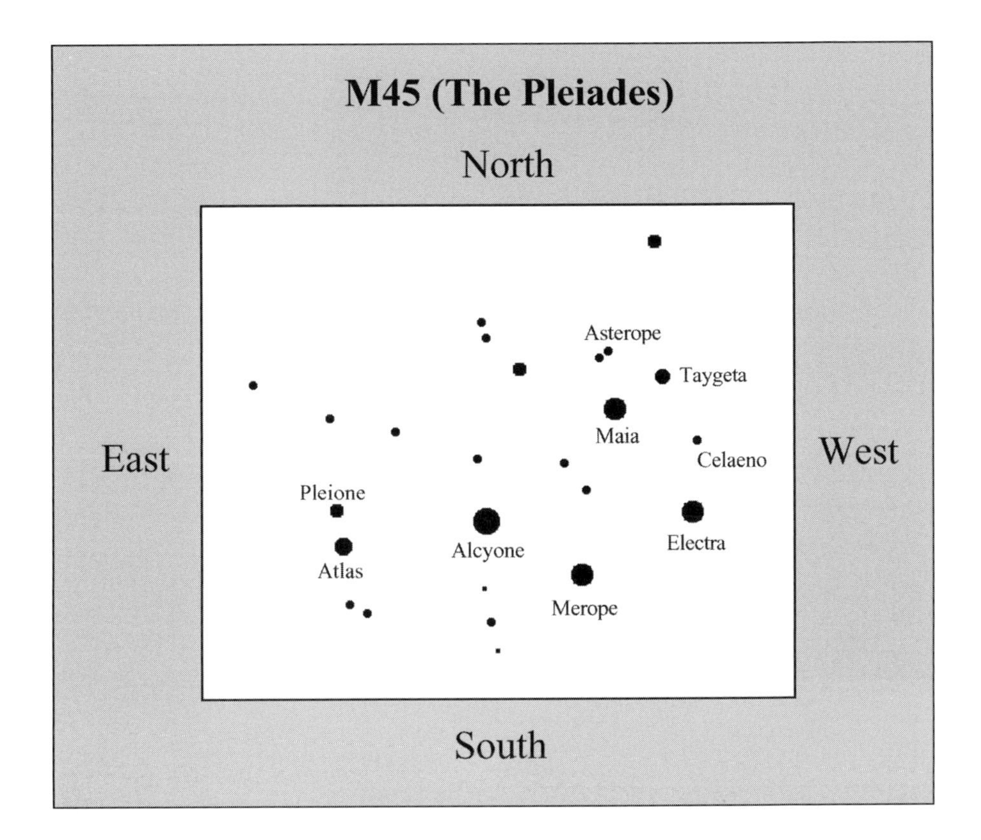

myths oft say, / And poets feign, that one has passed away. . . ." But, as I have discussed, the number of Pleiads one sees is a matter of atmospheric darkness and clarity, one's eyesight (only six Pleiads are considered "easy" to see) as well as the power of the mystical mind over the logical mind.

Early Eskimos regarded the Pleiades as a team of dogs in pursuit of a bear (the cluster really does look like a dogsled!). In the Baltic regions and ancient Russia, people saw the Pleiades as Baba Yaga, a gruesome witch who flies through the heavens on her fiery broom, scaring forest dwellers and eating children who trespass her lands; in keeping with the long tradition of the Pleiades in many cultures, Yaga is also a harvest goddess. In ancient Egypt the Pleiades were the stars of Hathor, and the Great Pyramid of Cheops may have had a chamber aligned to the cluster. The stars are the famous Makalii (little eyes) of Hawaiian legend, and may be associated with the demigod Maui, who used a mighty fishhook (the tail of Scorpius) to raise the Hawaiian islands from the sea. As the mountains rose from the depths, Maui intended to raise all the lands until all were united. But, because of the wrongful action of one of his brothers, the line snaps and the fishhook flies into the sky, where we see it today as the tail of Scorpius; and there the hook lies opposite the Pleiades (shown at right), which represents the scattered pearls of the Hawaiian islands.

During the Middle Ages, a time that marked the zenith of man's superstition, the Pleiades had a significant association with Samhain (our Halloween). On All Hallows' Eve, when the Pleiades were highest in the south at the stroke of midnight, witches celebrated their main sabbat. Fittingly, in his classic 1851 book *The House of the Seven*

Gables (Boston, MA: Ticknor, Reed, and Fields) – long a Halloween favorite Nathaniel Hawthorne tacitly alludes to the association of that haunted manse (see photo on page 114), which was built over the grave of an executed witch, with the cool, integrated light of the Seven Sisters:

> On every side the seven gables pointed sharply towards the sky, and presented the aspect of a whole sisterhood of edifices, ...

The occult side of the Seven Sisters may have a basis in fact. The eruption of the Santorini volcano in Greece in 1650 BC – one of the largest in the last 10,000 years – is said to have occurred during their midnight culmination. That eruption, believed to be the basis for the Legend of Atlantis (remember, the Pleiades were the Atlantides), erased a Minoan culture in the Santorini Island group. For these people, the midnight culmination of the Pleiades marked the end of their world. Apparently, ever since that

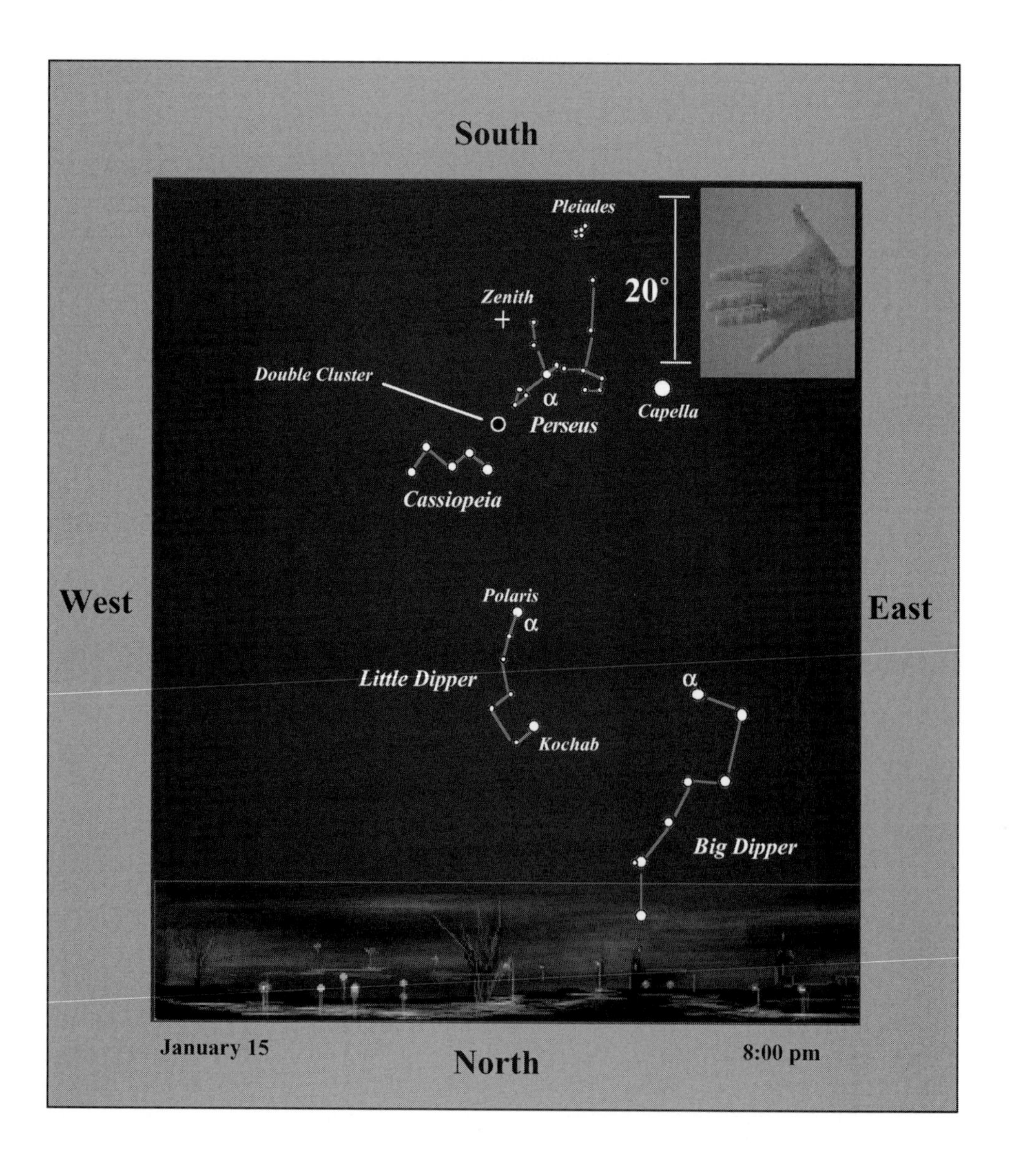
South
Pleiades
20°
Zenith
Double Cluster
α
Perseus
Capella
Cassiopeia
West
East
Polaris
α
Little Dipper
α
Kochab
Big Dipper
January 15
North
8:00 pm

event, the Pleiades have worn a Scarlet Letter in the eyes of the superstitious. Then again, the stars have been held in good standing by the Japanese, whose word "Subaru" translates into Pleiades; indeed the insignia of the Subaru car is a representation of this star cluster; it is also the name of one of the world's most powerful telescopes atop Mauna Kea in Hawaii.

Finally, compare the Hyades' appearance with that of the Pleiades, using both the naked eye and binoculars. The Hyades with their profusion of yellowish stars should look older than the Pleiades, which appear crisp and white. In fact, the Hyades are about 10 times the senior of the Pleiades.

Perseus, the Hero

If you go outside at 8:00 pm around January 15, turn to face north, then look high overhead. There you should see the unmistakable form of Perseus, the Hero (see the chart on page 114). I say unmistakable, because the constellation's core is a prominent spine of bright stars known as the Segment of Perseus, which is lodged comfortably between the M of Cassiopeia and brilliant Capella, the Alpha star of Auriga, the Charioteer. This graceful arc follows a dynamic section of Milky Way that slants across the heavens toward the west at a 45° angle. The brightest part of the Segment measures about a fist across in length and consists of the stars Eta (η), Gamma (γ), Alpha (α), Sigma (σ), Phi (φ), and Delta (δ) Persei. A fainter extension curves to the northeast from Delta and includes the stars Mu (μ), Lambda (λ), and 48 Persei. So the overall shape of the spine is a J or fishhook.

The brightest star in the Segment, and in the constellation, is Alpha (α) Persei (Mirfak). Mirfak is a spectacular supergiant star of spectral-type F. It is also a borderline Cepheid variable with a titillating color that has been described as brilliant lilac and ashy. That said, I see it as sunbeam yellow in color. Mirfak belongs to the glittering gems associated with the Alpha Persei Moving Cluster – a large scattered cluster of stars whose brightest members lie between Alpha and Delta Persei, the very heart of the famous Segment. Like the Hyades in Taurus, and the parts of the Big Dipper in Ursa Major, those in Perseus are part of a true physical system. The cluster consists of some 50 members splashed across 47 light years of space. The cluster's stars are principally luminous O- and B-type supergiants and comprise what is known as the Alpha Persei OB3 Association. The group is estimated to be about 80 million years young (about as old as the Pleiades). The stars are moving together at about 10 miles per second toward Beta Tauri. But don't start plotting its motion just yet, because it will take the group about 90,000 years to change its present position by 1°.

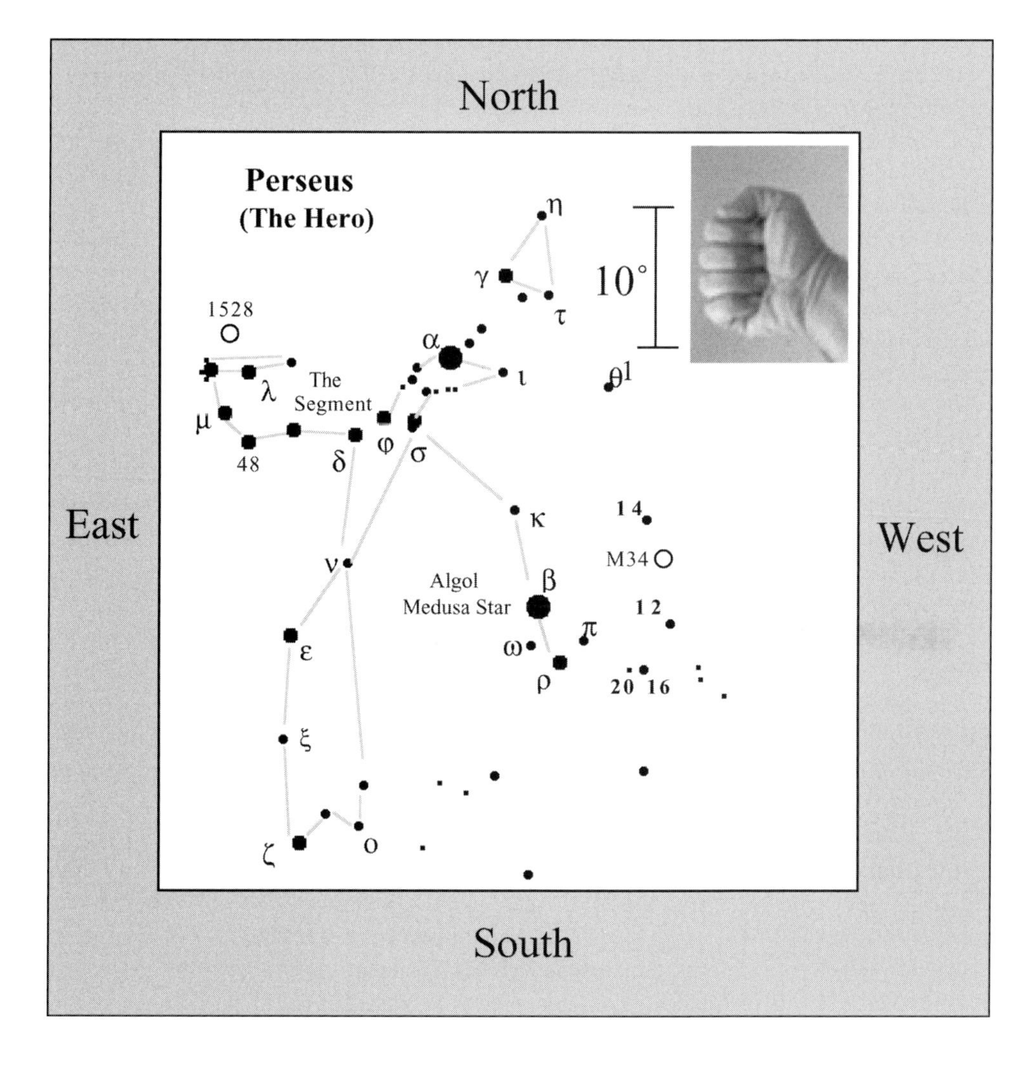

About a dozen of the cluster members lie within 2° of Alpha Persei and shine at magnitude 6 or brighter. So, as with the Pleiades, the region appears misty with direct vision and resolved with averted vision. A look at this region today in even the smallest of binoculars is one of abject wonder. The field is one of the finest in the sky, with stunning swirls and streams of starlight that, with imagination, are reminiscent of an explosion in a spaghetti factory.

In 7 × 50 binoculars, the cluster has an exquisite serpentine body, one befitting of the Sea Monster that mythical Perseus had turned to stone. Seen with southwest "up," the beast's hideous head is a conspicuous 1°-wide circlet of 6th-magnitude suns with Alpha Persei marking its gaping jaws. Sigma (σ) Persei marks the rising hump of the monster's charging body, while Delta (δ) Persei is the tip of its tail. But look closely at Sigma. What color do you see? While most of the stars in the vicinity are liquid white, some have pretty warm hues. Sigma is a smoldering fire-orange, Eta (η) is monarch orange, Tau (τ) is marigold, and Gamma (γ) is imperial yellow. To cool down, sneak another peek at Delta Persei; this ice blue B-type giant is 530 light years distant – the distance to the Alpha Persei Moving Group.

From Alpha Persei, drop your gaze almost 10° south and slightly west to Beta (β) Persei (Algol). Here is the famed Demon Star – the one that marks the position of the severed head of the gorgon Medusa (see page 91). Algol is one of the most monitored stars in the night sky. Bright and predictable, this prototype eclipsing binary star varies between magnitude 2.1 to 3.4 every 69 hours. The binary pair consists of a bright B-type star and a dimmer, but larger, K-type star, which orbit a common center of mass. The stars' orbital plane is inclined to our line of sight, so we see one star pass in front of the other, like clockwork. The drama for naked-eye and binocular observers, however, occurs only when the larger K-type star eclipses the bright companion (which it does by 79 percent), causing an obvious fading by more than a magnitude. The minimum lasts for about five hours before the bright star emerges from eclipse and we see the star gradually return to normal brightness. The whole event takes only 9.6 hours to complete the cycle. When the bright star eclipses the dimmer one, only a slight dip in brightness occurs and is not of interest to the binocular observer. You can use the chart on page 117 to monitor the brightness variations of this magnificent "winking" star; stellar magnitudes are given in parentheses.

In the same binocular field, about 5° to the northwest is the bright open star cluster M34. Shining at 6th magnitude, this loose aggregation of some 60 stars – all about 100 million years old – is visible to the unaided eye under a dark sky. The cluster equals nearly a full Moon in apparent diameter, which at a distance of 1,450 light years is equivalent to a true physical extent of about 10 light years. The cluster contains about a dozen suns brighter than 9th magnitude, all of which can be resolved in 7 × 35 binoculars, you can double that number of stars in 10 × 50 binoculars; several of these stars are white giants. The brightest star near M34 – a crisp 7th-magnitude sun to the southeast – is not a true cluster member.

If you use the chart on page 115 and swing your binoculars up to 4th-magnitude Lambda (λ) Persei, you'll find another treat: open star cluster NGC 1528. But unlike M34, NGC 1528 is twice as distant and not easily resolved in ordinary binoculars. It does form a beautiful 18′-round glow, one that is bright enough to be seen in bright

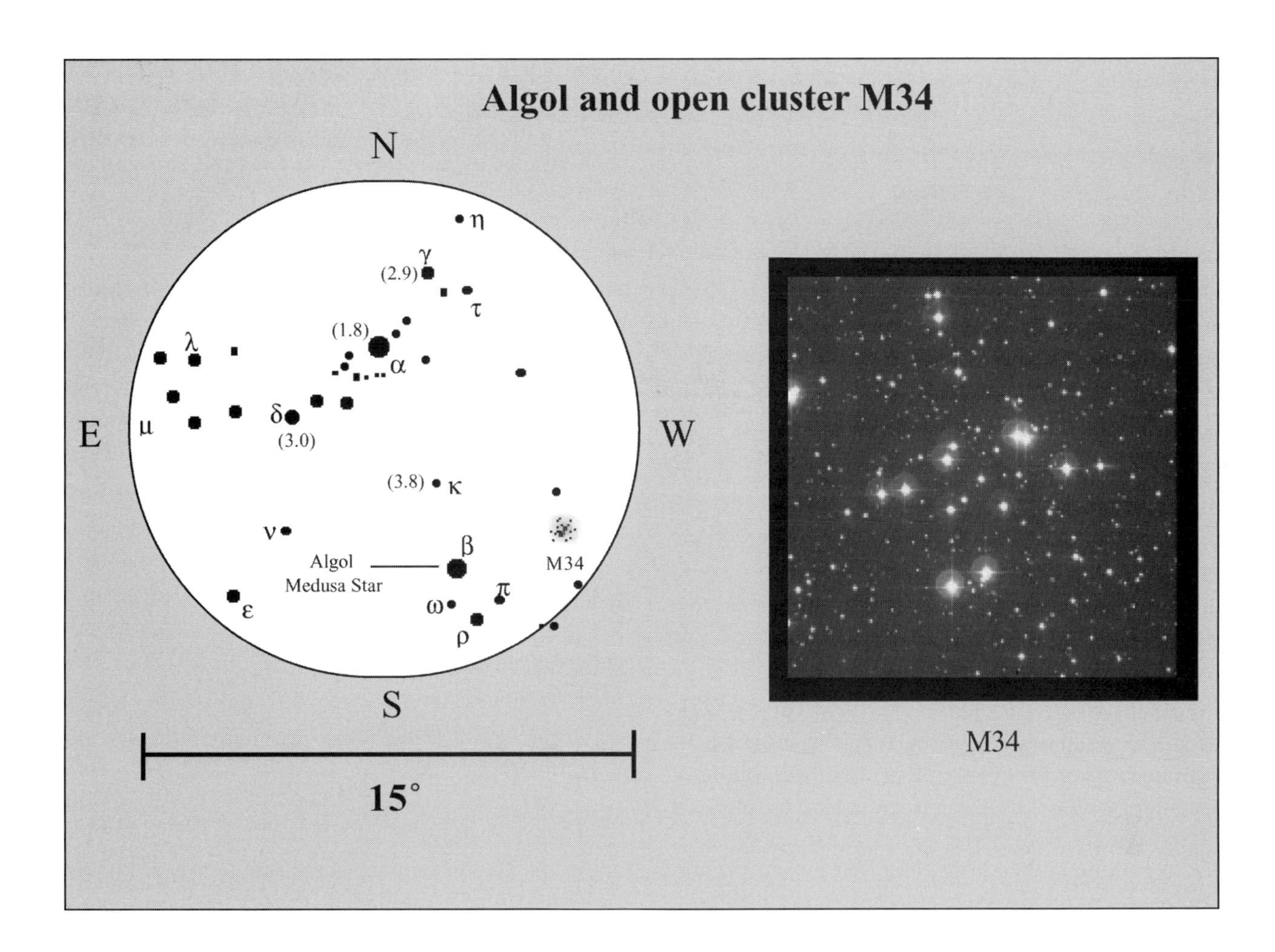

moonlight. With an estimated age of 370 million years, the cluster is three times older than M34, so its faded glory seems in line with reality.

Our tour of Perseus ends with a not-so-subtle finality, the Double Cluster of Perseus. This dynamic duo of the starry firmament is rivaled only by the Pleiades in grandeur and scope. Here are two open star clusters, each 7,300 light years distant, appearing as a 4th-magnitude knot in the gentle folds of the Milky Way. To the naked eye, these clusters are a visual punctuation mark – a fuzzy colon – just one binocular field due east of Delta (δ) Cassiopeiae in the W or M asterism; the Double Cluster announces a curious end to the great naked-eye wonders of the Milky Way – at least from mid-northern latitudes. Once the eye leaves Perseus and heads south, the fabric of the winter Milky Way becomes a bleak curtain of light – a gossamer ghost of its summer counterpart. It's as if some unseen hand has tied a knot at the position of the Double Cluster and wetted it with tears; indeed, near here, each August we see the radiance of the wonderful Perseid meteor shower – otherwise known as the tears of Saint Lawrence.

The glow from this pair of stellar islands stretches across two Moon diameters; their bright cores are separated by 25′ – nearly a full Moon diameter. So intense is the starlight that each component is visible in binoculars in deep twilight. As the sky darkens, more and more stars begin to pepper the field until the true wonder of the spectacle is revealed.

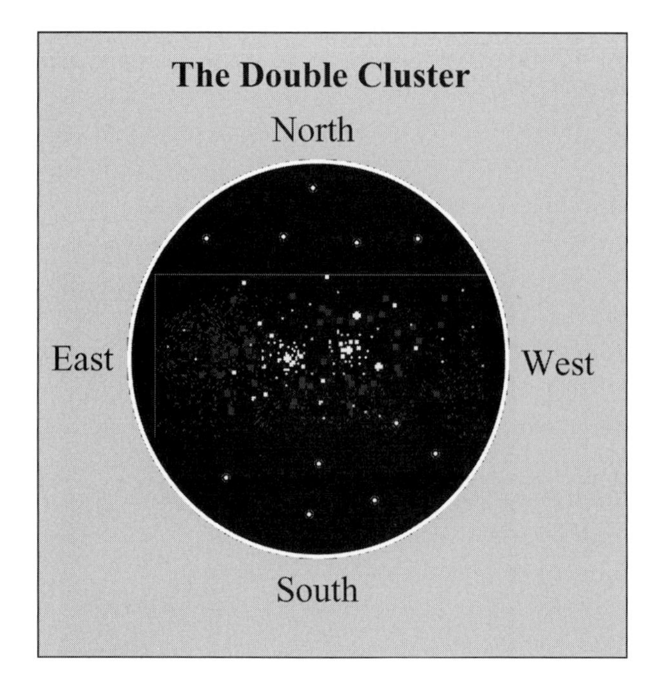

I cannot forget one childhood view of the pair under dark New Hampshire skies. A friend showed me the twin treat in a pair of zoom binoculars that brought the sparkling diamonds into sharp focus against the black velvet of the night. Each cluster had its own visual signature. The easternmost component (NGC 884) was a shattered crystal sphere caught at the moment of impact under a soft moonlit night. Its western companion (NGC 869) was a glistening diamond broach, a treasure displayed under glass in a museum of stellar wonders. The view was hypnotic, like staring into the eyes of a vampire, causing the mind to plummet into the world of the dark unknown. When I "awakened" and lowered the binoculars, in the place of the Double Cluster, I saw two puncture wounds, like pinpricks, in the pale and bloodless neck of the Milky Way.

By the way, on a more saintly note, more than a thousand years ago, when a collective of theologians set out to restructure the constellations with their own divine thoughts, Perseus became St Paul, Cassiopeia morphed into Mary Magdalene, and radiant Andromeda became a faceless sepulcher!

11 February

Down fell the red skin of the lion
Into the river at his feet.
His mighty club no longer beat
The forehead of the bull;

Henry Wadsworth Longfellow,
The Occultation of Orion (1845)

Orion, the Hunter dominates the southern horizon. He has beaten back Taurus the Bull, though the horned beast

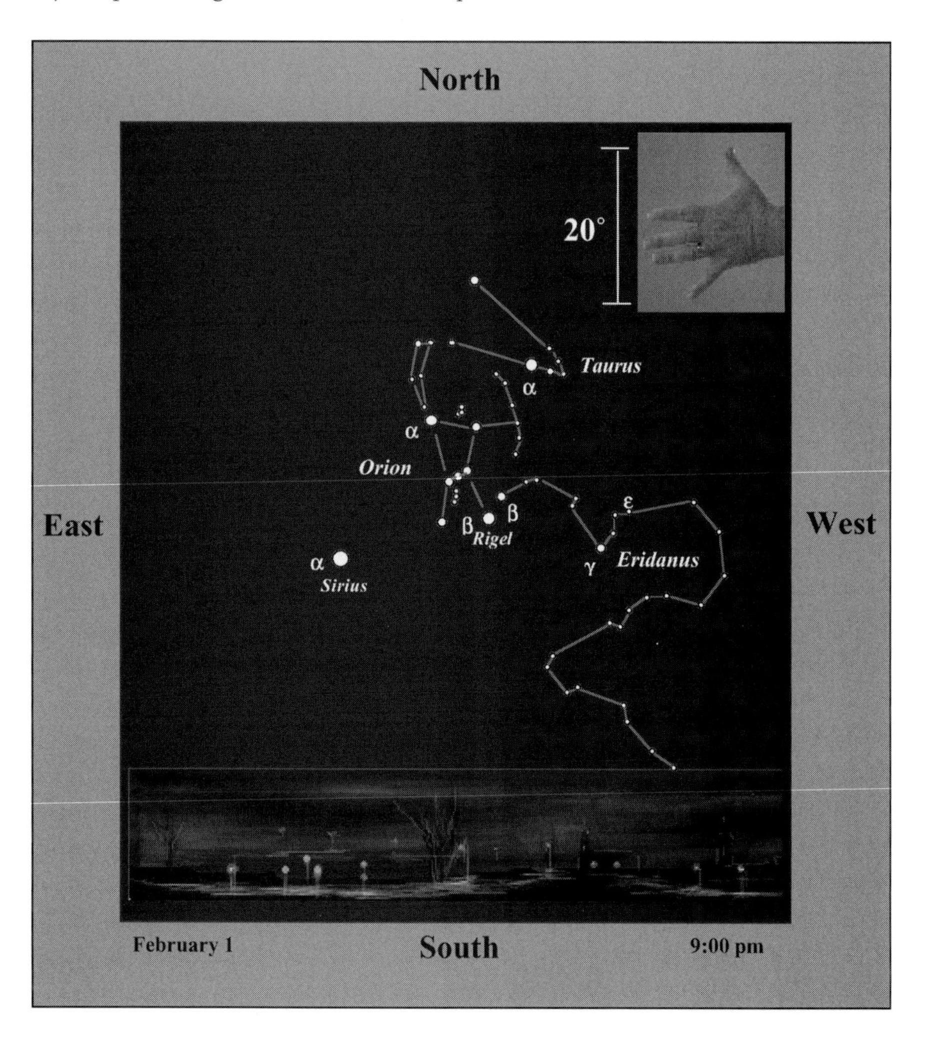

refuses to let him at the Seven Sisters. Cetus is also making a tailward plunge into the west. And one of the Fishes has already made the dive. It's as if the animal kingdom is retreating from the mighty Hunter, not daring to take their eyes off his mighty club. But before we journey to this dynamic constellation, we'll take a look at the celestial scene by his feet.

Eridanus, the River

Eridanus is a large and inconspicuous constellation whose Alpha star is below the horizon as seen from mid-northern latitudes. If you imagine Alpha as the River's headwaters, the constellation winds its way northward, bursts through the dam of the southern horizon and meanders twice before ending at the 3rd-magnitude, *A*-type giant, Beta (β) Eridani (Cursa). Cursa is easy to spot $3\frac{1}{2}°$ northwest of sparkling white Beta Orionis (Rigel). Now turn around and look north at the Big Dipper. Imagine that Cursa *may* be part of the Ursa Major Moving Group, which includes the five central stars of the Big Dipper! At a distance of 89 light years, Cursa is at the right distance to be a Group member. The only problem is Cursa's age, which is older than the 300 million years of the Group; Cursa is on its way to becoming an orange giant.

In Greek mythology, Eridanus is the river into which the reckless Phaeton "crashed and burned" the Chariot of the Sun. The Romans saw Eridanus as the River Po – the longest river in Italy. But the Greek astronomer Eratosthenes (276–194 BC) and Latin author Gaius Julius Hyginus (*c.* 64 BC–AD 17) finger it as the Nile, the only river on Earth that flows south to north, which the celestial river appears to do when rising above the horizon. In the aerial photograph of the Nile shown below, note how the vegetation bordering the river looks like a path of scorched terrain against the brilliant sands of the surrounding deserts – Phaeton's burning skid marks.

With our Northern Hemisphere bias, and without the ability to see the imaginary river Eridanus flow, we naturally assume that gravity is pulling its waters toward the horizon from the foot of Orion. (Southern Hemisphere observers, however, would see just the opposite.) No matter, start at Beta Eridani and run your binoculars "down" the river, until you get to 3rd-magnitude Gamma (γ) Eridani (Zaurak) – a pretty saffron star with a little trapezoid of dim suns next to it. At 220 light years distant, Zaurak is one of the few M-type giants visible to the unaided eye. It is 860 times brighter than our Sun and more than 130 times its diameter.

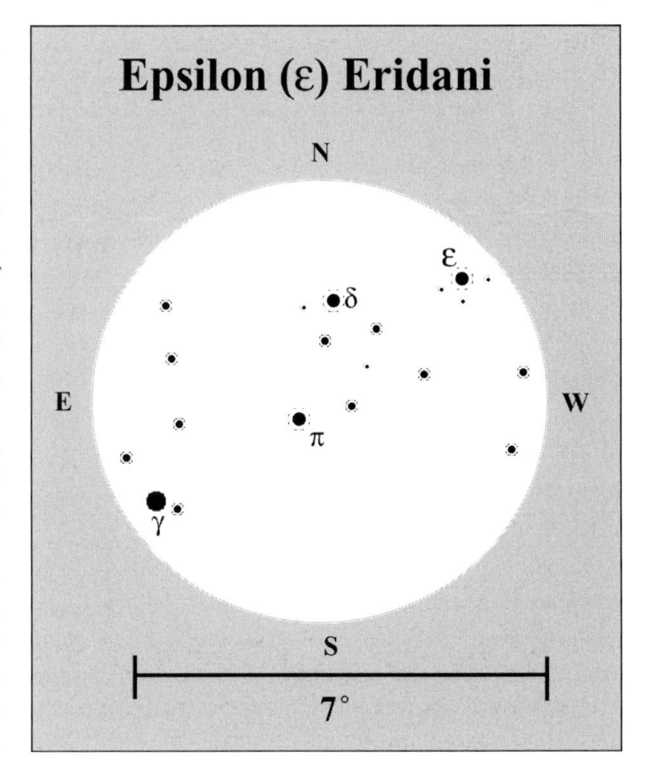

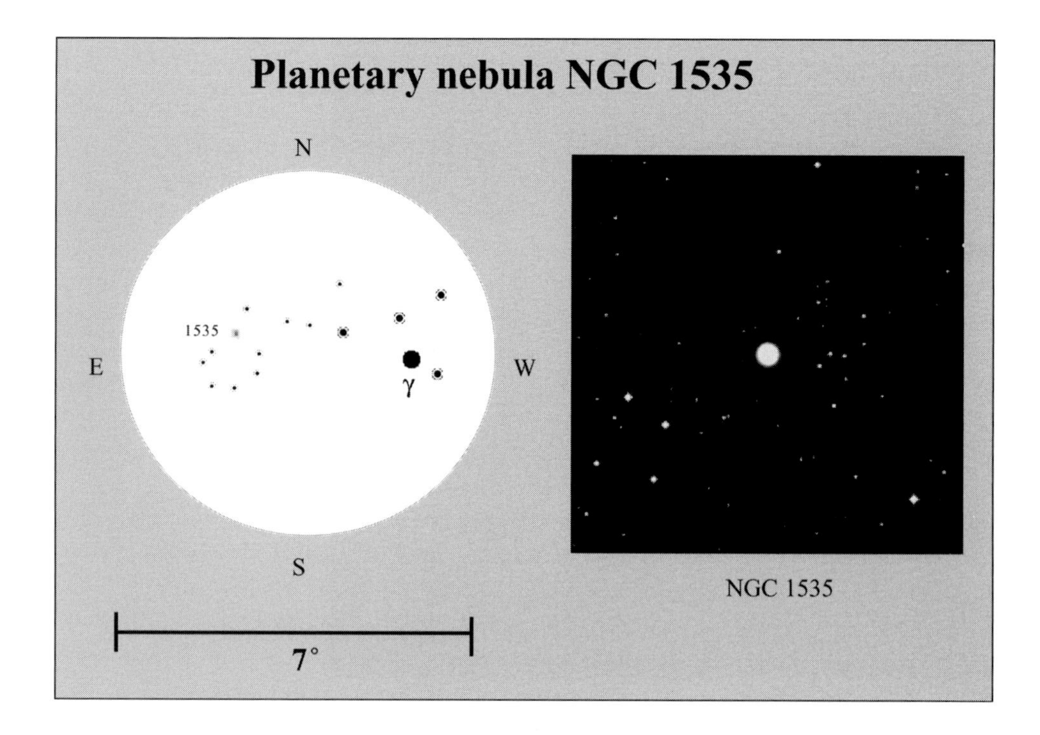

Now place Gamma Eridani in your binoculars as shown in the chart on page 119, then move your eye over to 4th-magnitude Epsilon (ε) Eridani – a deep orange dwarf only 10.5 light years distant. We have already discussed this star's involvement in Project Ozma (see page 106). But since that time, we have learned that the star not only has a cold dust disk surrounding it, but at least one, and possibly two, Jupiter-sized planets orbiting it. If true, Epsilon Eridani is the closest star to the Sun with a known solar system. And while the Jupiter-sized planets discussed here do not have life in human form, perhaps Epsilon harbors a world that does. Certainly, Epsilon's solar system will be the target of future searches for extraterrestrial intelligence.

Return your gaze to Gamma Eridani, but place it in the west side of the field. If you're under a dark sky, use the chart above and challenge yourself to see the star-like object labeled NGC 1535. This nearly stellar 9th-magnitude object requires averted vision to see, but if you catch enough photons, you'll be looking into Cleopatra's Eye – a lovely planetary nebula some 5,000 light years distant.

Orion, the Hunter (Hercules in Disguise), and Lepus, the Hare

The time has come to recognize Orion the Hunter, the greatest constellation of winter, and one of the most recognizable star patterns in the entire celestial sphere. The greatness of a constellation is measured in three ways: the brightness of its principal stars, the number of its bright stars, and the proximity of these stars to one another. Orion wins on all accounts. Its main form is a distinct hourglass 15° in length comprised of seven stellar beacons – two of which are among the most glorious stars in the night sky. Three gems of equal brightness form the pinched waist of the hourglass; formally known as Orion's Belt, these three stars stretch only 3° across the sky and cause an immediate spectacle to the unaided eye on crisp winter nights.

But before we dive into Orion's wealth of celestial treasures, use the chart on page 121 to find the little stars of Lepus, the Hare. Lepus is of interest because of its curious association with the great hunter who brandishes a mighty club in one hand, upholds a lion's hide as a shield in his other, wears a sword, and travels the sky with two hunting dogs (the star Sirius is part of the greater dog). Why would such a mighty hunter be chasing rabbits? Apparently, in ancient Sicily, hares were a menace to the island's crops. To them, the hare was a little devil.

But really, do you need someone as great as Orion to be bashing bunnies? Apparently, at least for the ancient Greeks and Romans, the true mythological significance of Lepus has been lost, though it's possible that the Hare is a metaphor for Orion being swift-footed, like a boxer, or virile.

Of course, not all cultures saw Lepus as a Hare. In ancient Persia, the bunny's body was seen as four thirsty camels near the Milky Way, which sweeps vertically past it and Orion to the east. The Chinese saw the stars as a shed. But once again, to understand the relationship between Orion, Eridanus, and Lepus, we perhaps need not look any further than to ancient Egypt, where the stars of Lepus formed the boat of the great god Osiris, who stands boldly before us as Orion.

Osiris not only presided over the life and death of his people, but also the fertile flooding of the Nile. This fact also explains the proximity of Orion (Osiris) to Eridanus (the Nile). As we have seen, the Bull was also revered by Egyptians and was part of a grand ceremony to celebrate

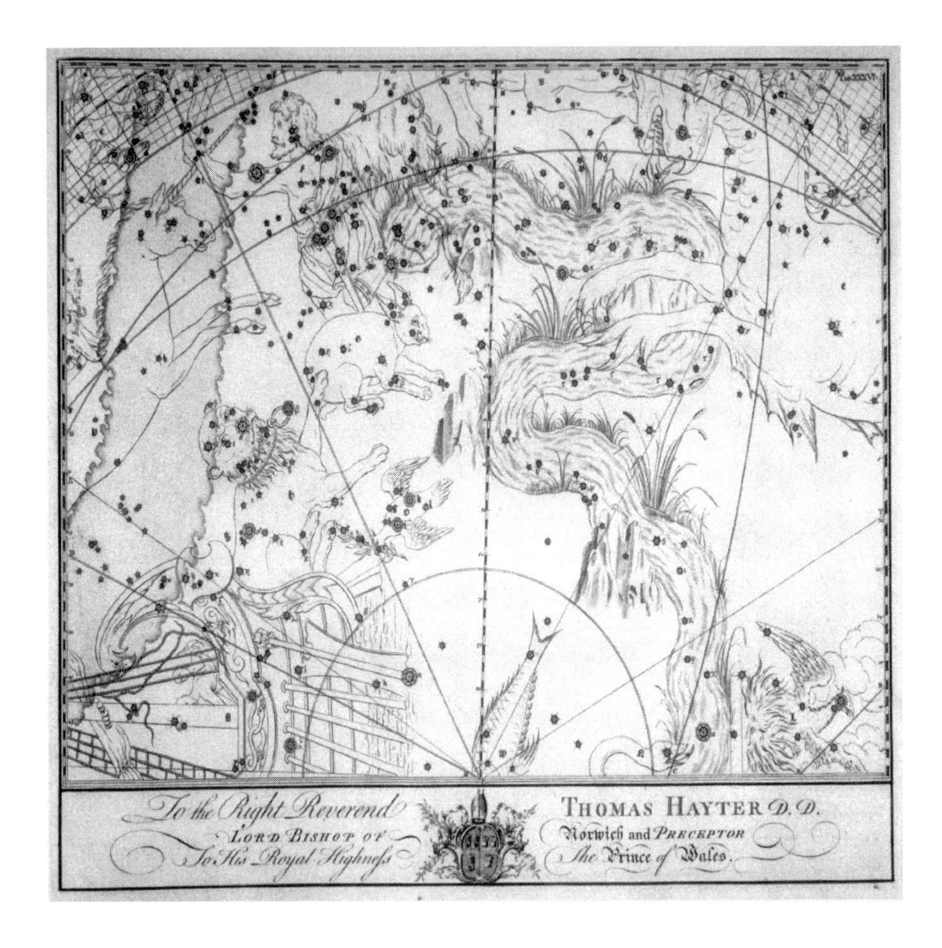
To the Right Reverend THOMAS HAYTER D.D.
LORD BISHOP OF Norwich and PRECEPTOR
To His Royal Highness the Prince of Wales.

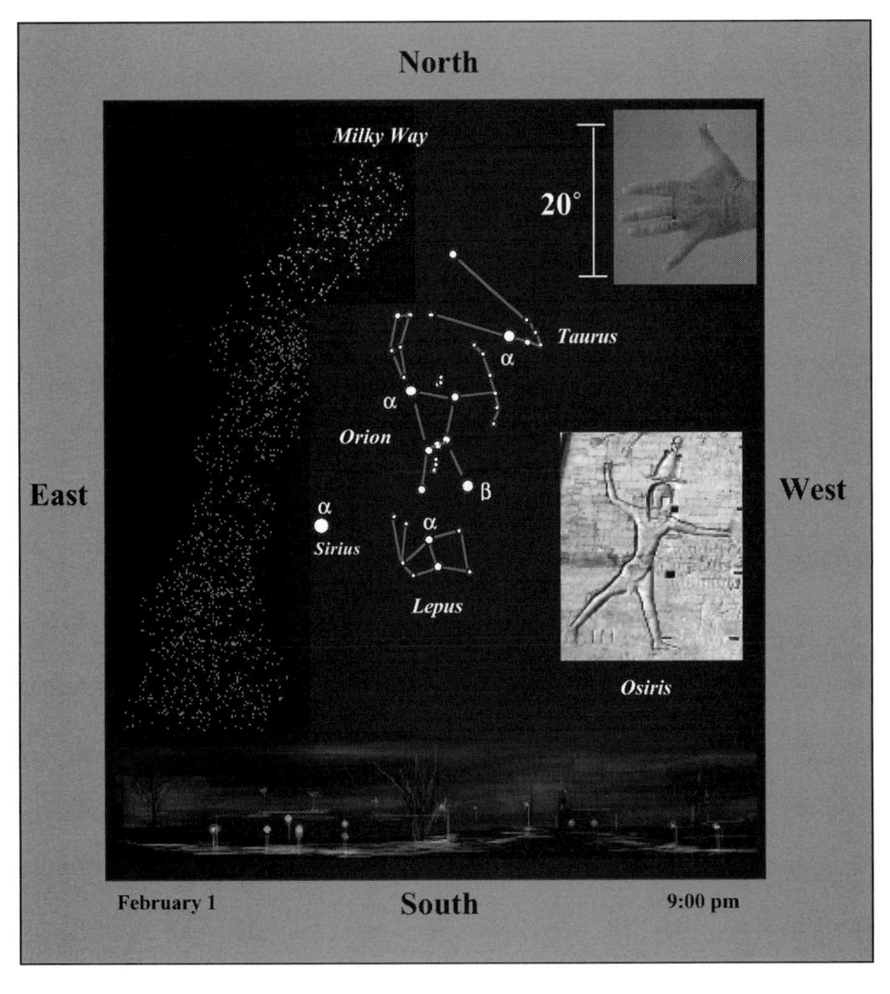
North
Milky Way
20°
Taurus
α
α
Orion
East
West
β
α
Sirius
α
Lepus
Osiris
February 1
South
9:00 pm

the rising of the Nile and the return of fertility. The Bull was placed in a vessel elaborately decorated and conveyed down the Nile to Memphis, where a grand festival was held to celebrate his birthday.

In Egyptian hieroglyphs, the Hare too is a lunar sign that symbolizes fertility; just as the Moon waxes and wanes – mirroring the cycle of death and rebirth – so too do the waters of the Nile. Osiris is sometimes portrayed with a hare's head, because each year, the great god was sacrificed to the Nile in the form of a hare to guarantee the river's annual flooding. When depicted as a hare on top of a single blue ripple, the hare becomes a spiritual symbol for the very essence of life itself.

Little Lepus is not devoid of celestial wonders. One is globular star cluster M79. Shining at about 7.5 magnitude, M79 is an easy binocular target under clear, dark skies, especially since its disk is only 6′ across. To see it, visualize a line from Alpha (α) to Beta (β) Leporis, and extend it southwest about $4\frac{1}{2}°$. There you will see a 5th-magnitude star. Messier 79 is only 30′ northeast of that star. The globular lies 43,000 light years distant and in true physical extent spans 75 light years of space.

Now tilt your binoculars back to Alpha Leporis, then over to Mu (μ). If you place Mu in the lower left of the field of view, you will be centered on R Leporis, Hind's Crimson Star. This Mira-type variable is a prime example of a carbon star – one of the reddest stars known. It has been called "a gleaming crimson jewel," but that's when it's seen through a telescope and appears as red as a vampire's lips. R Leporis varies between magnitude 7.3 and 9.8 every 420 days.

It's time to turn our attention to Orion. In classical mythology, Orion was a mighty giant, the son of Neptune. His countenance was striking and, like Christ, he could walk on water. Orion quite suspiciously wields a giant club and has a lion cloak. I say suspiciously because, as Ian Ridpath explains in his *Star Tales*, Orion apparently originated with the Sumerians and was envisioned as their great hero Gilgamesh fighting the Bull of Heaven. "Gilgamesh," Ridpath says, "was the Sumerian equivalent of Heracles [Hercules], which brings us to another puzzle. Being the greatest hero of Greek mythology, [Hercules] deserves a magnificent constellation such as this one, but in fact is assigned a much more obscure area of sky. So is Orion really [Hercules] in another guise?" Indeed, one of the labors of Hercules was to fight the Cretan Bull. Hercules carried a enormous club, and he wore the pelt of the Nemean lion. "Despite these facts," Ridpath continues, "no mythologist hints at a connection between this constellation and [Hercules]."

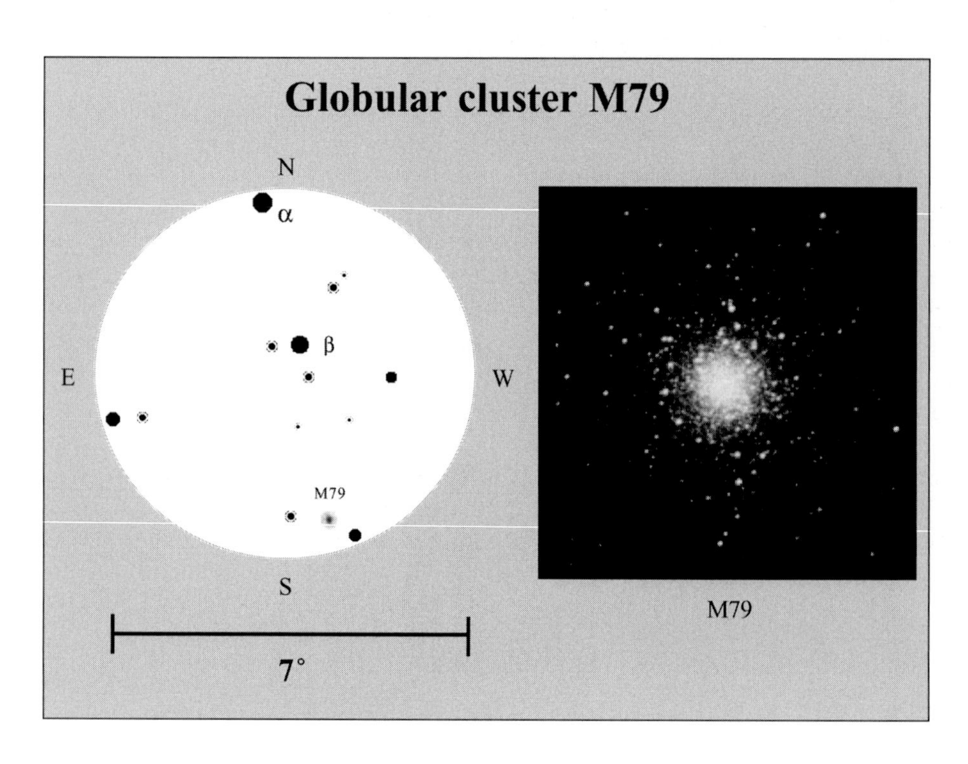

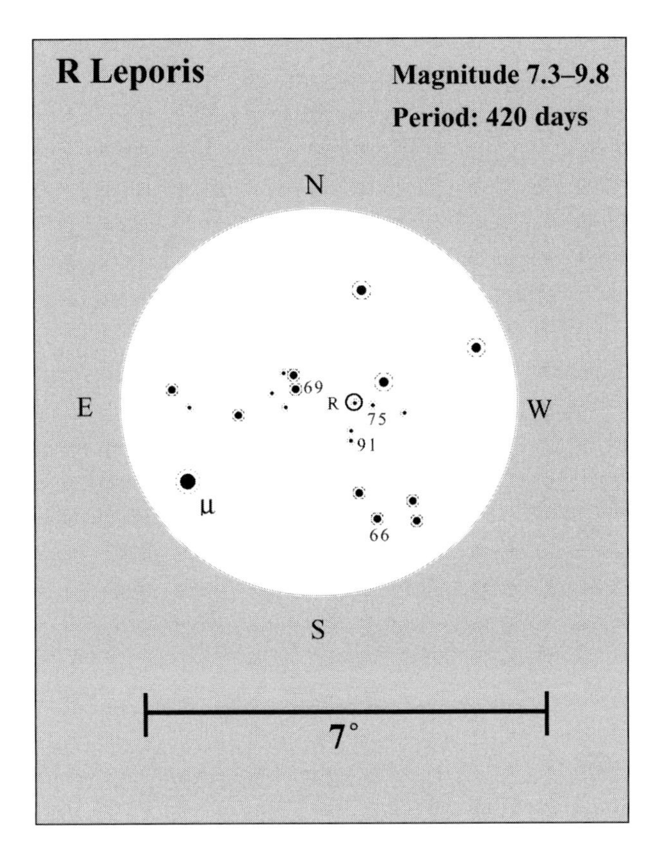

In his poem "The Occultation of Orion," Longfellow comments on the brilliance of Orion and his associated myth:

> And, slow ascending one by one,
> The kindling constellations shone.
> Begirt with many a blazing star,
> Stood the great giant Algebar,
> Orion, hunter of the beast!
> His sword hung gleaming by his side,
> And on his arm, the lion's hide . . .

Orion is not so much battling the Bull as the Bull is protecting the Pleiades from Orion, who lusted after the Seven Sisters. In one version of the tale, Orion was most attracted to Merope, daughter of King Oenopion. Failing to learn his lesson, the giant tried to take her by force. When the King learned of this treachery, he got Orion drunk, poked out his eyes, then tossed him onto the seashore. When the blinded hunter awakened, he heard the sound of a Cyclops' hammer (either that or he had a hangover) and followed the noise until he reached the forge of Vulcan. Seeing Orion's despair, the great Roman god helped Orion to the water's edge and turned him to the east to face the direction of the rising Sun. When the giant opened his eyes to the Sun's burning rays, his eyesight was immediately restored.

Still Orion did not learn his lesson, because he then tried to ravish the virgin Diana. To stop the dimwitted brute, Diana opened the Earth and unleashed Scorpius, the Scorpion, who stung the hunter in the heel, killing him instantly with its poison. Grieving over her deed, Diana had Jupiter place Orion among the constellations, so she could look upon his beautiful form without him ever laying a hand on her. As for the Scorpion, it too was placed in the sky opposite Orion. To this day, whenever Orion rises, the Scorpion sets, and vice versa; never shall the two meet again.

Alpha (α) Orionis (Betelgeuse) is not the constellation's brightest star – that honor goes to Beta (β) Orionis (Rigel) – but it is a remarkable star. Like Antares, the Scorpion's red pulsing heart, Betelgeuse is a 1st-magnitude M-type supergiant. Its light varies from 0.2 to 1.5 over multiple periods between roughly half a year and six years. When most luminous, it shines 100,000 times more brightly than our Sun; at minimum light, it is still 40,000 times more luminous than our star. That's why, despite Betelgeuse's great distance (some 425 light years), it appears so obvious in our night sky, burning with an intense sunburst orange color. As with all red giants, the star is nearing the end of its life and is probably in the process of fusing helium into carbon and oxygen in its core. One day, like Antares, Betelgeuse will extinguish itself in a supernova explosion. And like the Crab supernova event witnessed in AD 1054 (page 111), the blast will be bright enough to be seen in the daylight and intense enough to cast dim shadows at night.

While Betelgeuse shows us a red supergiant on the verge of extinction, its bright counterpart, Rigel, shows us the same fate at the other end of the spectrum. Lying at a distance of 775 light years, Rigel, like Betelgeuse, is in the process of dying and is expected to go supernova. Unlike Betelgeuse, however, Rigel was born together with many of the hot blue stars, including those that comprise Orion's Belt and Sword.

If you center your binoculars on the three 2nd-magnitude stars that comprise Orion's Belt (known to the Arabians of old as the String of Pearls), you'll see several lovely arrangements of little stars threading their way across, or looping around, each member of the Belt (see the charts on page 124). From west to east, the Belt stars are the OB-type giant Delta (δ) Orionis (Mintaka), B-type supergiant Epsilon (ε) Orionis (Alnilam), and O-type supergiant Zeta (ζ) Orionis (Alnitak) – the brightest O-type star in the sky. Again, these stars lie in the same region of space as Rigel and are 915, 1,340 and 800 light years distant, respectively. All are destined to die in a fiery supernova explosion at some unknown date in the future. Orion is like a ticking time bomb, so the Hourglass asterism is a decent metaphor, because for many of Orion's stars, time is running out.

But not all of Orion's stars are dying. In fact, many are in the process of being born. So Orion is a yin–yang constellation. If you are under a dark sky, look at Alnitak with keen averted vision, or better yet, try to cover the star with the edge of a distant object, like a roof or tree limb. Just 15′ east–northeast is the bright but elusive emission nebula NGC 2024, a pale, sepulchral glow that could easily be mistaken for an optical reflection or the ghost of Alnitak. The nebula NGC 2024 is an active star-forming region

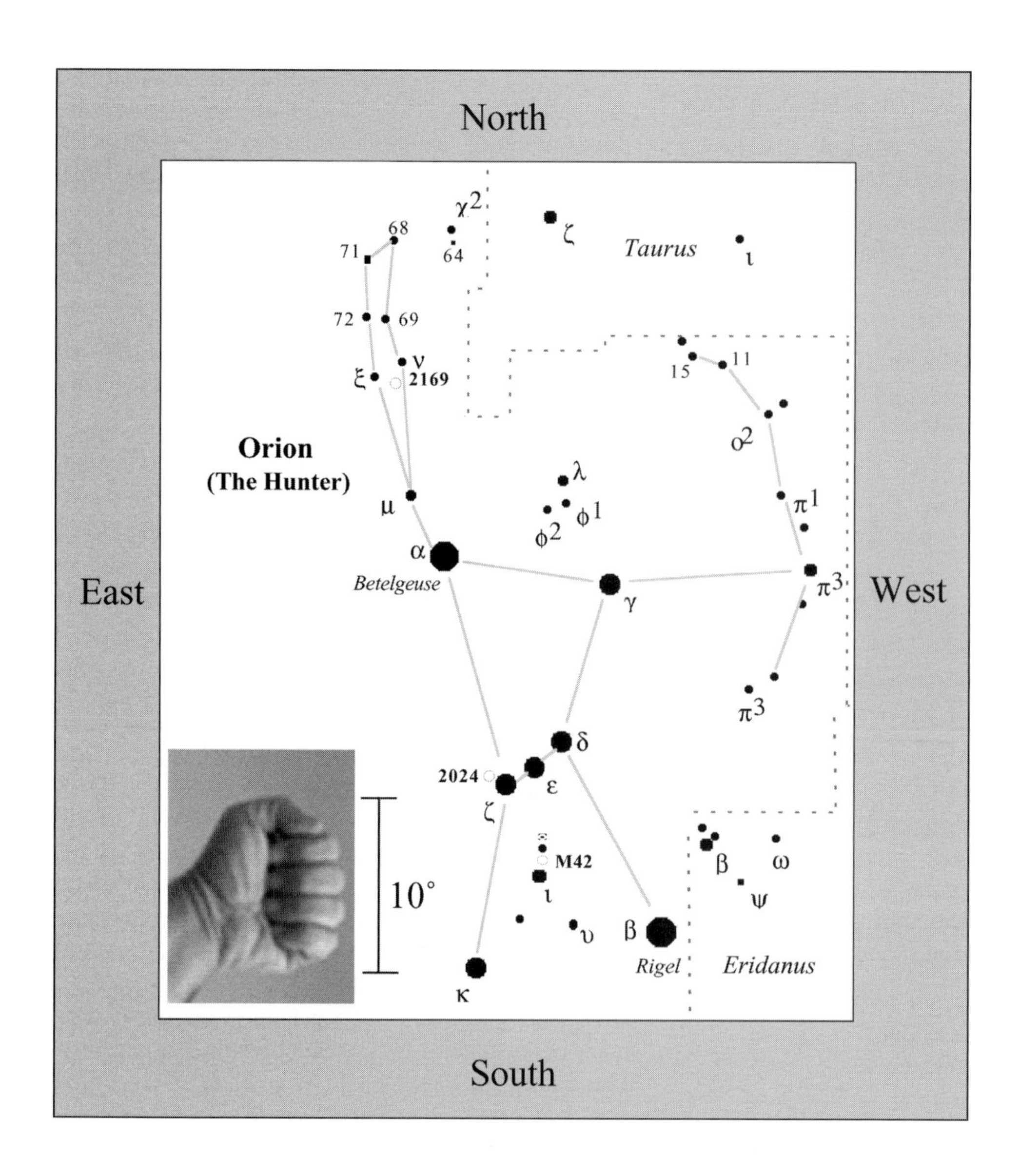
North
South
East
West
Orion
(The Hunter)
Taurus
Eridanus
Betelgeuse
Rigel
2169
2024
M42
10°

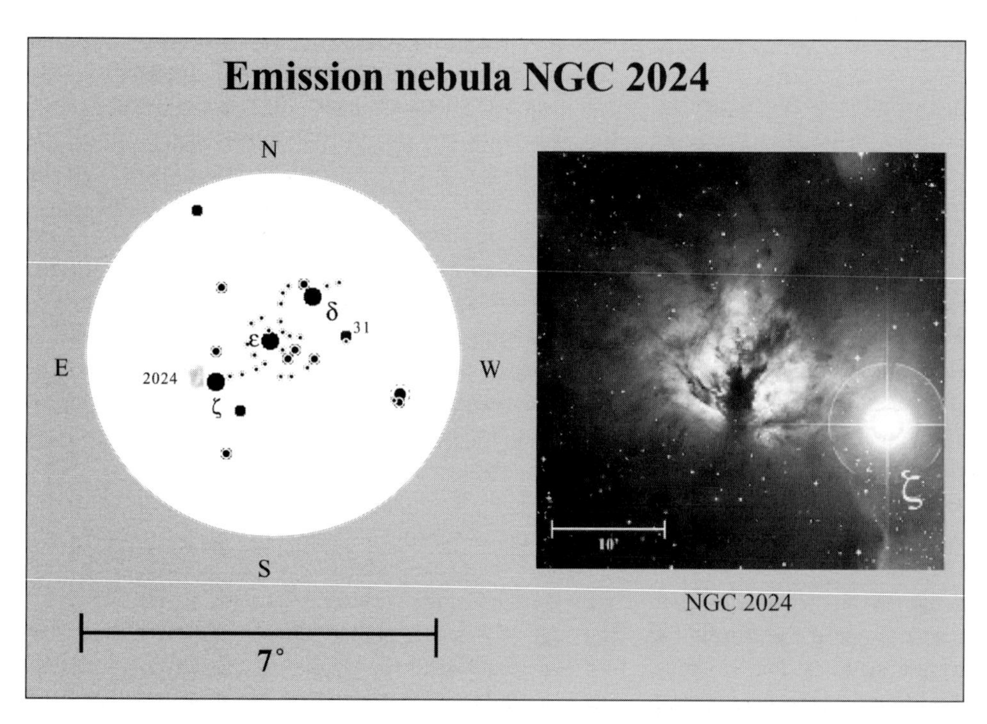
Emission nebula NGC 2024
N
S
E
W
2024
31
7°
10'
NGC 2024

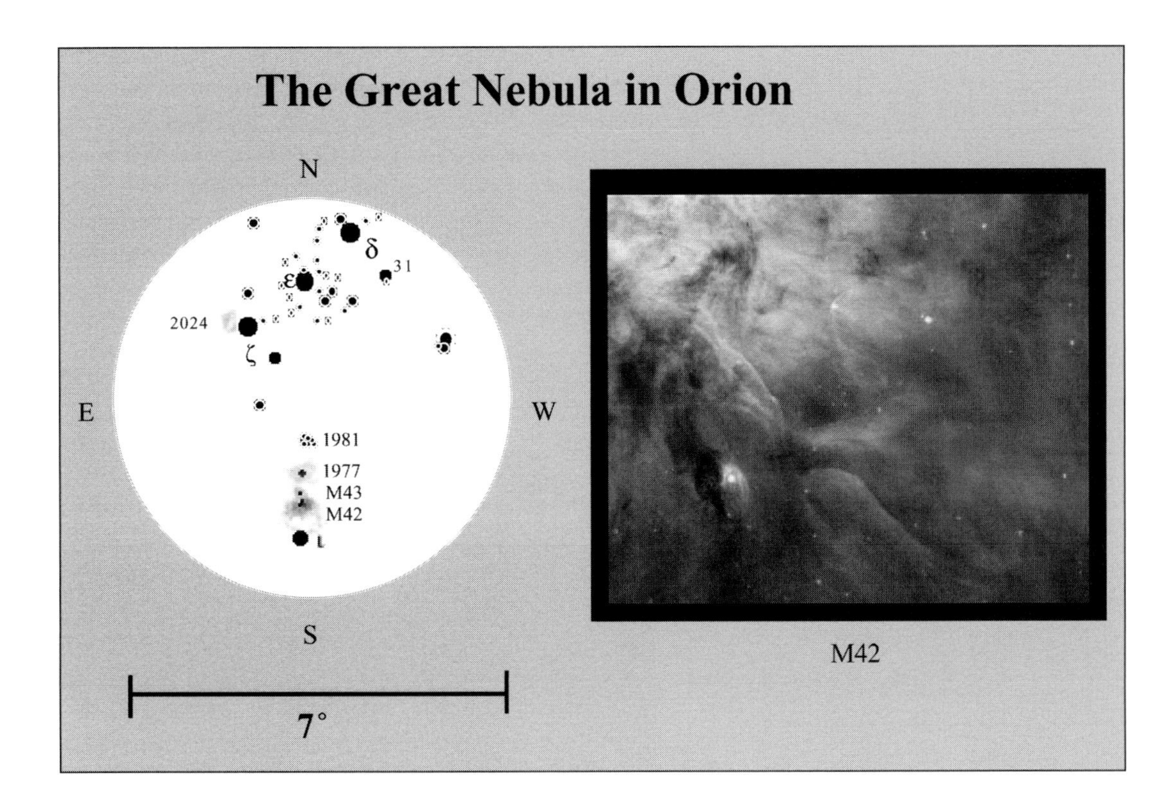

some 1,300 light years distant. Energetic photons from young embedded stars strip the surrounding hydrogen atoms of their electrons, causing the nebula to glow. And though you cannot see it, NGC 2024 contains a cluster of stars 300,000 years young.

Hanging obliquely from Orion's Belt is his Sword – a short row of three naked-eye stars, the middle one of which, if you look at it with averted vision should appear fuzzy – like a puff of fairy dust under the moonlight. That fuzziness is Theta (θ) Orionis or M42, the Great Orion Nebula – an enormous cloud of fluorescent gas 40 light years in diameter and some 1,500 light years distant. Few sights in the night sky excite beginners more than being able to detect this, the most glorious of all nebulae, without optical aid. Its beauty and glory is justly magnified in binoculars and telescopes, which also reveal its little companion nebula, M43 kissing it to the north, and the famous Trapezium star cluster at its core. The Trapezium began shining only about a million years ago and there are about a thousand unseen members hiding in its dense cloudscapes. The Hubble Space Telescope image of its core (see above) shows tumultuous swirls of churning gas, 40-billion-mile-long "comets" of dust and gas, whose comas enshroud newborn stars, and dark protoplanetary disks silhouetted against the nebula's nascent glow. Here are all the ingredients for solar and planetary creation. The Orion Nebula is, in fact, a stellar nursery.

Now look a little above the starlike speck of M43, about only $\frac{1}{2}^\circ$ north where you'll see (under a dark sky) the breath-like glow of NGC 1977, which is also part of the Orion Nebula complex. Look for a close pair of 5th-magnitude stars inside the nebula's brightest section; the two 5th-magnitude stars look like car headlights emerging from a fog. Some observers have reported seeing the nebula with the unaided eye, claiming it was "easy." See if you can detect it too.

Finally, note that Orion's Sword actually has four naked-eye stars. It's just that the lower three are the more obvious with a quick glance. But if you turn your binoculars to that fourth, northernmost star in Orion's Sword, you'll find that it's actually a 4th-magnitude open star cluster. Known as NGC 1981, this loose gathering of suns is spread across an area of sky comparable to that of the full Moon. Eight of these scintillating jewels can be seen quite comfortably with 7 × 50 binoculars: look for a tight arc of three 6th-magnitude suns (oriented north–south) set between a solitary 7th-magnitude star to the east and four roughly 7th-magnitude suns, in a Y-shaped pattern, to the west. Before leaving this region, be sure to also check it for several pretty pairings of double stars, especially around bright Iota (ι).

Raise your binoculars to the three stars marking Orion's head: 3rd-magnitude Lambda (λ) Orionis and its two, 4th-magnitude attendants: Phi^1 (ϕ^1), and Phi^2 (ϕ^2) Orionis. These stars form a distinct isosceles triangle midway between, and a little north of the Hunter's shoulders: Alpha (α) Orionis (Betelgeuse) and Gamma (γ) Orionis (Bellatrix). To the naked eye, these stars, tightly knit, might appear nebulous at first. But through your binoculars they are an attractive grouping of suns, with a pretty chain of a half-dozen 5th- to 6th-magnitude suns running between Lambda and Phi^1. Rays of fainter stars stream eastward from this line like the fins of a robinfish. Indeed, Lambda and Phi^1 are part of a true cluster of some 20 stars spanning an area of sky of about 70′ in apparent diameter. Known as Collinder 69, the cluster lies 1,600 light years distant and has been known since historical times as a "nebula."

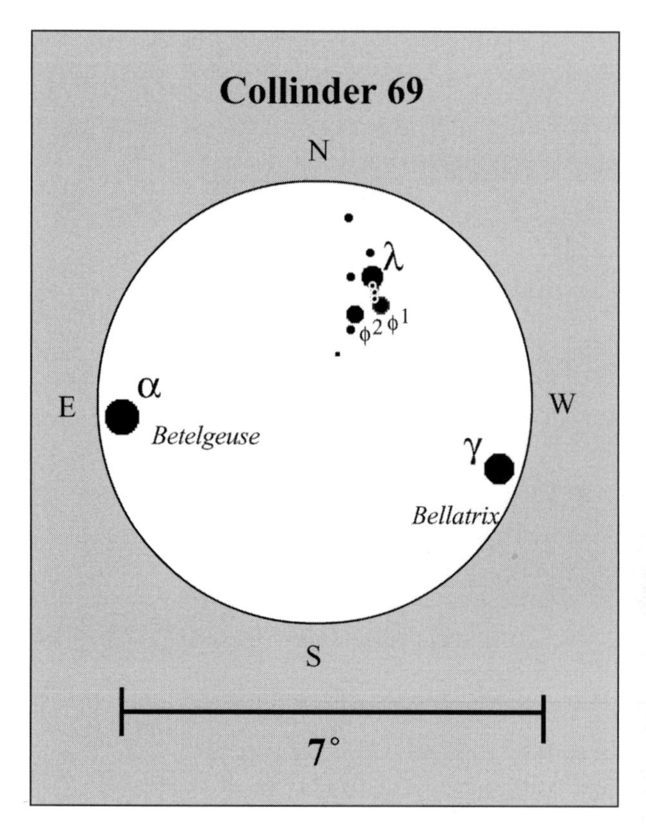

If you return to the Belt stars and position them in the field of view as shown below, you can find 6th-magnitude HR 1988, a Sun-type star 138 light years distant with one, and maybe two, planets. The more certain planet is nearly as massive as Jupiter and orbits the star a mere 12 million miles away; a year on this planet is only 14 days long! The second planet takes about five years to complete an orbit. But it is 13 times as massive as Jupiter and may, in fact, be a brown dwarf star.

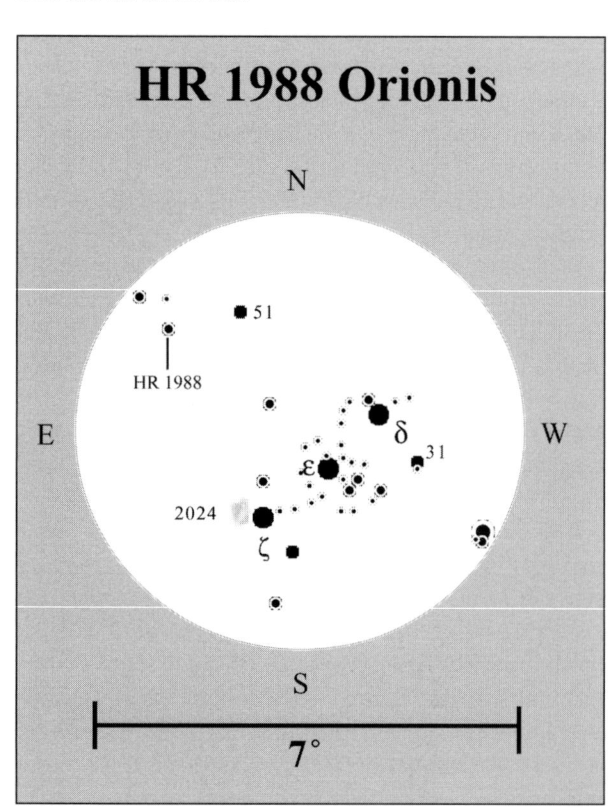

We end our tour of Orion with a little glitter. Use the chart on page 124 to look for the 6th-magnitude open star cluster NGC 2169, which is easy to find. Just raise your binoculars about 35′ southwest of the midpoint between Xi (ξ) and Nu (ν) Orionis. You should immediately see a tight double star centered in a fine mist. This 6′-wide mist is the unresolved light of the 15-million-year-old cluster, whose light has traveled 3,000 light years to reach your eyes. The cluster has a wonderful nickname, the Shopping Cart Cluster, whose form is apparent in the photograph below. NGC 2169 lies very close to the middle radiant of the Orionid meteor-shower, which occurs around October 18; these meteors are debris left behind by periodic Comet Halley.

Auriga the Charioteer

Auriga is another beautiful constellation whose brilliant Alpha (α) star (Capella) visually "sings" from the high heavens. Go outside at 7:30 pm around February 20, and you will see this glorious saffron star shining directly overhead at zero magnitude. The star is about two fully outstretched hands directly above Rigel and a little more than one hand north and slightly east of Alpha Tauri (Aldebaran). It is the brightest star in a 15°-wide pentagon of stars that strikes the eye. Capella is immediately obvious, not only because of its brightness and tantalizing hue, but also because of its proximity to the Kids – three roughly 3rd-magnitude stars forming a tight acute triangle 5° to the southwest.

Actually Capella and the Kids are supposed to be a mother goat (Capella) and her kid(s) resting in the protective arms of a man whose identity remains unclear. Ridpath says that the most popular interpretation is that he is Erichthonius, a legendary King of Athens who became the first person to harness horses to a chariot; Jupiter admired

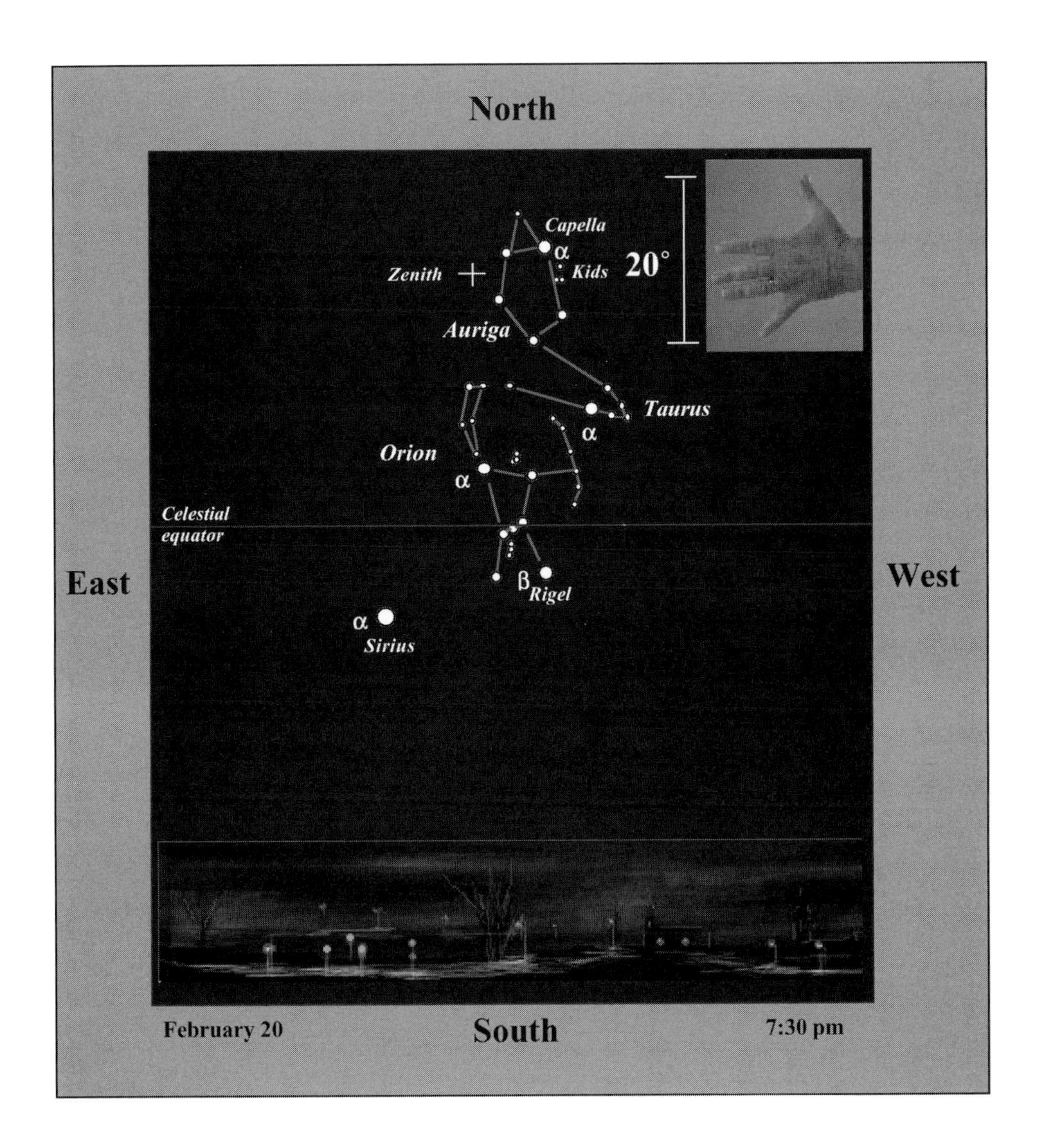

the ingenuity of Erichthonius so much that he placed the King in the sky.

But Auriga was patterned long before the Greeks. Its earliest depiction may be that of Christ as the Good Shepherd. In the time of Christ, Capella would have been seated high on the throne of heaven during the night around the time of Christ's ascension into heaven (Easter). According to the *Catholic Encyclopedia*, one of the few Christian symbols dating from the first century is that of the Good Shepherd carrying on his shoulders a lamb or a sheep, with two other sheep at his side. This interpretation, the Encyclopedia says, is in harmony with an ancient liturgical prayer for the dead: "We pray God . . . to be merciful to him in judgment, having redeemed him by His death, freed him from sin, and reconciled him with the Father. May He be to him the Good Shepherd and carry him on His shoulders [to the fold]." The association of Auriga with Christ is magnified when you consider its stately position in the pentagram of Auriga; early Christians attributed the pentagram to the "five wounds of Christ."

Although you cannot resolve the stars in your binoculars, Capella is actually two yellow giants orbiting one another only 60 million miles apart. One companion is 50 times more luminous than our Sun, while the other is nearly 70 percent brighter. Now compare the glow of Capella in your binoculars to that of 2nd-magnitude Beta (β) Aurigae (Menlalinan) – one binocular field to the east and twice as distant. It has a crisp lemon–lime tint, which is nicely offset by the golden glow of Pi (π) Aurigae immediately to its north. The color of Beta Aurigae is curious because the star is an *A*-type dwarf with a surface temperature of 9,200 Kelvin. Like Capella it is most likely two twin stars in a tight gravitational embrace.

If you want to jar your senses, place both Capella and the Kids in the same field of view. Besides being a beautiful sight, one worthy of a few moments of poetic thought, all these stars lie at greatly different distances. Capella is the closest at 42 light years distant. Eta (η) Aurigae at 220 light years. Zeta (ζ) is 850 light years away, while yellow Epsilon (ε) Aurigae (the brightest Kid) is a rare F-type supergiant 2,000 light years distant! Like Algol, the Demon Star in Perseus, Epsilon Aurigae is an eclipsing binary star that has a noticeable dip of 1 magnitude. But unlike Algol, the eclipse does not occur over a period of a few days but every 27 years. And the eclipse lasts for two years! University of Illinois astronomer James Kaler notes that the prevailing model for this system is that Epsilon is in mutual orbit with a star that is surrounded by a thick

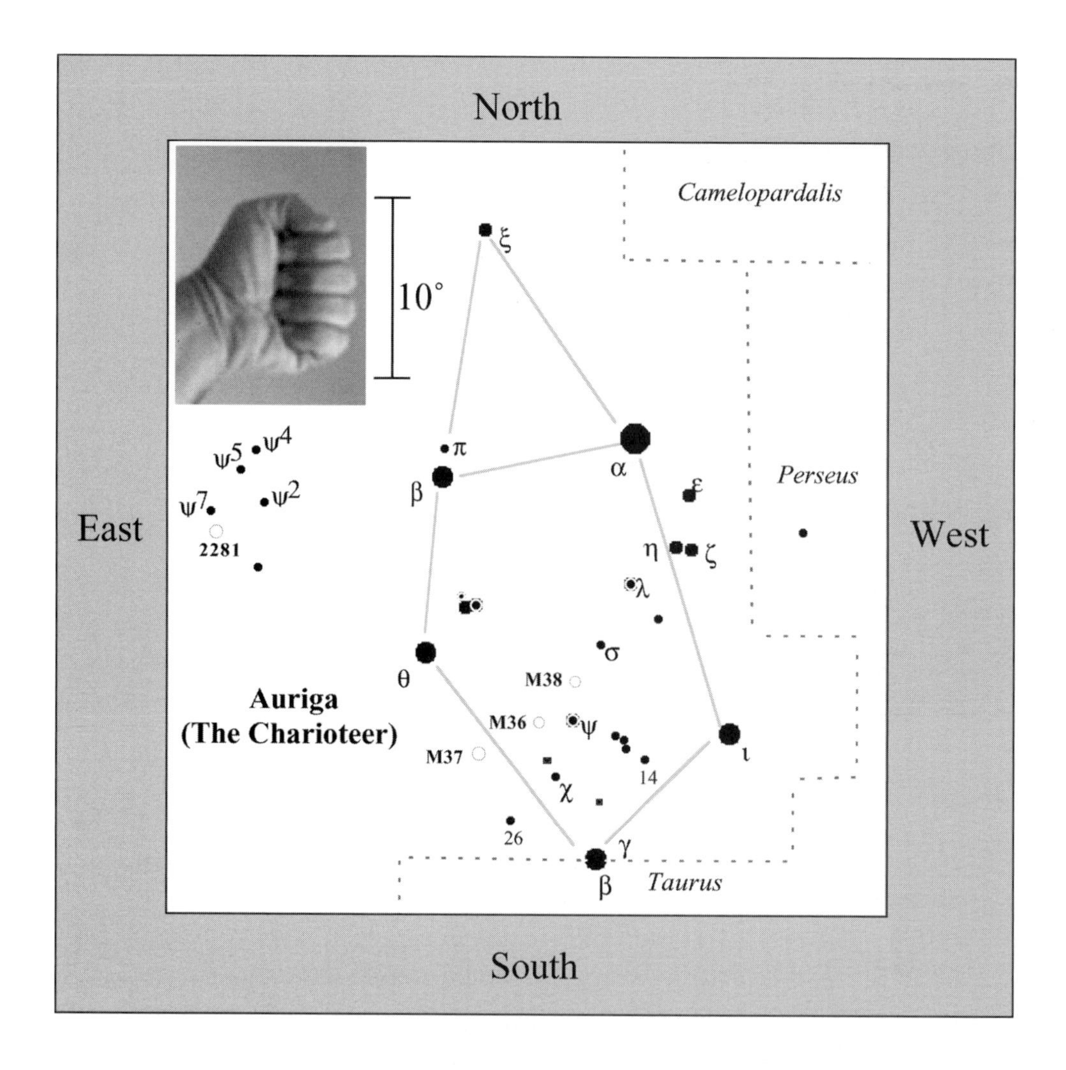

North
South
East
West
10°
Camelopardalis
Perseus
Taurus
Auriga
(The Charioteer)
ξ
π
α
β
ε
η
ζ
λ
σ
θ
ι
γ
ψ
χ
ψ5
ψ4
ψ7
ψ2
2281
M38
M36
M37
14
26

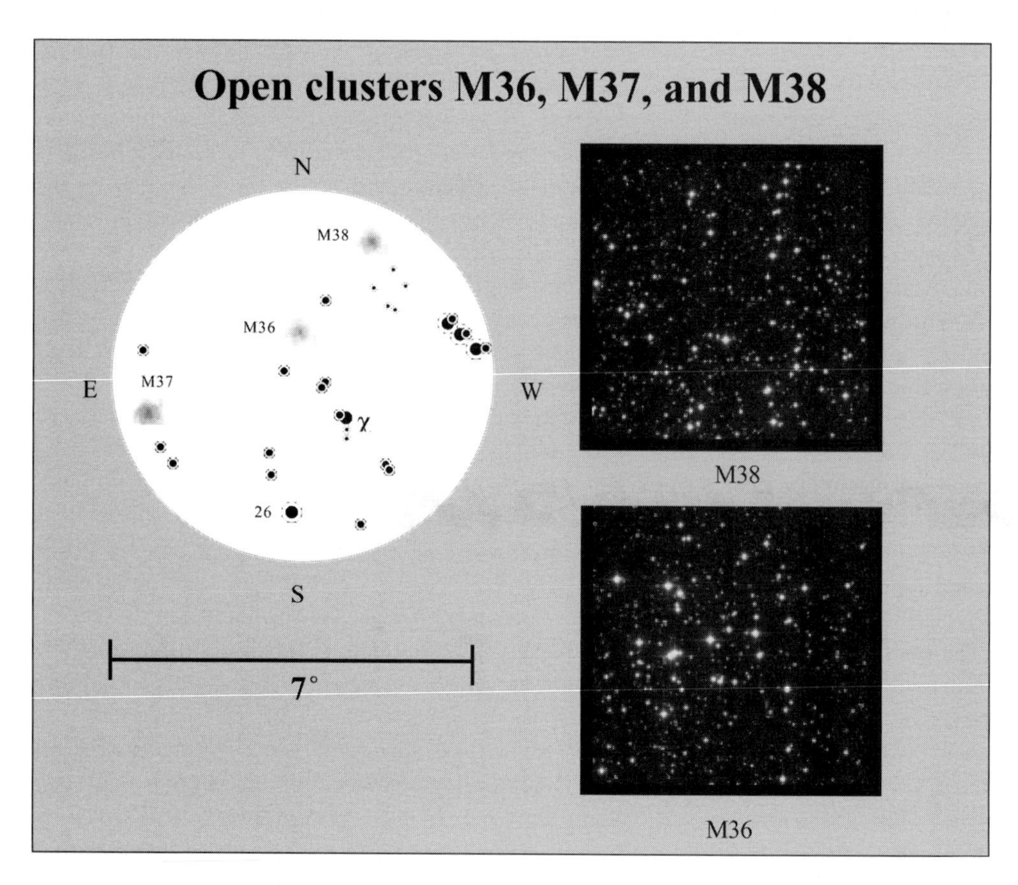

Open clusters M36, M37, and M38
N
S
E
W
M38
M36
M37
χ
26
7°
M38
M36

ring of obscuring dust set nearly edge on. The next eclipse will be in 2009–11.

The Milky Way pours through the constellation from northwest to the southeast. Sweep your binoculars around the pentagram's interior because it's a visual feast. The most plentiful harvest is one binocular field north of Gamma (γ) Aurigae (better known as Beta Tauri – a star with dual citizenship) where you'll find the bright (6th-magnitude) open star clusters M36 and Messier 38. Both clusters lie about 4,000 light years distant, but while M36 is some 20 to 30 million years young, M38 is ten times older. Messier 38 is about as old as yet another bright open star cluster, M37, which is about 6° to the southeast. All three clusters fit in the same field of view – a real triple treat.

Finally, use the top chart on page 128 to find the bright (5th-magnitude) but lonely open star cluster NGC 2281, which is in the remote eastern corridor of Auriga. In 7 × 50 binoculars, the cluster appears as a large diffuse glow nestled between two 7th- and 8th-magnitude stars; both of these stars have dramatic golden hues.

12 March

I'm a poor underdog,
But to-night I will bark
With the great Overdog
That romps through the dark.

Robert Frost, *Canis Major* (1928)

The end of winter is upon us. And though the stars still sparkle like cold crystal, the Earth has nearly

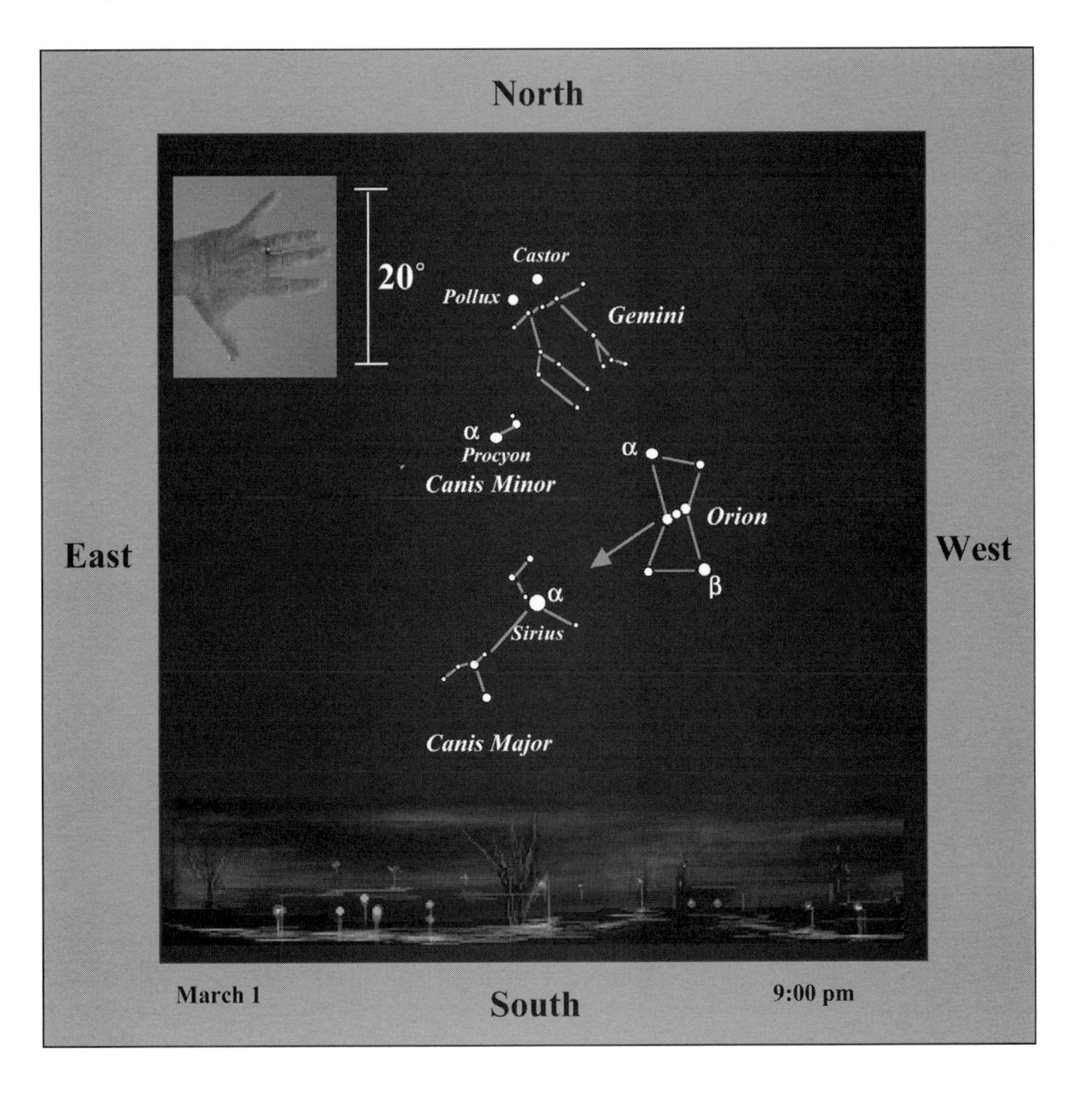

completed its orbit around the Sun. By month's end the stars will have come full circle. If you go out at 8:00 pm on March 1 and look north, you'll see the Great Bear poised high in the northeast, ready to claim its crowning position above the north celestial pole, just as it did when we started our stellar journey. The flowers too, like the stars, are returning. Spring is on the horizon, but our final stars are waiting for us in the south.

Canis Major, the Greater Dog

If March is for the dogs, it's because the night sky is now ruled by two of them: Canis Major and Canis Minor, the hunting dogs of Orion (see chart on page 129). We'll start with Canis Major rearing up in eagerness at Orion's feet. The constellation's Alpha star, Sirius, is the brightest star in the heavens. Shining at magnitude −1.5, this hot *A*-type dwarf (only 9 light years distant) is a molten white gem, like liquid starlight. Orion's Belt points right to it, though there's no mistaking this solitary luminous spectacle, known simply as the Dog Star. In ancient Egypt it was seen as the soul of Isis, and its helical rising was connected to the annual flooding of the Nile. The helical rising also announced the coming of the dog-days of summer.

If you look at Sirius in your binoculars, especially when it is near the horizon, you'll be treated to a spectacular display of flashing colors, like sunlight glinting through a swinging crystal. In fact, on some nights, when the white star is rising or setting through a contaminated atmosphere, it can appear red, just as the Sun can appear red at sunrise or sunset. This effect, combined with the radiant spectral flashes, may be responsible for an ancient myth – that Sirius was once a red star. Many believe that the myth is a misinterpretation of the words "searing flame," which appear in the astronomical poem *Phaenomena* by Aratus (c. 315–240 BC): "The tip of his terrible jaw is marked by a star that keenest of all blazes with a searing flame and him men call Seirius [*sic*]."

Sirius is a remarkable telescopic double star. It's remarkable because the companion, Sirius B (the Pup), is an 8th-magnitude white dwarf 10,000 times fainter than Sirius. The little star completes an orbit every 50 years. The two stars have such a great brightness difference that it's quite a challenge to separate the two in a telescope. It's possible that the Pup was once a giant B-type star (with a mass perhaps seven times that of the Sun) but managed to shed some 80 percent of its mass into space over the last 250 million years, probably via a strong stellar wind. Hubble Space Telescope observations of the Pup (see the image on this page) reveal that despite it being Earth-sized, the Pup has a mass that is 98 percent that of our Sun. Its powerful gravitational field is 350,000 times greater than Earth's, meaning that a 150 pound person would weigh 55 million pounds standing on its surface.

Voltaire included Sirius in his 1752 short story *Micromegas* – obviously inspired by Jonathan Swift's *Gulliver's Travels*. In the story, Micromegas, who lives on an enormous planet orbiting Sirius, stands 20 miles tall. When he travels to our Solar System, he is amused to find that the Saturnians are only about a mile tall. And when he arrives on Earth, he is further amazed at how puny its inhabitants are. Micromegas is able to scoop up a bunch of Earthlings with his thumb nail and hold philosophical conversations with them with the aid of a microscope. In the end, what Micromegas discovers is that though Earthlings are small their pride is infinitely great.

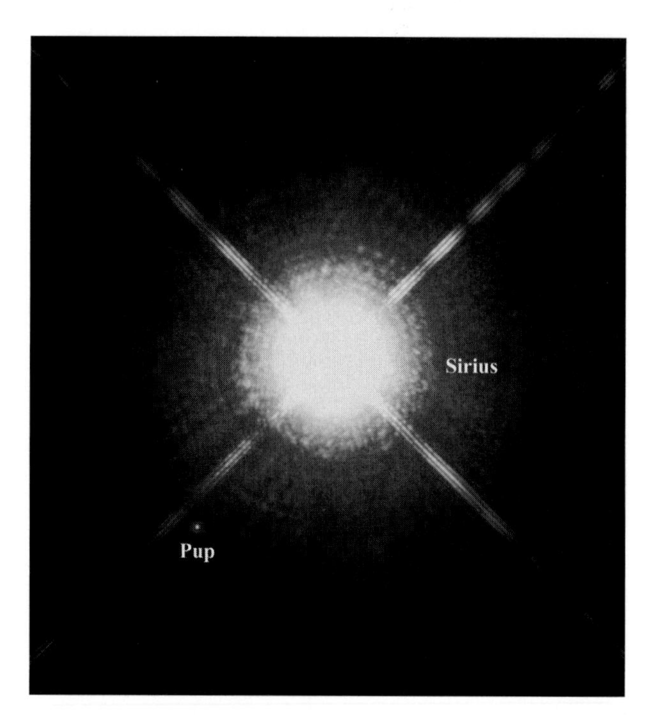

While we know of no planet around Sirius or its Pup, we do know that two planets – with masses five and seven times that of Jupiter – orbit the Dubhe-like K-type orange giant HR 2447 Canis Majoris, which lies 395 light years away in the lonely southwestern quadrant of the constellation. The planets orbit the 5th-magnitude star with periods of 430 and 2,500 days, respectively. The farthest planet orbits at a distance that equals that of Mars's orbit around our Sun. Seen from that distance, HR 2447 would appear 5° across in the sky – large enough to fill the gap between the Pointer Stars in the Big Dipper. The other planet is in an Earthlike orbit.

In case you hadn't noticed, Zeta Canis Majoris is a very pretty binocular double. The primary is a blue B-type dwarf 335 light years distant; its companion requires averted vision in 10 × 50 binoculars but it looks like a tiny 8th-magnitude fleck of warm light 3′ away to the north. Compare the fresh blue light of Zeta with the lemon-peel yellow hue of 2nd-magnitude Delta (δ) Canis Majoris (Wezen) in the dog's rump. Wezen is a remarkable F-type yellow supergiant 1,800 light years distant! One can only imagine how luminous the star would appear if it were at the distance of Sirius! The star is 50,000 times as luminous as the Sun and has a diameter some 400 times larger! And while Wezen is only about 10 million years old, it has ceased fusing hydrogen at its core and is on its way to

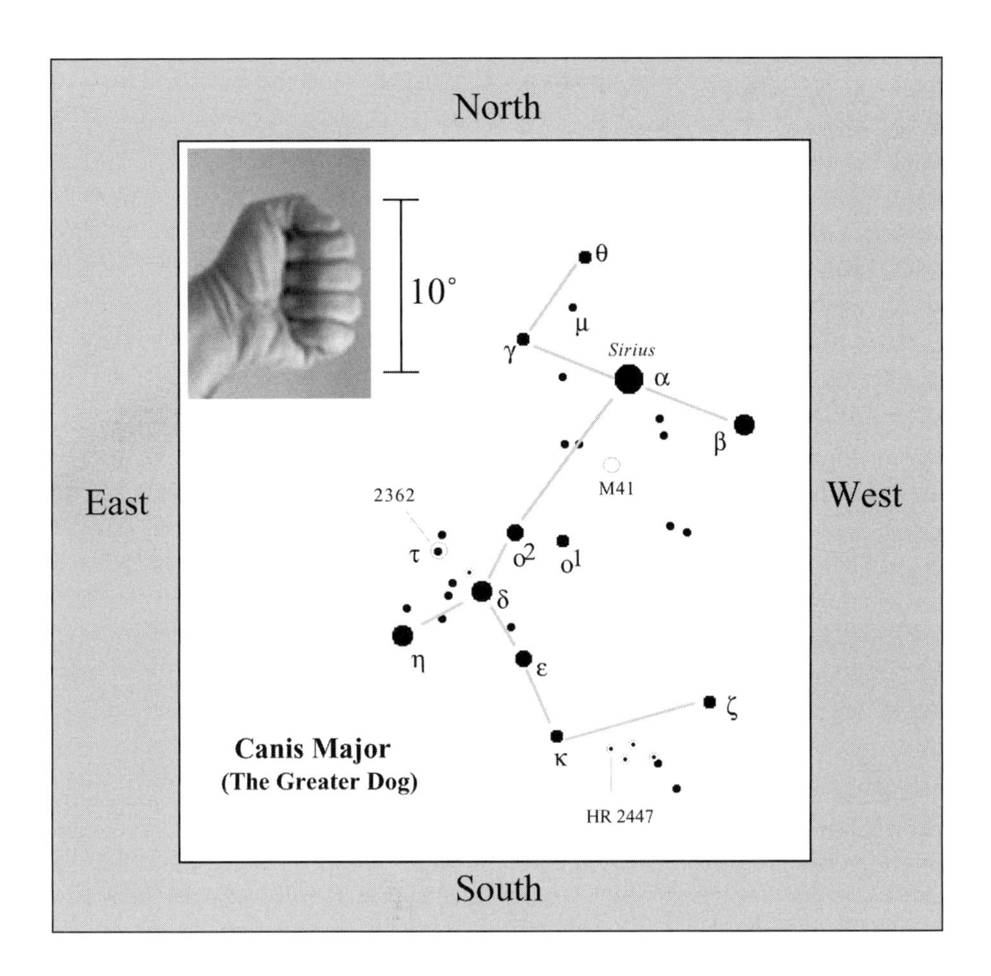

becoming a brilliant supernova. The same fate awaits the 2nd-magnitude blue supergiant star Eta (η) Canis Majoris (Aludra) about $4\frac{1}{2}°$ to the southeast. Aludra may be 3,200 light years distant, but that estimate might be in error by a factor of two, placing it at around the same distance as Wezen. In binoculars, I find Aludra's color pale Iris; look for a 7th-magnitude partner about 2′ to the west.

In the same binocular field as Delta and Eta Canis Majoris is the seemingly inconspicuous 4th-magnitude star Tau Canis Majoris. But don't be fooled. If you're under a dark sky and can hold your binoculars steady, you'll see that Tau is surrounded by a narrow halo of light only 6′ across. This halo is actually a tight gathering of suns comprising the rich and young Tau Canis Majoris Cluster, or NGC 2362. Although the cluster looks only like a teasing haze in binoculars (and only that with keen averted vision), it's a vivid reminder of how well the night sky can hide its treasures from the curious gaze of skywatchers. Equally impressive under a dark sky is the binocular Milky Way surrounding the cluster, which is laced with ribbons of dark clouds.

For a special treat, if you place Sirius in the top of your field of view, the dazzling, 4th-magnitude open star cluster M41 should be near the field's center (see the chart on page 132). Although the cluster can easily go unnoticed – because of the visual attraction of its bright luminary to the north – it has been known throughout antiquity as a fuzzy star; Aristotle recorded it with his unaided eye as a star with a tail. Lying 2,100 light years distant, M41 fills an area of sky 30 percent larger than the full Moon. Many of the cluster's stars can be resolved in simple binoculars. At first M41 will appear as a sharp round glow north of a nice double star. In time, many of the resolvable stars also appear in pairs that surround a bright arc of three suns. The 100-million-year-old cluster has about 80 stars, about a dozen of which are easily seen in binoculars. The stars are physically spread across 25 light years of space.

We will now slip across the northeastern border of Canis Major into the large and vague constellation Puppis, the Stern of the ship Argo in which Jason and his Argonauts set sail on in their epic adventure to find the Golden

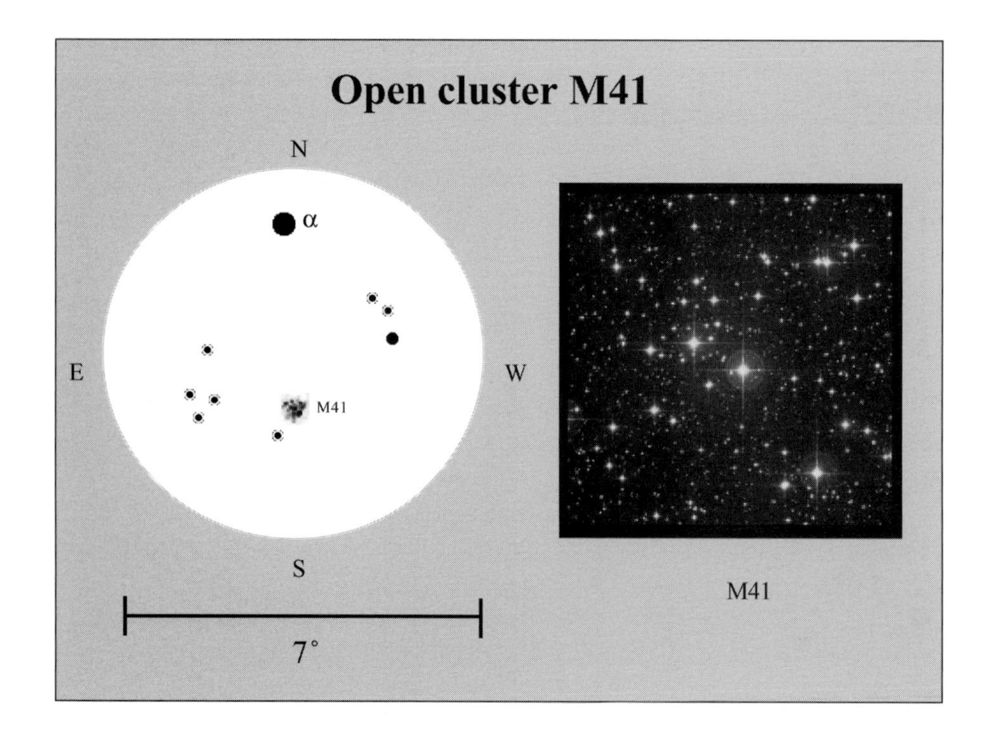

Fleece. We'll just be looking at the inconspicuous northern end of that vast and mighty vessel, where we'll find two delightful celestial treasures worthy of a binocular hunt: the glorious open star clusters M46 and M47. All you have to do is center Sirius in your binoculars, then set a course two binocular fields due east. You'll suddenly collide upon the shores of these two stellar islands – first M47, then M46 – which are separated by a mere $1\frac{1}{2}°$. (You could also start by finding 4th-magnitude Alpha (α) Monocerotis in the celestial Unicorn; if you place that star in the top of your binocular field of view, the two clusters will stand out boldly at the bottom.) Seen together they are truly one of the binocular highlights in the heavens.

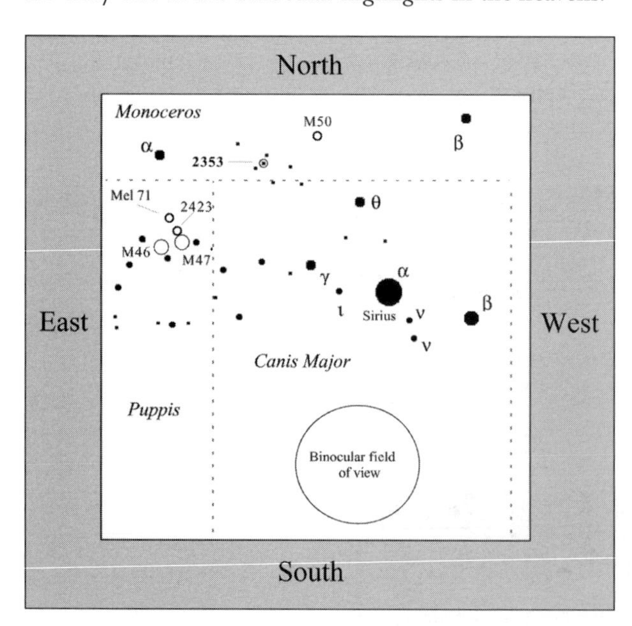

Messier 47 is a random shuffling of stars of various brightnesses, tossed against the Puppis Milky Way. Very bright and beautiful in binoculars, the cluster sports an obvious double star at its core, which is part of a sharp line of equally bright suns. This line is surrounded by about a dozen other fainter stars scattered haphazardly about like scattered jewels. In contrast, M46 is a round and uniform 6th-magnitude sphere of snowy starlight. If you relax your gaze and use averted vision, you may also see another knot of starlight above and slightly east of M47. This is yet another star cluster, 7th-magnitude NGC 2423. And if you're really astute, you may see an even smaller and dimmer knot further to the northeast; this is open star cluster Melotte 71. Both of these subsidiary clusters appear to be attached to M47 by a chain of 8th- to 9th-magnitude suns. Seen together, all these clusters look like a sprinkling of treasures tossed wildly asunder by some celestial pirate – a scene that disturbs the tranquility of the quiet neighborhood in this corner of the Milky Way.

Despite their proximity in the sky, M46 and M47 have nothing in common. Messier 47 is a loose and sparse

open cluster 1,500 light years distant and measures 14 light years in true physical extent. Messier 46, on the other hand is 5,300 light years distant. And while M46 is a relatively young cluster (about 300 million years), M47 is a baby by comparison, being a mere 55 million years old – placing it in the same age category as the Pleiades star cluster in Taurus (page 112). Messier 46 has yet another curiosity, actually an illusion – one that is visible only through a telescope. That's the superimposition of a planetary nebula against the cluster. Actually, the ring-shaped planetary nebula, which you can see in the photograph at left, is some 1,200 light years in the background.

From M46 and M47, swing up to Alpha Monocerotis, and place that star in the left side of the field. The 6th-magnitude open star cluster NGC 2353 will be at the opposite side of the field. Look for three 6th-magnitude stars in an arc only $\frac{3}{4}^{\circ}$ long and oriented north–northwest to south–southeast. The middle of these three suns is NGC 2353. The cluster lies 3,400 light years distant, on the edge of a bright association of starlight whose formation was triggered by a single supernova that detonated about 1 million years ago. This event occurred at the edge of a dense cloud of gas and dust that measured about 3 light years wide and had a mass of about 1,000 times that of our Sun. Then about 100,000 years ago, the propagating blast wave slammed into another cloud of dust and gas, triggering star formation that continues to this day in this region. With an age of about 76 million years, NGC 2353 severely predates these events, but is still a young cluster. By the way, the 6th-magnitude orange star on the cluster's southern edge is an unrelated star.

The Milky Way around NGC 2353 is peppered with interesting clusters and elegant starfields; photographs will add swaths of nebulosity. But there's one other outstanding open star cluster – lonely and obscure, M50 in Monoceros. If you place NGC 2353 in the lower left of the field, M50 will be at center, looking like a sesame seed of tightly packed suns nestled in an elliptical haze just north

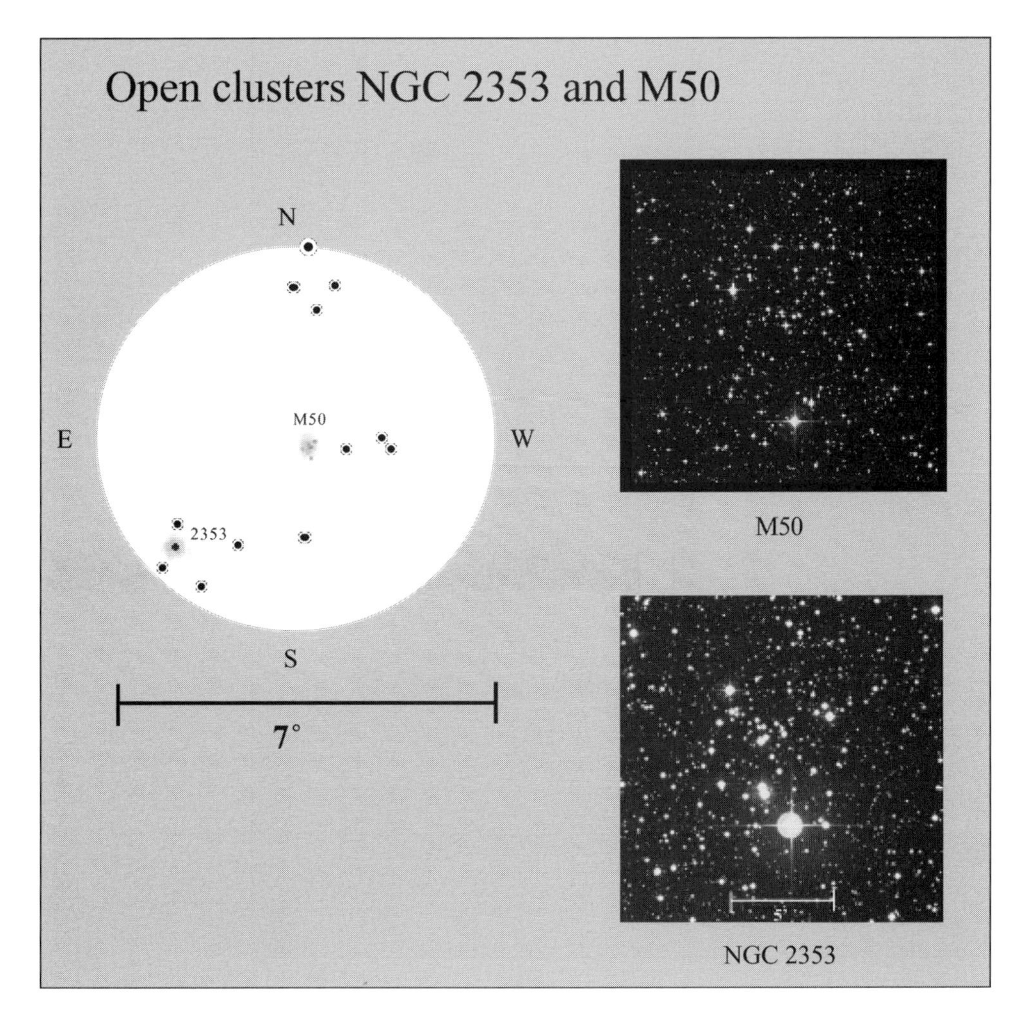

of a faint star. When you are done admiring the cluster, use your binoculars to look for a double helix of stars running 2° south of the cluster.

Canis Minor, the Lesser Dog

Use the chart on page 129 to find 0-magnitude Alpha (α) Canis Minoris (Procyon), which is about one fully outstretched hand to the north–northeast of Sirius. Procyon is a splendid sunkiss-yellow dwarf 11 light years distant. Through binoculars it is attended by a pale-green companion to its east, which, when viewed with averted vision, has two fainter stars attending it. So this singular yellow sun is really a meeting place for little wonders.

Like the greater Dog Star, Sirius, Procyon has an 11th-magnitude, white-dwarf companion (Procyon B) much too close (5″) to separate in binoculars. The stars circle a common center of mass, completing an orbit once every 41 years. The density of Procyon B, which is nearly 1.5 times the diameter of the Earth, is just 20 percent that of Sirius B.

The mythology of the Lesser Dog is very murky. Commonly seen as one of Orion's hunting dogs, it has also been identified as a dog belonging to the huntress queen, Diana, and Anubis, the jackal-headed Egyptian god of cemeteries and embalming.

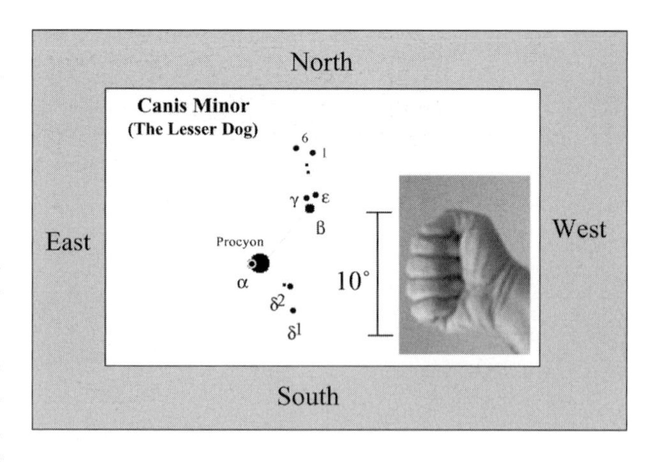

The main stars of Canis Minor all fit in the same binocular field, allowing you to compare the subtle colors of its stars. I have to wonder if the intense yellow of Procyon does not interfere with the perceived color of 3rd-magnitude Beta (β) Canis Minoris (Gomeisa). Beta is a blue, B-type dwarf that has a curious pale-olive sheen, a color that seems to play off of yellow Procyon, orange Gamma (γ), and peach-colored Epsilon (ε) Canis Minoris.

From Procyon, look far to the west for Alpha Orionis (Betelgeuse) and center it in your binoculars. Now use the chart below to move a little more than one binocular field to the east where you will see the dynamic open star

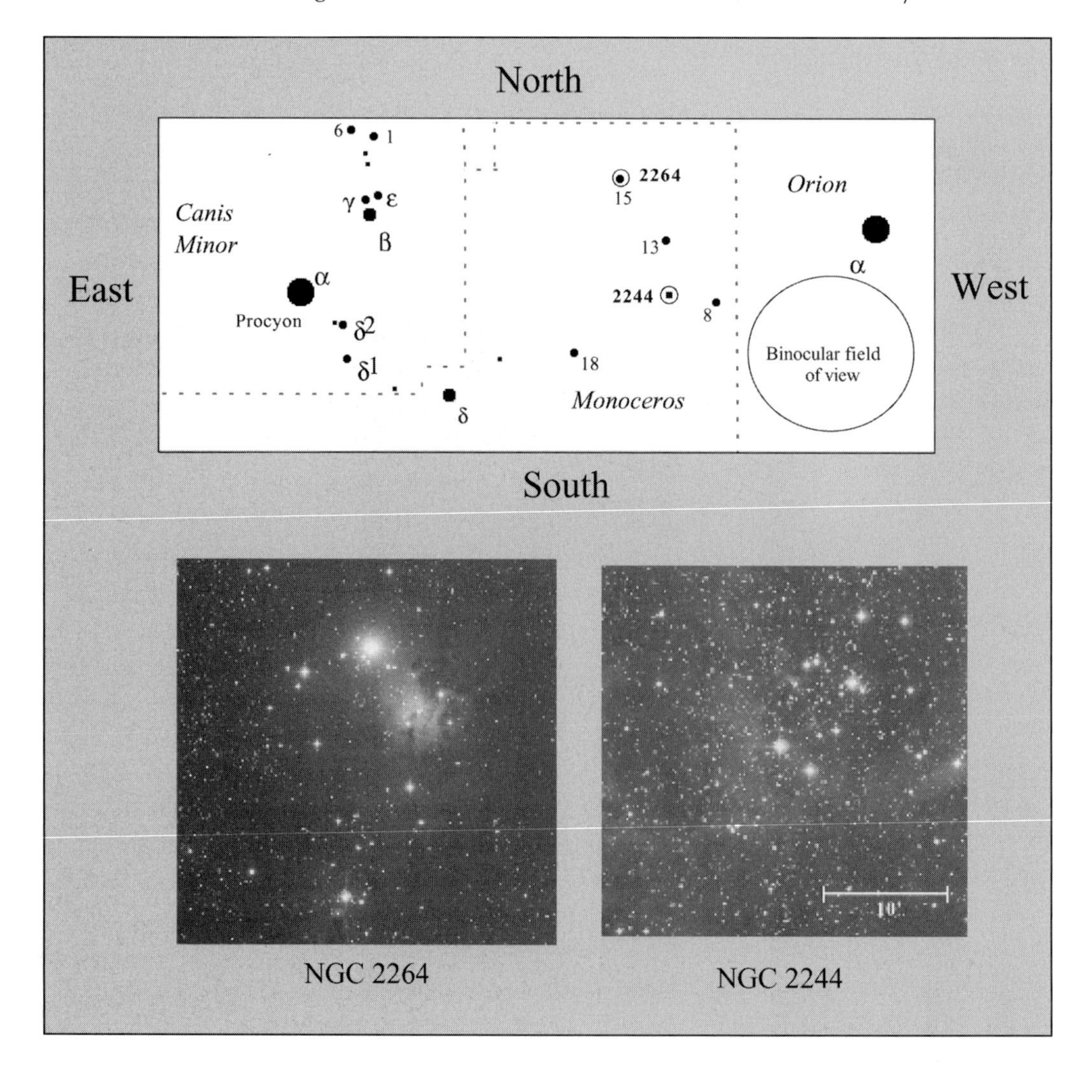

NGC 2264

NGC 2244

cluster NGC 2244. Nestled in the center of the famous wreathlike Rosette Nebula, NGC 2244 is a bright and beautiful open star cluster just southeast of the midpoint between 8 and 13 Monocerotis, respectively. Under dark skies, the cluster is immediately recognizable as a 30′-wide rectangle of six binocular suns, which is surrounded by the extensive, though dim, Rosette Nebula. The brightest star in the rectangle is 6th-magnitude 12 Monocerotis, and the two middle stars in the rectangle are nice doubles. The cluster lies nearly 5,000 light years distant and spans about 115 light years of space.

Now place NGC 2244 at the bottom of the field and look at the top for NGC 2264 – another bright and obvious open star cluster in Monoceros called the Christmas Tree Cluster (which is seen upside down, the 5th-magnitude star 15 Monocerotis marks the tree's base). Under a dark sky, 15 stars are immediately apparent in 7 × 50 binoculars. The Christmas Tree Cluster, in fact, looks like the cluster in the Rosette Nebula (without the nebulosity), only slightly warped, or dented. Look for two loops of stars – one on either side of the cluster – resembling butterfly wings. Today we know the Christmas Tree Cluster and the "faint milky nebulosity" is only part of a much larger star-forming region some 3 to 30 million years young. The Christmas Tree Cluster alone extends across more than 30 light years of space. By the way, the tree's base, 15 Monocerotis, is a famous semiregular variable star that also goes by the name S Monocerotis. It has a pure blue sheen.

Gemini, the Twins

Our final constellation of the season, Gemini, the Twins, the most northerly of all the zodiacal constellations, stands with its feet in the river Milky Way. Go out on March 20 at 7:30 pm and you will see its two sparkling jewels, Castor and Pollux, high overhead. Castor and Pollux mark the heads of the Twins – sons of Jupiter born to Leda after Jupiter disguised himself as a Swan. The Twins were hatched from an egg. And their sister was Helen of Troy. Each had a special talent. Castor was good at keeping horses and Pollux was a pugilist (a boxer). Like many twins, they were inseparable in blood and affection. And that's the way we see them in the sky, shoulder to shoulder, arms around one another, standing side by side. In classical mythology the Twins joined Jason in his search for the Golden Fleece. On that voyage, after a fantastic storm had passed, stars appeared on the heads of the brothers; the "stars" were a special twin manifestation of St. Elmo's fire – an atmospheric plasma that appears during electrical storms. In his *Natural History* Pliny explains the connection between the two phenomena:

> Stars become visible at sea and on land; I have seen them with the appearance of lightning clinging to the spears of soldiers on guard-duty at night in front of the rampart. At sea I have observed St. Elmo's fire on the yard-arms and other parts of a ship, jumping about with a sound like a voice, just as birds hop from perch to perch. When these stars come singly they are heavy and sink ships, and, if they fall down into the bottom of the hold, they destroy them by fire. When they come in pairs, they signify safety and herald a successful voyage. Men say that the terrifying star called Helena is put to flight by their approach. This is the reason the two stars are named Castor and Pollux and men at sea call on them as gods.

Castor is Gemini's Alpha star. Shining at a clean 1.6 magnitude, it is actually slightly dimmer than Pollux. But diamond-crisp Castor, a hot *A*-type star (like Vega), 52 light years distant, is one of the sky's most celebrated multiple star systems. When you look at the light of Castor, you are seeing the combined light of six stars all swirling around a common center of mass – though you will need a telescope to see the brightest members, which are a couple of arc seconds apart. See if you don't detect a kiss of yellow light to Castor's glow in your binoculars. By comparison, 34-light-year-distant Pollux, Gemini's Beta (β) star, shines at 1st magnitude and has a distinct canary-yellow tint that seems to warm up when compared to Castor's virginal purity in binoculars. Pollux really is a dynamite star – a *K*-type giant 10 times the diameter of our Sun. It also has a planet with a mass at least three times that of Jupiter and an orbital period of about $1\frac{1}{2}$ years. Now compare the light of Pollux to that of neighboring Sigma (σ) Geminorum – a 4th-magnitude *K*-type giant 122 light years distant with a molten-gold hue. Also look for two near-mirror asterisms that I call the calla lilies; one is roughly northwest of Upsilon (υ) Geminorum and the other is east of Pollux (see chart below).

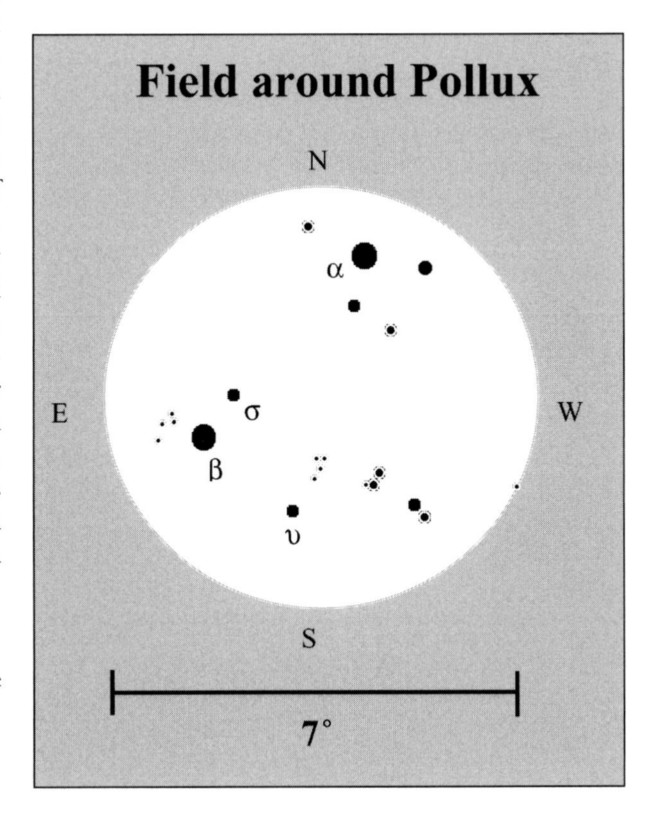

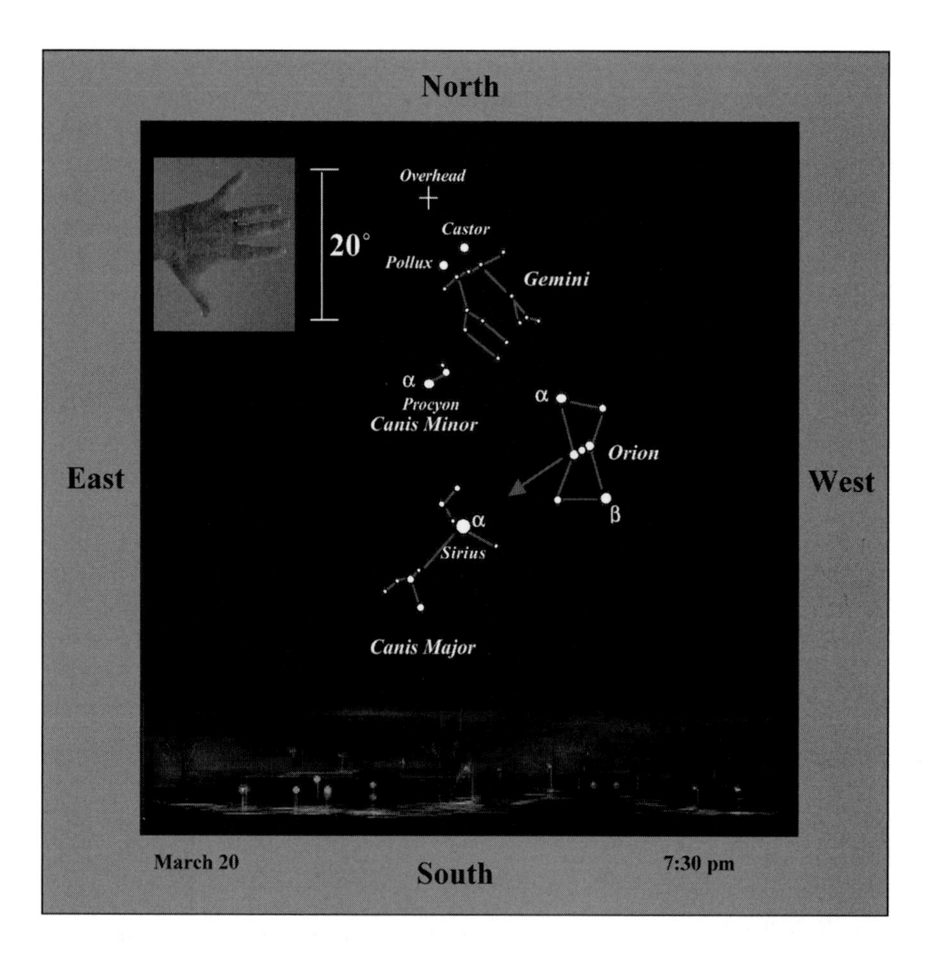

North
Overhead
20°
Castor
Pollux
Gemini
α
Procyon
Canis Minor
α
Orion
East
West
β
α
Sirius
Canis Major
March 20
South
7:30 pm

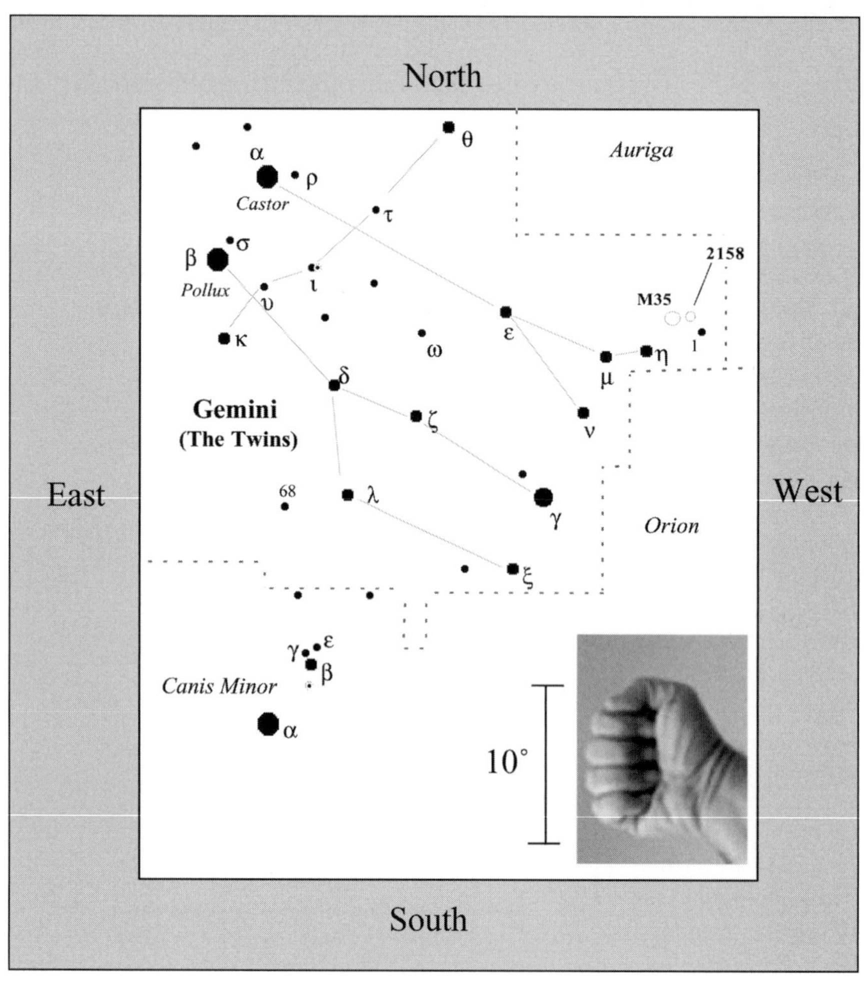

North
Auriga
Castor
Pollux
2158
M35
Gemini
(The Twins)
East
West
68
Orion
Canis Minor
10°
South

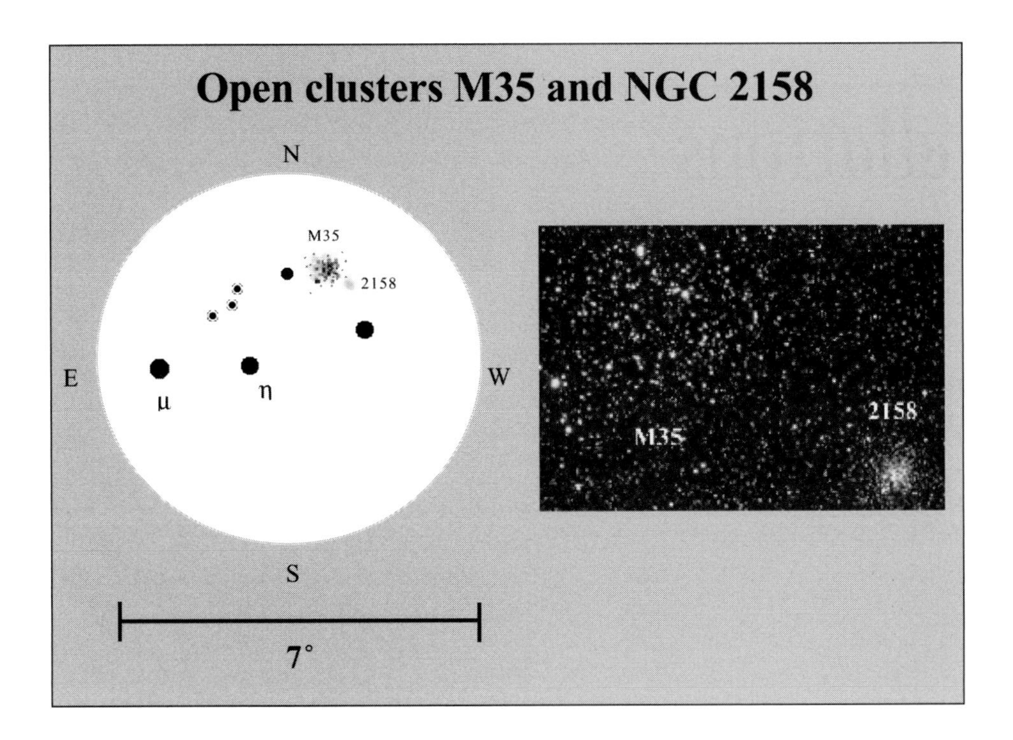

Use your binoculars and the bottom chart on page 136 to sweep around the borders of the Twins and note the wide variety of star colors. Be sure to look at Zeta (ζ) Geminorum – a pretty yellow supergiant with a faint companion that might require averted vision to see. Zeta is also a Delta Cepheid variable star 1,200 light years distant; it varies between 3.7 and 4.2 magnitude every 10.2 days. Mu (μ) Geminorum is a dynamic blood-gold star, an M-type red giant 232 light years away. And 3rd-magnitude Eta (η) Geminorum (Propus) is jet-fire orange; it's another M-type giant 350 light years distant.

Stop at Eta Geminorum, because if you haven't already noticed, it leads the way to the M35 – a bright 5th-magnitude open star cluster nearly 3,000 light years distant in the plane of the Milky Way. Under a dark sky, the cluster is visible to the unaided eye as a mottled swath of foamy light with a diameter equal to that of the full Moon. In 10 × 50 binoculars the cluster is immediately resolved as an elliptical splash of resolvable suns, all nervous starlight as the eye tries to capture each little flash in such a small area.

Messier 35 is one of the subtle splendors of our galaxy, a depiction of the wonderment of stellar birth and family. These stars – some 200 or more in total – were all born together from the same cloud of dust and gas in space. For the last 110 million years or more they have been traveling the gentle curves of our spiral galaxy as a unified whole. It will take countless years for the cluster to lose a hold on its members. If you are very astute, you might also see M35's "companion" cluster, NGC 2158 immediately to its west and a tad south. I place companion in quotes because this dynamic pairing, like so many others in our seemingly two-dimensional sky, is nothing but a wonderful illusion. The cluster NGC 2158 is an older star cluster six times more distant than M35. Messier 35 is also near the point in the sky where the Sun resides during its summer solstice, where it reaches its zenith in our sky on the longest day of the year.

It's time to put down your binoculars, and sweep the celestial vault with the broad brushstrokes of your eyes. Try concentrating on the arrangement of the sky's brightest stars. Do you see anything special? The bright stars, Sirius, Procyon, Pollux, Castor, Capella, Aldebaran, and Rigel, are part of the great Winter Hexagon, or Circle, which I call the Winter Wreath. If you add Betelgeuse, you can rearrange the pattern so that it is a celestial G.

At this point, we must now depart. Our journey has ended. At month's end we will have come full circle. Winter will have ended, and the spring stars will return to their posts. It is appropriate that I leave you with the stars of Gemini to ponder. Notice how the northerly twin is pointing to the heavens while the southerly one is pointing to the Earth. The stars are speaking to us. They are telling us of the importance of the union between Heaven and Earth. "The greatest delight which the fields and woods minister," Emerson said (and I would add the stars), "is the suggestion of an occult relation between man and the vegetable. I am not alone and unacknowledged. They nod to me, and I to them." So Look up and wonder at the stars. What we see may be a mirror and a measure of ourselves.

And we two lovers shall not sit afar,
Critics of nature, but the joyous sea
Shall be our raiment, and the bearded star
Shoot arrows at our pleasure! We shall be
Part of the mighty universal whole,
And through all Aeons mix and mingle with the Kosmic Soul!

Oscar Wilde, *We Are Made One with What We Touch and See* (1917)

The constellations

Constellation refers to the grouping of stars within a region of sky that has been divided by international agreement. There are 88 official constellations.

Abbrev.	Constellation	Latin genitive	Abbrev.	Constellation	Latin genitive
And	Andromeda	Andromedae	Lac	Lacerta	Lacertae
Ant	Antlia	Antliae	Leo	Leo	Leonis
Aps	Apus	Apodis	LMi	Leo Minor	Leo Minoris
Aqr	Aquarius	Aquarii	Lep	Lepus	Leporis
Aql	Aquila	Aquilae	Lib	Libra	Librae
Ara	Ara	Arae	Lup	Lupus	Lupi
Ari	Aries	Arietis	Lyn	Lynx	Lyncis
Aur	Auriga	Aurigae	Lyr	Lyra	Lyrae
Boo	Bootes	Bootis	Men	Mensa	Mensae
Cae	Caelum	Caeli	Mic	Microscopium	Microscopii
Cam	Camelopardalis	Camelopardalis	Mon	Monoceros	Monocerotis
Cnc	Cancer	Cancri	Mus	Musca	Muscae
CVn	Canes Venatici	Canum Venaticorum	Nor	Norma	Normae
CMa	Canis Major	Canis Majoris	Oct	Octans	Octantis
CMi	Canis Minor	Canis Minoris	Oph	Ophiuchus	Ophiuchi
Cap	Capricornus	Capricorni	Ori	Orion	Orionis
Car	Carina	Carinae	Pav	Pavo	Pavonis
Cas	Cassiopeia	Cassiopeiae	Peg	Pegasus	Pegasi
Cen	Centaurus	Centauri	Per	Perseus	Persei
Cep	Cepheus	Cephei	Phe	Phoenix	Phoenicis
Cet	Cetus	Ceti	Pic	Pictor	Pictoris
Cha	Chamaeleon	Chamaeleontis	Psc	Pisces	Piscium
Cir	Circinus	Circini	PsA	Pisces Austrinus	Piscis Austrini
Col	Columba	Columbae	Pup	Puppis	Puppis
Com	Coma Berenices	Comae Berenices	Pyx	Pyxis	Pyxidis
CrA	Corona Australis	Coronae Australis	Ret	Reticulum	Reticuli
CrB	Corona Borealis	Coronae Borealis	Sge	Sagitta	Sagittae
Crv	Corvus	Corvi	Sgr	Sagittarius	Sagittarii
Crt	Crater	Crateris	Sco	Scorpius	Scorpii
Cru	Crux	Crucis	Scl	Sculptor	Sculptoris
Cyg	Cygnus	Cygni	Sct	Scutum	Scuti
Del	Delphinus	Delphini	Ser	Serpens	Serpentis
Dor	Dorado	Doradus	Sex	Sextans	Sextantis
Dra	Draco	Draconis	Tau	Taurus	Tauri
Equ	Equuleus	Equulei	Tel	Telescopium	Telescopii
Eri	Eridanus	Eridani	Tri	Triangulum	Trianguli
For	Fornax	Fornacis	TrA	Triangulum Australe	Triangulum Australis
Gem	Gemini	Geminorum	Tuc	Tucana	Tucanae
Gru	Grus	Gruis	UMa	Ursa Major	Ursa Majoris
Her	Hercules	Herculis	UMi	Ursa Minor	Ursae Minoris
Hor	Horologium	Horologii	Vel	Vela	Velorum
Hya	Hydra	Hydrae	Vir	Virgo	Virginis
Hyi	Hydrus	Hydri	Vol	Volans	Volantis
Ind	Indus	Indi	Vul	Vulpecula	Vulpeculae

APPENDIX B

Nova hunting with binoculars

I first learned of Tycho's Star in the mid-1960s. Anton Pannekoek summarized the discovery in his 1961 *A History of Astronomy* (New York: Dover Publications), and the account fascinated me. Like Tycho, I too "knew all the stars of the heavens perfectly" since boyhood. I also wondered if I, like Tycho, could discover a new star.[1] The desire had nothing to do with fame or glory; unlike with comets, which can carry a discoverer's name, novae receive a bland alphanumerical designation that only identifies the constellation in which it was discovered; for instance, V1500 Cygni, or Nova Cygni 1975. Most nova discoverers are destined to become lost in a fog of perpetual anonymity. No matter, there is something exciting about the fact that novae, like rebels, can shake the very foundations of an accepted framework. Novae are anti-Aristotlean manifestations – things that mar the purity and steadfastness of the ancient and reliable sky. Every nova is a blemish on the heavens' Mona Lisa. It is a flash of mystery, a visual enigma.

I started nova hunting by using my father's binoculars to survey the brightest stars in a region of the Milky Way centered on Cassiopeia. To establish a foundation – a source for comparison – I plotted the stars visible at a glance through the binoculars, then taped the chart into a small notebook. The faded star chart shown here is the original one I created in the 1960s for my first nova hunts.

Whenever Cassiopeia was visible, I would sit outside in a chair and place this star chart on a little table by my side. I would then train my binoculars on a particular spot in Cassiopeia and check to see if anything new had appeared. I suppose for inspiration, or perhaps in the belief that lightning could strike twice in the same place, I plotted on this chart the position of Tycho's Star (look for a small star with a circle around it).

I continued to search for novae in this way until the mid-1970s – until I met Margaret Harwood (1885–1979), who was the former director of the Maria Mitchell Observatory in Nantucket, a small island off the coast of Massachusetts. In a memorable exchange of ideas, she passed on to me these words of wisdom: "If you want to find a nova, look in Scutum."

Her words turned out to be prophetic. In June 1975 Paul Wild of the Astronomical Institute at Berne University discovered photographically an 8th-magnitude nova in Scutum. But a search of pre-discovery photographs revealed that the nova had actually peaked around 6th magnitude about 35 days prior to Wild's discovery . . . but no one had noticed. Then, in August of that year a brilliant nova burst forth near Alpha (α) Cygni (Deneb), disrupting the pattern of both the Northern Cross and the ancient Swan. The new star battled its way to prominence in the summer Milky Way. Seeing this star gave me an appreciation for what it must have been like for Tycho to have seen his new star burning so brightly in an esteemed constellation.

Then, in October, 1976, the renowned comet and nova discoverer George Alcock of Peterborough, England, found a new binocular star that disrupted the famous Coathanger asterism Vulpecula. That discovery not only inspired the formation of the United Kingdom's Nova Patrol but also the visual discovery in January, 1977 of yet

[1] Although both novae and supernovae suddenly appear in the night sky as "new stars," they are not. A supernova is, in fact, the death of a single massive star, while a nova is a star that brightens spectacularly – by up to a factor of a million or so – before dimming to its pre-nova state. The brightening of a nova is caused by mass transfer in a close binary system. A white-dwarf primary drains hydrogen gas from an orange or white-dwarf companion. As hydrogen builds up on the white dwarf's surface, it rapidly ignites, creating a large nuclear explosion on the surface of the star. As the ejected shell of gas rapidly expands, we see it as a surge in the star's brightness. When the expanding gas diffuses, we see the new star fade to normal light.

another nova, Nova Sagitta 1977, which John Hosty of Huddersfield, England, discovered visually with a pair of broken 10 × 50 binoculars.

With this rash of nova discoveries, I decided to start a new nova patrol. With Harwood's words haunting my mind, I decided to start hunting in Scutum. So, in the summer of 1977, I went to the Harvard College Observatory library, checked out a box that contained charts produced by the American Association of Variable Star Observers, pulled out the wide-field chart for R Scuti, and used it as the foundation for my new nova-search program. Star by star, I began to survey the R-Scuti field with the 3-inch finderscope attached to the Observatory's 9-inch Clark refractor in hopes of finding an obvious discrepancy.

In time, though, I felt the search was inadequate. There had to be a better way, and there was. Peter Collins (a friend and observing partner at Harvard College Observatory) suggested that perhaps we should use our binoculars to memorize the Milky Way to 8th magnitude as Alcock had. Collins was convinced of the need to search because he had just attended Cecelia Payne-Gaposchkin's Henry Norris Russell lecture on novae, in which she noted that another Harvard astronomer, Solon Bailey, had estimated that some 50 novae 7th magnitude or brighter should erupt each year in our galaxy. (Collins later said that he had initially misunderstood the reference and was under the impression that some 50 novae erupted each year in the night sky – many of which were obviously going undetected.) Regardless, a critical spark was ignited and by the fall of 1977, both Collins and I began to search for novae by creating and memorizing "mini-constellations" as seen through our binoculars. To help us in our searches, comet discoverer Michael Rudenko created a special set of star charts that I still use today.

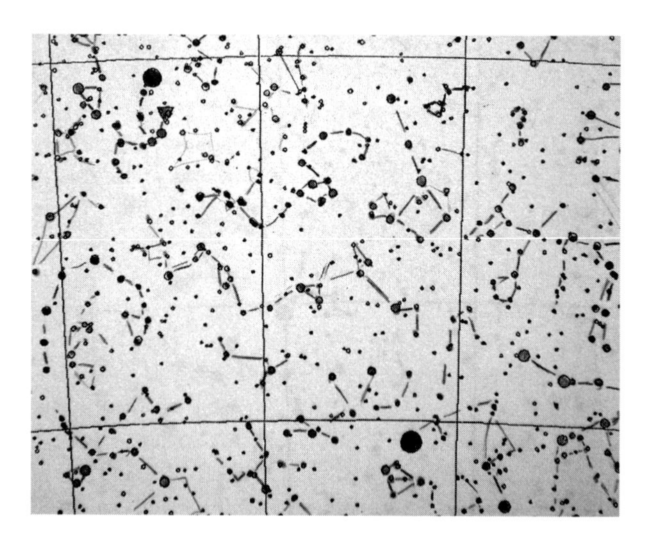

Memorizing the Milky Way in binoculars is not as hard to do as it sounds. In fact, I found it therapeutic, especially since I could let my imagination run free when I created my constellations. Each clear night, I created and memorized about a half-dozen of them. On the next clear night, I would review the constellations I had just learned and look for changes. If everything checked out okay, I would add new constellations, then repeat the entire process on the next clear night. Memorizing new star patterns took only a few minutes.

It would seem that the more constellations I learned, the more time I would need to hunt – but it doesn't work that way. In fact, by repeating the sweeps each clear night, the patterns I learned became rapidly more familiar. After a number of years of training, I could, with confidence, make a binocular sweep from Cassiopeia to Sagittarius in a matter of minutes and be confident (to a degree) that no novae 7th-magnitude or brighter were visible in those regions of the sky.

Since memory is not perfect, a sweep of the heavens would usually rake up several suspects, all of which needed to be checked against Rudenko's nova-hunt charts; since I traced the outlines of the mini-constellations I created on these charts, locating a suspect was simple. But the method is not perfect. It relies on the chance that a nova will occur in a location that alters the shape of one of the memorized star patterns. If a nova occurs, for instance, within a line of stars that forms part of one of the mini-constellations, it might go undetected in a quick sweep.

There is also a certain speed that is most beneficial; the more time I spend looking at a specific region of sky, the more suspects I acquire, and the more checking I have to do. Also, because I memorized the Milky Way to a fainter magnitude than Collins, my area of search was not as large as his. And this may have been an important factor in the end, because Collins went on to discover five novae through his binoculars between 1977 and 1992, while, to this date, I have not had a credited success.

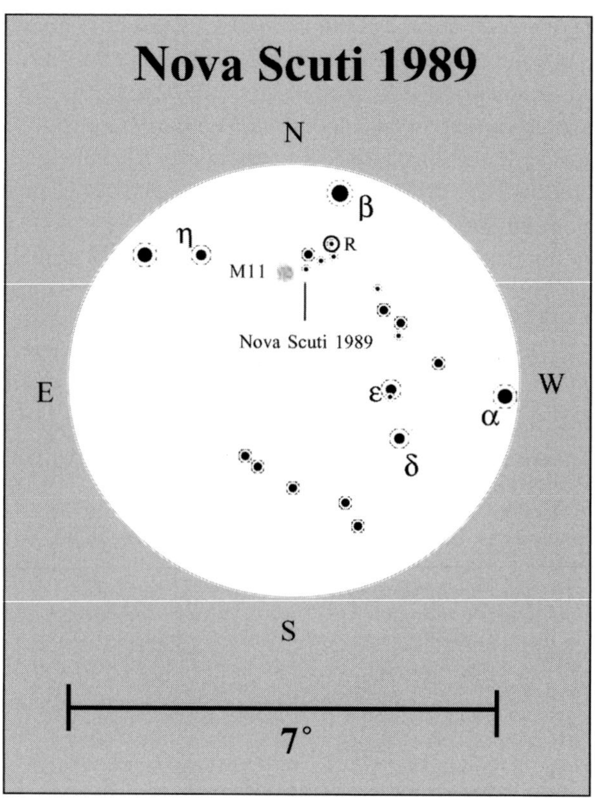

I say credited success because I did discover a nova in Scutum. On the evening of September 16, 1989, I made a search for novae through holes in clouds with my 10 × 50 binoculars from my home in West Roxbury, Massachusetts. Here is the excerpt from my diary:

> Clouds broke enough for me to nova hunt [between] rolls of altocumulus. Sometimes large holes would appear and remain long enough for me to make good coverage. The scan in Sagittarius was desultory because of clouds, but the remainder – from Scutum to western Cassiopeia – was thoroughly monitored. One suspect near M11.

The last sentence is the most important one. After I completed my sweep from Scutum to Cassiopeia, I returned my gaze to Scutum, to check the suspect, but clouds had, by then, completely covered the region. The skies did not clear for another 11 days, so I could not confirm my observation. Meanwhile, on the morning of September 23, I received a phone call from Daniel Green at the Central Bureau for Astronomical Telegrams, who asked me if I had heard about the new nova. I immediately said, "Dan, don't tell me. It's in Scutum, isn't it?" He confirmed my suspicion. The new nova was discovered photographically once again by Paul Wild on September 20. It was shining at about magnitude 10.5 just 20′ west of M11 and about 45′ southeast of R Scuti. I had estimated the new star's magnitude was about 8. Indeed, pre-discovery images taken by Rob McNaught at Siding Spring Observatory, show that the photographic magnitude of the nova was 7.6 on September 12 and at magnitude 8.5 on September 17 (Universal time).

Margaret Harwood was right. My first nova discovery was in Scutum. The question is, do you want to carry her torch? Do you want to discover a nova? Perhaps you'd like to concentrate on an area in Scutum.

APPENDIX C

Photo credits

Chapter 1
Stephen James O'Meara
Stephen James O'Meara
Stephen James O'Meara
Stephen James O'Meara
Stephen James O'Meara
M97: Digitized Sky Survey[1]
Stephen James O'Meara
Three galaxies: Digitized Sky Survey[1]
Digitized Sky Survey[1]
NASA and The Hubble Heritage Team (AURA/STScI)
Digitized Sky Survey[1]
Digitized Sky Survey[1]
Digitized Sky Survey[1]
Digitized Sky Survey[1]
All galaxies: Digitized Sky Survey[1]
M81 and M82: Digitized Sky Survey[1]
M101: Digitized Sky Survey[1]

April
Stephen James O'Meara
Stephen James O'Meara
Stephen James O'Meara
Stephen James O'Meara
Leo Triplet: Digitized Sky Survey[1]
Stephen James O'Meara collection
Stephen James O'Meara
Stephen James O'Meara
Stephen James O'Meara
Stephen James O'Meara
Stephen James O'Meara collection
Stephen James O'Meara
M67: Digitized Sky Survey[1]

May
Stephen James O'Meara
Coma galaxy cluster: Hubble Space Telescope WFPC Team and NASA
M64: NASA and The Hubble Heritage Team (AURA/STScI)
M53: Digitized Sky Survey[1]
Stephen James O'Meara
Stephen James O'Meara
M51: Digitized Sky Survey[1]
M3: Digitized Sky Survey[1]
Stephen James O'Meara
M104: NASA and The Hubble Heritage Team (STScI/AURA)

June
Dr. Andrew Wood
Stephen James O'Meara
Stephen James O'Meara

July
M13: Digitized Sky Survey[1]
Stephen James O'Meara
M92: Digitized Sky Survey[1]
Supernova diagram: NASA/CXC/M.Weiss
M4: Digitized Sky Survey[2]
Stephen James O'Meara
Stephen James O'Meara
Caduceus and painting of Adam & Eve: with permission from: www.drblayney.com/Asclepius.html
Ophiuchus: Michael Oates/Manchester Astronomical Society
Thoth: Stephen James O'Meara
Egyptian solar symbol: Stephen James O'Meara
Stephen James O'Meara
Stephen James O'Meara

August
Stephen James O'Meara
M22: Nigel A. Sharp, REU program/AURA/NOAO/NSF
Stephen James O'Meara
M8: A. Caulet (ST-ECF, ESA) and NASA
M17: NASA, H. Ford (JHU), G. Illingworth (UCSC/LO), M. Clampin (STScI), G. Hartig (STScI), the ACS Science Team, and ESA
M16: NASA, ESA, and The Hubble Heritage Team (STScI/AURA)
Stephen James O'Meara
M11: Digitized Sky Survey[2]
Stephen James O'Meara
M57: The Hubble Heritage Team (AURA/STScI/NASA/ESA)
Stephen James O'Meara
M29: Digitized Sky Survey[1]
ankh: Stephen James O'Meara

September
Stephen James O'Meara
Stephen James O'Meara
M71: Digitized Sky Survey[1]
M27: Digitized Sky Survey[1]

October
Stephen James O'Meara

Stephen James O'Meara
M30: Digitized Sky Survey
NGC 7009: WFPC2, HST, NASA
M2: Digitized Sky Survey[2]
Stephen James O'Meara

November

Stephen James O'Meara
Stephen James O'Meara
Stephen James O'Meara
Stephen James O'Meara
Stephen James O'Meara
NGC 281: Digitized Sky Survey[1]
M103, NGC 663, and NGC 457: Digitized Sky Survey[1]
M52 & NGC 7789: Digitized Sky Survey[1]
Stephen James O'Meara
Stephen James O'Meara
Cassiopeia A: NASA, ESA, and The Hubble Heritage (STScI/AURA)-ESA/Hubble Collaboration
M15: Digitized Sky Survey[1]
O'Meara 1: Digitized Sky Survey[1]

December

M31, M32, and M110: Digitized Sky Survey[1]
Stephen James O'Meara collection
NGC 752: Digitized Sky Survey[1]
M33: Digitized Sky Survey[1]
Cetus: Michael Oates/Manchester Astronomical Society
NGC 253: Digitized Sky Survey[2]
NGC 288: Digitized Sky Survey[2]
M77: KPNO and NASA

January

Stephen James O'Meara
NGC 1647: Digitized Sky Survey[1]
M1: Digitized Sky Survey[1]
Pleiades: NASA/JPL/Space Science Institute
Stephen James O'Meara
Stephen James O'Meara
Stephen James O'Meara
M34: Digitized Sky Survey[1]
Stephen James O'Meara

February

Stephen James O'Meara
NGC 1535: Digitized Sky Survey[2]
Orion: Michael Oates/Manchester Astronomical Society
Osiris: Stephen James O'Meara
Stephen James O'Meara
M79: Digitized Sky Survey[2]
NGC 2024: Digitized Sky Survey[2]
M42: HST, Hubble Heritage
NGC 2169: Digitized Sky Survey
M36 & M38: Digitized Sky Survey[1]
NGC 2281: Digitized Sky Survey[1]

March

Sirius: NASA, ESA, H. Bond (STScI), and M. Barstow (University of Leicester)
Tau Canis Majoris Cluster: Digitized Sky Survey[2]
M41: Digitized Sky Survey[2]
M46 and 47: Digitized Sky Survey[2]
M50: Digitized Sky Survey[2]
NGC 2353: Digitized Sky Survey[2]
NGC 2264 and NGC 2244: Digitized Sky Survey[1]
M35 and NGC 2158: Digitized Sky Survey[1]

Appendix B

Stephen James O'Meara
Stephen James O'Meara

[1] All deep-sky photographs with northern declinations are from the Digitized Sky Survey, Northern Hemisphere, courtesy the Palomar Observatory and NASA/AURA/STScI.

Note: The Digitized Sky Survey was produced at the Space Telescope Science Insitute (STScI) under US Government grant NAG W-2166. The images of these surveys are based on photographic data obtained using the Oschin Schmidt Telescope on Palomar Mountain and the UK Schmidt Telescope. The plates were processed into the present compressed digital form with the permission of these institutions.

The National Geographic Society Palomar Observatory Sky Atlas (POSS-I) was made by the California Institute of Technology with grants from the National Geographic Society.

The Second Palomar Observatory Sky Survey (POSS-II) was made by the California Institute of Technology with funds from the National Science Foundation, the National Aeronautics and Space Administration (NASA), the National Geographic Society, the Sloan Foundation, the Samuel Oschin Foundation, and the Eastman Kodak Corporation. The Oschin Schmidt Telescope is operated by the California Institute of Technology and Palomar Observatory.

[2] All deep-sky photographs with southern declinations are courtesy of the UK Schmidt Telescope (copyright in which is owned by the Particle Physics and Astronomy Research Council of the UK and the Anglo-Australian Telescope Board) and the Digitized Sky Survey created by the Space Telescope Science Institute, operated by AURA (Association of Universities for Research in Astronomy), Inc., for NASA, and are reproduced here with permission from the Royal Observatory Edinburgh.

Index

Other books by Stephen James O'Meara

Readers who have got the astronomy 'bug' and are intrigued to witness more celestial sights may be interested in Stephen James O'Meara's other books. His Deep-Sky Companions series brings together nearly 330 astronomical gems, all observed through a 4-inch refractor.

Deep-Sky Companions: The Messier Objects
Foreword by David H. Levy
9780521553322
This book provides new and experienced observers with a fresh perspective on the Messier Objects, the most widely observed celestial wonders in the heavens. It contains superb drawings, detailed finder charts, and new astronomical data on each object.

". . . there is no better guide than Stephen O'Meara's book."

Stephen P. Maran, Nature

Deep-Sky Companions: The Caldwell Objects
Foreword by Patrick Moore
9780521827966
Carefully compiled by Sir Patrick Moore, the Caldwell Catalog covers the entire celestial sphere, highlighting cosmic wonders for observers worldwide. This book presents O'Meara's beautiful sketches and detailed visual descriptions, and discusses each object's rich history and astrophysical significance.

". . . the definitive guide to the Caldwell catalog . . . "

Orion

Deep-Sky Companions: Hidden Treasures
9780521837040
These 'hidden treasures' have been carefully chosen based on their popularity and ease of observing. Stunning photographs and beautiful drawings accompany detailed visual descriptions of the objects, none of which are included in the Messier or Caldwell catalogs.

". . . a completely fresh look at the most sublime and distant objects in our night sky . . ."

David H. Levy
Co-discoverer of comet Shoemaker-Levy 9
Science Editor, Parade Magazine

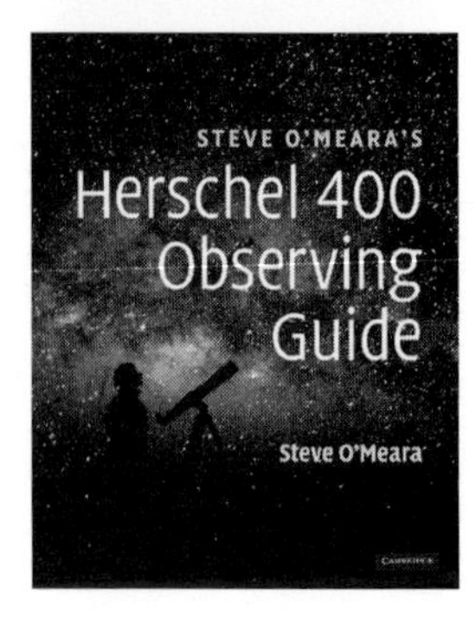

Steve O'Meara's Herschel 400 Observing Guide
9780521858939
This richly illustrated guide leads you through the galaxies, nebulae, and star clusters in the Herschel 400 list. It is ideal for intermediate or advanced astronomers who have tackled the Messier objects, or are working towards their Herschel badge.

"O'Meara takes a list of admittedly faint objects and injects new life into them."

Michael Bakich, Astronomy.com